Categories for Software Engineering

José Luiz Fiadeiro

Categories for Software Engineering

José Luiz Fiadeiro
University of Leicester
Department of Computer Science
University Road
Leicester LE1 7RH
United Kingdom

Library of Congress Control Number: 2004113132

ACM Computing Classification (1998):
D.2.11, F.3.1, D.2.1, D.2.4, D.1.3

ISBN 3-540-20909-3 Springer Berlin Heidelberg New York

Springer is a part of Springer Science+Business Media

springeronline.com

Printed in Germany

Cover design: KünkelLopka, Heidelberg
Production: LE-TeX Jelonek, Schmidt & Vöckler GbR, Leipzig
Typesetting by the author
Printed on acid-free paper 45/3142/YL - 5 4 3 2 1 0

From a loving father

Preface

Why Another Book on Category Theory?

In the past ten years, several books have been published on category theory either by computer scientists or having computer scientists as a target audience (e.g. [6, 12, 22, 89, 105], to which a precious collection of little gems [90] and the chapter cum book [91] should be added). Isn't the working computer scientist spoilt with choice?

Although each of the above mentioned books presents an approach of its own, there is one aspect in common in their view of computer science: the analogy between arrows (morphisms) and (classes of) computations. This "type-theoretic" or "functional" approach corresponds to a view of computer science as a science of computation, i.e. a discipline concerned with the study of computational phenomena where the focus is on the nature and organisation of computations.

However, there is another view of computer science where the focus is, instead, on the development of computer programs or systems. This is the approach that supports, for instance, software engineering. From this point of view, arrows do not capture computational phenomena, or abstractions thereof, but instead relationships between programs, or abstractions of programs, that arise in the development of computer systems, for instance, refinement of higher-level specifications into executable programs [100, 104], and superposition of new features over existing systems [72].

Not surprisingly, this same difference in the points of view can be found when logic is taken as a mathematical domain for formalising aspects of computer science. The "computations as proofs" paradigm is the one that corresponds to the "classical" application of category theory. Terms of the logic correspond to objects in a category of programs whose morphisms capture (partial) computations. From a logical point of view, the perspective that we take in this book is not centred on terms but on theories as system specifications. Morphisms then capture what in logic is known as "interpretations between theories", the cornerstone for the formalisation of refinement in program development and other operations on specifications and system designs [17, 82, 103, 104].

Category theory can also be presented as the branch of mathematics that, par excellence, addresses "structure". As the introduction will try to explain, this is because category theory causes structure to emerge from relationships between objects as captured by arrows, and not extensionally as in set theory. Indeed, the term *morphism*, often used for arrow in category theory, has in its etymology the notion of *preservation of form*. What these structures are, or mean, is up to the "user". Hence, in the "classical" approach, we find applications of category theory that address the structure of computations. In the approach that is taken in this book, the structures that are addressed are those that capture modularisation principles in software development, in particular, those that have been used for constructing distributed systems (e.g. as in [81]) and, more recently, emerging in the guise of what has become known as software architectures [50].

The practical difference between the two approaches in what concerns category theory in general, and this book in particular, is that the reader will not find as many references to algebraic topology or related fields of mathematics as applied, for instance, to domain theory. Although this book is still "mathematical", the software engineer will find the mathematics applied to objects of its day-to-day concerns: programs, object classes, specifications, designs, and so on.

This approach can be also situated as belonging to the class of applications of category theory to general systems theory, namely in the tradition initiated in [52, 53, 64], an area of science that, as the name indicates and the introduction elucidates, encompasses more than computational systems in the traditional sense. Through books aimed at wider audiences like [71], a unifying view of complex systems as they arise in disparate areas like physics, biology, social sciences, economics and, yes, informatics, has started to emerge (pun intended), which is a clear indication of new levels of maturity in science in general and informatics in particular. Hence, one of the purposes of this book is to help computing scientists and software engineers acquire formal tools that will enable them to follow and participate in this "new" culture.

A trait that is common to all these areas is a view of complex systems as communities of interacting, simpler, autonomous entities. Whereas, in areas like biology or social sciences, the notion of "community" is intrinsic, its use in areas like software engineering is more artificial and is normally identified with methods and development techniques that, in the past few years, have attempted to tackle complexity by borrowing the organisational principles that can be recognised in such "natural" communities. Object-oriented modelling, agent-based programming and component-based development all make use, in one way or another and with different emphasis, of this analogy. This brings us to the application area covered in the third part of this book.

CommUnity is the name of a language for parallel program design that is similar to Unity [19] but adopts instead an interaction model that places it in the realm of these more general and unifying approaches to systems.

It addresses in particular the most recent trend, *service*-oriented software development, an (r)evolution of the popular object-oriented modelling techniques for the "Internet-age" or what is becoming known as the "real-time" or "now" economy. The distinctive feature of this new trend is in the emphasis that it puts in the externalisation and explicit modelling of interactions as first-class citizens so that systems can be more easily reconfigured, in run-time, and without interruption of vital services. These characteristics match, precisely, features that are intrinsic to category theory, namely those that distinguish it from set theory.

That is why, even if a substantial part of this book is illustrated with examples borrowed from software engineering practice, we decided to devote three chapters to the application of category theory to CommUnity and its relationship to software architectures. This material will provide an opportunity for the reader to see concepts and techniques of category theory applied in an integrated and systematic way. At the same time, the reader will be able to appreciate how far one can go in formalising software development methods and techniques in mathematical frameworks, which is essential for a mature engineering discipline and, in my opinion, is our responsibility as computing scientists.

This Book and Its Many Authors

Mentioning the connections between category theory and general systems theory is a good opportunity to give due credit to Joseph Goguen for the profound inspiration that his work has instilled, a sentiment that I know is shared by many other researchers in computing science. He has expressed his own views on the applications of category theory to computing in several publications, most notably in [57], which include detailed summaries of technical results that we all have found very useful when categorical approaches were still regarded, at best, as "exotic" [60, 61, 66, 102]. All of us regret that this material has never found its way to a textbook. Because it is not our aim to fill this gap, the reader is strongly encouraged to consult this rich legacy at his or her own pace, bearing in mind that the list of references that is provided at the end is far from being complete.

Completeness is, in fact, a concern that has remained largely alien to my research agenda. (This observation is intended to make some readers smile, but you can take it literally.) This book is more about a personal experience than the output of a rational process of identifying "the" or "a" complete categorical kernel that software engineers can use as a toolbox. The only justification for the inclusion of many concepts and constructions is that they were of help to me, either technically or aesthetically, making it likely that they will be directly useful for other people "like me". The exclusion of many other, even very basic concepts,[1] can be justified by the officious disclaimer that "the line has to be drawn somewhere", but, most of the time, the reason is that I never stumbled upon them in my daily routine or simply that I have not developed an understanding about them that is deep enough to add any value to what can be found in other books.

This personal experience has gone through well identifiable periods, each of which is associated with a different focus of interest in computing and a group of people with whom I worked directly and whose contributions I would like to acknowledge. My first contact with category theory was when I was studying mathematics as an undergraduate at the University of Lisbon, and Prof. Furtado Coelho challenged the wrath of my fellow students, and his fellow staff, by including this most exotic, difficult and useless of subjects in the curriculum of Algebra II. Applications to computing science came a year later through the study of Goguen and Burstall's Theory of Institutions as a means of formalising conceptual modelling and knowledge representation approaches, under the supervision and in collaboration with Amílcar and Cristina Sernadas [46]. This is when things started to get serious.

In 1988, I started what has been a very rewarding collaboration with Tom Maibaum. During the three years I spent at Imperial College, we developed a categorical approach to object-oriented development based on temporal logic specifications [39], a marriage between my previous work with institutions and the ideas of Tom Maibaum and Paulo Veloso on the nature of specifications in system development [103]. Their contribution permeates the material that is exposed in a way that cannot be referenced in the same way as a technical result. I have been very fortunate to be able to keep exchanging ideas and experiences with them; there are always hidden subtleties that only come to the surface when you are challenged by people like them and required to scratch the innermost levels of your understanding to satisfy their curiosity.

During this same time, Félix Costa explored the categorical semantics of objects from the point of view of algebraic models of concurrency [21]. In 1992, we brought it all together [32]! My collaboration with Félix provided much of the inspiration that led to my own understanding of the application of category theory to systems modelling. Although specific contributions are acknowledged with references to his work, it would be unfair to reduce his contribution to this book to those occasions.

The next phase is devoted to the (then) emerging field of software architecture. It is centred on a language – CommUnity – that I developed together with Georg Reichwein and Tom Maibaum in an initial period, and later on with my students Antónia Lopes and Michel Wermelinger. It

[1] Yes, I know that I will not be forgiven for having left out "must-haves" such as the Yoneda lemma, Cartesian-closed categories, topoi, monads, and so on.

started as a proof of concept, showing that Goguen's categorical approach could be applied to parallel program design in the style of Unity [19] and Interacting Processes [48]. Later, it evolved into a prototype language for architectural modelling [34], a process that led me to understand many concepts that, until then, were blurred by the poor expressive power of the formalisms with which we had been working: non-determinism versus under-specification [77], refinement versus composition [78], and the role of "signatures" in separating computation and coordination [35]. Some of this is revealed in Part III of this book, but you will have to wait for another book to have the full story!

Although this "architectural" period is still very much alive (which does not mean that the others are already dead), another step in this evolution process has occurred: the realization that category theory provides a perfect fit to support service-oriented software development, for instance in the sense of Web-services. But this step is so recent that, in fairness, I cannot acknowledge/blame anybody in particular for it. Nevertheless, it is unlikely that it would have happened so soon, or at all, if I had not accepted the challenge that Luís Andrade presented me with for working with ATX Software SA in putting these "theories" into "practice". This has been a very rewarding process that has given me the opportunity to understand the implications of many of the structures and mechanisms that are intrinsic to category theory. I hope that I have managed to permeate this understanding in the way the material is exposed in the book.

This is probably why this book is being finalised now and only now: during each of the periods I mentioned, a book was planned and parts were sketched. It is only now that the work of so many people has contributed to the contents that I can safely write it on my own without feeling guilty for excluding anybody in particular from coauthoring it.

It so happens that the last thrust in writing this book was made during my first year at the University of Leicester, a renowned address for research in category theory and its applications to computer science. Although I can honestly assure the reader that the decision to join Leicester was not for the advantages of promoting this book, it is certainly a privilege for the book to bear this affiliation!

Finally, I should thank all the colleagues and students who have trod with me the paths that you can choose to follow in this book. The opportunity to discuss and lecture on many of the topics that are covered contributed decisively in helping me reach the level of maturity that made me decide that this book could be written. The feedback I received from tutorials presented at events such as ECOOP, ETAPS, FME, OOPSLA and TOOLS also helped me decide that this book should be written. The interest and encouragement of people like Ira Forman and Desmond D'Souza have also reassured me that the message could perfectly go beyond the walls of academia.

Special Acknowledgements

Although the previous paragraphs have given me the opportunity to acknowledge the contributions of a number of people and institutions, there are some specific colleagues to whom I would like to express my deepest gratitude for direct contributions to this book:

- *Félix Costa*: A significant part of the material covered in Part II was developed jointly with him as reported in [32, 33]. As already mentioned, much of my own understanding of category theory and its role in computing science grew from discussions with him.
- *Antónia Lopes* and *Michel Wermelinger*: The fact that Part III of this book was essentially extracted from [37, 79] is a good indication of how important and extensive their contribution has been. CommUnity as we know it today is as much theirs as it is mine.
- *Tom Maibaum*: His encouragement and support in the earlier phases of the production of the book were decisive.
- *Uwe Wolter*: He had the courage to follow an early draft in a course that he gave in 2002/2003 (repeated in 2003/2004) at the University of Bergen (Norway). As a result, I received precious amounts of feedback, which was invaluable for the final tuning of the material and the way it is now presented.

Finally, I would like to thank the EPSRC for an eight-month visiting fellowship at King's College London in 1999, which gave me the opportunity to make a significant advance in the writing of the book, to Janet Maibaum for her help in setting Microsoft Word up to the job[1] and to the team at Springer for their enthusiasm, support and advice.

Gesse,[2] April 2004 *José Luiz Fiadeiro*

[1] Yes, this book is a proof that writing about category theory is not reserved to users of a well-known typesetting system that I will not name... And this remark is not meant as a recommendation for the products developed by a company that I have already named...

[2] The little village in the French Pyrenees, by the river Aude, where this book was written and revised from 2000 until completion.

Contents

1 Introduction **1**
1.1 The Social Life of Objects 1
1.2 Categories Versus Sets 3
1.3 Overview of Typical Application Areas 5
1.4 What Can Be Found in This Book 9

Part I Basics

2 Introducing Categories **15**
2.1 Graphs 15
2.2 Categories 20
2.3 Distinguished Kinds of Morphisms 27

3 Building Categories **31**
3.1 Some Elementary Operations 31
3.2 "Adding Structure" 33
3.3 Subcategories 37
3.4 Eiffel Class Specifications 43
3.5 Temporal Specifications 46
3.6 Closure Systems 53

4 Universal Constructions **57**
4.1 Initial and Terminal Objects 58
4.2 Sums and Products 61
4.3 Pushouts and Pullbacks 67
4.4 Limits and Colimits 75

5 Functors **83**
5.1 The Social Life of Categories 83
5.2 Universal Constructions Versus Functors 89

Part II Advanced Topics

6 Functor-Based Constructions **95**
6.1 Functor-Distinguished Kinds of Categories 95
6.2 Structured Objects and Morphisms 110
6.3 Functor-Structured Categories 117
6.4 The Grothendieck Construction 124
6.5 Institutions 128

7 Adjunctions **141**
7.1 The Social Life of Functors 141
7.2 Reflective Functors 145
7.3 Adjunctions 151
7.4 Adjunctions in Institutions 160
7.5 Coordinated Categories 167

Part III Applications

8 CommUnity **177**
8.1 A Language for Program Design 177
8.2 Interconnecting Designs 182
8.3 Refining Designs 191

9 Architectural Description **197**
9.1 Motivation 197
9.2 Connectors in CommUnity 199
9.3 Examples 204
9.4 An ADL-Independent Notion of Connector 211
9.5 Adding Abstraction to Connectors 214

10 An Algebra of Connectors **221**
10.1 Three Operations on Connectors 223
10.2 Higher-Order Connectors 227

References **237**

Index **245**

1 Introduction

1.1 The Social Life of Objects

Questions that we are frequently asked are: What is category theory good for? Why should I use category theory? These questions usually indicate a genuine and healthy reaction to the proposal of a new piece of mathematics that one is invited to learn, similar to the reaction we have each time we are asked to change our eating habits and, say, replace butter with olive oil when cooking... How is this going to make us happier? Or healthier? When is the change justified?

The question also indicates that the nature and role of category theory is not completely clear to many people. The way we like to present category theory is as a toolbox similar to set theory: as a kind of mathematical lingua franca in the sense that it can be used for formalising concepts that arise in our day-to-day activity. It constitutes, however, a richer toolbox in the sense that the instruments that it provides are more sophisticated and thus make it easier to model situations that are more complex and that involve structured objects. On the other hand, because these instruments are more sophisticated than those of set theory, they require a dedicated learning effort. Briefly, in category theory one can do as much as in set theory, in an easier way when it comes to formalising and relating different notions of "structure", but at the cost of learning a few more concepts and techniques.

So, I would like to reformulate the original question as: Having been brought up to think about the world in terms of sets, why should I now change and use another frame of reference? The purpose of this book is to convince you that this change of reference is worth making, that is, that this book will have been worth reading and that the concepts and techniques that it introduces belong in the mathematical toolbox of the software engineer.

For many of "us", the key factor that justifies this change is related to the fact that, whereas concepts in set theory are typically formalised extensionally, in the sense that a set is defined by its elements, category theory provides a more implicit way of characterising objects. It does so in terms of the relationships that each object exhibits to the other objects in the universe of discourse. The best summary of the essence of category theory that I know is from the logician Jean-Yves Girard.[1] For him, category theory characterises objects in terms of their "social lives".

In my opinion, this focus on "social" aspects of object lives is exactly the reason for the applicability of category theory to computing in general, and software engineering in particular. To realise why this is so, one just needs to think that current software development methods, namely object-oriented ones, typically model the universe as a *society* of interacting objects. Agent-oriented methods are based on the same societal metaphor. This focus on interaction is not accidental; it is an attempt at tackling the increasing complexity of modern software systems. In this context, complexity does not necessarily arise from the computational or algorithmic nature of systems, but results from the fact that their behaviour can only be explained as emerging from the interconnections that are established between their components. By promoting interactions as a focal point in the definition of the structure of a system, one also brings software to the realm of natural, physical and social systems, something that seems to be essential for the development of well-integrated systems. Category theory is advocated as a good mathematical structure for this integration precisely because it focuses on relationships and interactions! The work of Goguen on general systems theory [52, 53, 64], and recent books like [94] show exactly that.

This is why many of the examples that are used throughout this book address what has become known in software engineering as software architecture [50], i.e. precisely the study of the gross modularisation principles that should allow us to design systems as possibly standard structures of smaller components and the interconnections between them. The focus that category theory puts on morphisms as structure-preserving mappings is paramount for software architectures because it is the morphisms that determine the nature of the interconnections that can be established between objects (system components). Hence, the choice of a particular category can be seen to reflect, in some sense, the choice of a particular "architectural style". Moreover, category theory provides techniques for manipulating and reasoning about system configurations represented as diagrams. As a consequence, it becomes possible to establish hierarchies of system complexity, allowing systems to be used as components of even more complex systems (i.e. to use diagrams as objects), and for inferring properties of systems from their configurations.

The ultimate conclusion that we would like the reader to draw is that high-school education could well evolve in a way that equips our future generations with tools that are more adequate for the kind of systems that they are likely to have to develop and interact with. We believe that the

[1] What stronger evidence would one need of the close relationship between logic and category theory...

teaching of mathematics could progressively shift from the set-theoretical approach that made it "modern" some decades ago to one that is centred on interactions. This is, of course, a big challenge, mainly because it is not enough for the mathematical theory to be there; the right way needs to be found for it to be transmitted. We believe that recent books such as [75] are putting us on a path towards meeting this challenge, and we hope this book makes a further contribution. However, we will be satisfied if the rationale for such a shift can somehow emerge from the way we motivate and present the basics of category theory.

1.2 Categories Versus Sets

Let us consider an example in order to make clear the difference in approaches between set theory and category theory. For this purpose, there is nothing better than showing how certain mundane set-theoretic constructions are modelled in category theory.

Consider, for instance, the characterisation of the empty set. In set theory, the empty set is characterised precisely by the property of not having any elements. There are several ways in which we can say this using a formal notation. Here is one:

$$(\forall x)x \notin \emptyset.$$

The important point here is that any characterisation requires the use of the membership relation $\in$, i.e. the characterisation is made with respect to the elements that belong to the set.

Consider now the characterisation of the empty set in category theory. As discussed above, such a characterisation involves the definition of a "social life" of sets, i.e. it has to be made relative to the way sets interact with one another. Hence, the obvious question to ask first is "What is the social life of sets"?

The first important thing to understand is that category theory does not provide an answer to questions like this one; it is up to whoever is formalising a particular domain of discourse to come up with a definition of "social life" that is convenient. Convenience here has to be measured against the formalisation activity that is being undertaken, possibly as an abstraction of real-world phenomena. Hence, it is not subject to mathematical proof. Category theory requires some basic properties of such a "social life" so that the whole mathematical machinery that we are about to describe can be applied successfully. Such basic properties are defined in Sect. 2.2. In a nutshell, they prescribe ways in which one is related to oneself and the way one's relations' relations are our own relations.

Each time, during both classes and industry-oriented tutorials, our audiences were first confronted with the need for defining a social life for sets, the immediate answer was:

Set A is related to set B iff $A \subseteq B$.

Discussing the reason why this answer comes up spontaneously is well beyond the scope of this book.[1] The characterisation of the empty set based on this social life of sets is quite easy: the empty set is the only set that is related to every other set by this particular relationship.

Prompted for another definition of social life, the following answer was given several times:

Set A is related to set B iff $A \cap B \neq \emptyset$.

According to this definition, the empty set is the only set that is not related to any set; all non-empty sets relate at least to themselves. Incidentally, we shall see in Sect. 2.2 that this definition of social life does not define a category, one of the reasons being that, in a category, every object is at least related to itself in a canonical way. Nevertheless, the example is useful for showing that changing the definition of social life may lead to quite different characterisations of the same objects.

We typically have to force the audience to come up with the "standard" definition of social life between sets:

A social relationship between a set A and a set B is given as a total function $f{:}A \to B$.

The characterisation of the empty set in this case is very similar to the first one: the empty set is such that, given any other set A, there is one, and only one, function to A – the empty function. Notice that, although concepts such as "contains", "intersection" and "function" are defined via the membership relation, the characterisation of the empty set given in the three cases does not involve it directly.

For further evidence that the nature of objects changes according to the "social life" that is of interest, consider the characterisation of singletons. According to the first definition, a singleton is such that only the empty set and the singleton itself relate to it. In the second case, a singleton is such that it relates to the sets that contain the element: $\{a\}$ relates to B iff $a \in B$. This is a good evidence for the fact that the second definition is not very "categorical": it reduces to set membership and makes the identity of the elements visible. This is also the case with the first definition: in both cases, two singletons are socially equivalent, in the sense that they relate to any other set in the same way, iff they are equal. Incidentally, a possible set-theoretic characterisation of a singleton set $\{a\}$ is:

$x \in \{a\}$ *iff* $x = a$.

[1] Nevertheless, the author is willing to collaborate in any research project that aims at understanding the psychology of modern formalism.

The use of equality between elements makes clear the need to look inside the set.

The third definition characterises singletons as sets into which there is one, and only one, total function from any other set. Indeed, the function being total, a singleton offers no choice for the mapping to be established: everything is mapped to this single element. According to this definition, two singletons cannot be distinguished because they relate in exactly the same way with the other sets. This is quite intuitive because, not being able to use set membership, we cannot look inside the singletons and notice that they have different elements. Hence this social life is a better abstraction from set membership than the other two.

The "social" way of characterising objects is, in fact, similar to the way, in object-oriented programming, the view of objects that matters for building systems is that of the methods that objects make available to the other objects to interact with. However, the view that matters for their implementation may be different, reflecting the fact that different uses reflect different structural views of the same concept (which in category theory means different categories).

Other examples could be given from branches of engineering or our day-to-day praxis. For instance, another example is the way we understand devices as mundane as hi-fi stereo systems. When assembling a hi-fi system from separate components it is important to know how each component can be connected to the others; the way they are implemented in terms of microcircuits probably explains the restrictions on the connections that can be established, but it is something that a user would like to see abstracted away. The characterisation of human behaviour can also be used as an example. There is probably a neuro-physiological justification for what we call a "shy" or "expansive" person, but the meaning that we normally attach to such features of human nature is derived from the way people interact with us. The final word, however, comes from Saint Exupéry: "The meaning of things lies not in the things themselves, but in our attitude towards them".

In brief, like all mathematics, category theory is about providing us with abstraction mechanisms. In our opinion, the fact that these mechanisms relate to interaction make category theory particularly suited for certain aspects of computing science, and software engineering in particular.

1.3 Overview of Typical Application Areas

In this section, we summarise some of the main applications areas of categorical techniques we know, with references to the literature, bearing in mind that our focus is software engineering and not computer science as a whole. We strongly suggest [26, 57] for excellent introductions to (and overviews of) the field.

1.3.1 General Systems Theory

Goguen started exploring category theory as a mathematical toolbox in the early 1970s, applying it to general systems theory [52, 53, 64]. The famous motto [57]

> "given a category of widgets, the operation of putting a system of widgets together to form a super-widget corresponds to taking a colimit of the diagram of widgets that shows how to interconnect them"

was first applied to mathematical models of system behaviour. We also find in this area the origins of latter applications to object-oriented modelling [58]. In fact, one of Goguen's early works is even called, very appropriately, "objects" [54].

But this is just one example of the many applications of category theory to the area of general systems. We would like to encourage the non-specialist to get acquainted with the field through one of the recent books that have appeared in the wider market of science, like [71, 94].

1.3.2 Algebraic Development Techniques

This is one of the most typical areas of the application of category theory to computer science. One of the earlier and most influential movements started in the second half of the 1970s with the work of a group of researchers at IBM (Joseph Goguen, Jim Thatcher, Eric Wagner and Jesse Wright), very aptly named ADJ (for *adj*unction), around the initial semantics of abstract data type specification [65]. The pioneering work of Rod Burstall and Joseph Goguen around the language CLEAR [17, 18] showed how simple universal constructions (e.g. colimits) could be used to give semantics to operations for structuring specifications (e.g. computing the sum of two specifications, parameterising a specification, etc.). See also [95, 96] for some more advanced examples and [101] for an example of tool support. The area produced textbooks such as [28, 29, 76, 92] as well as surveys [8, 26] that provide a good showcase for "categories at work", namely in what concerns elegance and economy of means.

One of the finest hours of this research programme was the development of the theory of institutions, a categorical formalisation of the notion of logic developed by Goguen and Burstall in the early 1980s [62]. The aim of this effort was to provide a mathematical framework in which the specification building operators defined for the language CLEAR could be made independent of the underlying logic and thus available for other specification formalisms. The theory of institutions is, in our opinion, one of the best examples of the usefulness of category theory as a unifier of concepts and techniques developed by different research teams in response to different or similar needs. As a science, computing is still very young and, hence, fragmented in that it is often difficult to find the "universal laws" that we recognise as the objects of study in other sciences. Through the theory of institutions, category theory was shown to be a powerful tool for abstracting from individual and uncoordinated efforts some universal laws (or "dogmas" as they are called in [57]) that apply to specifications in general [95]. This categorical framework was subsequently extended and put to use, for example, in applying algebraic specification techniques to other computational paradigms (e.g. object-oriented) and defining ways for specifications to be mapped from one formalism to another.

1.3.3 Concurrent and Object-Oriented Systems

Concurrency theory is another area to which categorical techniques have been applied in the "engineering" view that interests us in this book. This domain of application is dominated by the work of Glynn Winskel [106, 107, 108], who showed how models for concurrent system behaviour like transition systems, synchronisation trees and event structures can be formalised in category theory. The idea is that each process model is endowed with a notion of morphism that defines a category in which typical operations of process calculi are given as universal constructions. This work was continued in the 1990s in collaboration with Mogens Nielsen and Vladimiro Sassone [97], exploring the use of adjunctions as a means of translating between different models. The chapter [109] provides an excellent overview of this exciting application area.

This stream of work (including [15]) was used for giving semantics to object-oriented systems by Félix Costa, Hans-Dieter Ehrich, and Amílcar and Cristina Sernadas [20, 21, 25], also with contributions from Goguen [24]. In a nutshell, these authors showed how universal constructions can be used to express object-oriented features like encapsulation, inheritance and composition in concurrency models endowed with richer notions of state. The work of Goguen on sheaf-theoretic models [58] also establishes an interesting connection to the earlier work on general systems theory.

Applications of category theory to the (logical) specification of concurrent and object-oriented systems can also be found. These include our work in the late 1980s and early 1990s [38, 39, 40, 43]. Basically, we showed how the techniques developed for the modularisation of equational and first-order logic specifications of abstract data types could be applied to concurrent and object-oriented systems by using modal and temporal logics developed for the specification of reactive systems [13].

Briefly, the idea was to fit the specification of such systems in the "institutional" picture as set up by Goguen and Burstall [32, 33]. We later showed that our work is directly related to the General Systems tradition [31], in the sense that the module structure defined by our categorical formalisation is directly compositional over the run-time structure of the system. This led to applications of the categorical techniques to parallel program design [44], in which the notion of morphism captures what in the literature is known as superposition or superimposition [19, 48, 72].

There is also a substantial amount of work in the application to concurrent and object-oriented systems of the algebraic approach to abstract data type specification [7], most notably by Egidio Astesiano and his colleagues in Genova. Basically, these approaches present different alternative ways of bringing states and transitions to the universe of discourse of abstract data types.

1.3.4 Software Architectures

Applications of categorical techniques to the semantics of interface description and module interconnection languages were developed by Goguen in the early 1980s [55] and more recently recast in the context of the emerging interest in software architecture [59]. These applications are in the tradition of the algebraic approach to abstract data types, more specifically in what has become known as "parameterised programming" [56], in the sense that they capture functional dependencies between the modules that need to be linked to constitute a given program.

The view of architectures that is captured in this way is somewhat different from the one followed in the work of Dewayne Perry, David Garlan and other researchers who have launched software architecture as we know it today [14, 50, 99]. This more recent trend focuses instead on the organisation of the *behaviour* of systems as compositions of components ruled by protocols for communication and synchronisation. As explained in [2], this kind of organisation is founded on *interaction* in the behavioural sense, which explains why formalisms like the chemical abstract machine (or CHAM) [16] are preferred to the functional flavour of equational logic for the specification of architectural components.

This why, in the early 1990s, we proposed a categorical toolset for architectural description based on our work on the formalisation of parallel program design techniques [34, 42]. The idea is that, contrarily to most other formalisations of architectural concepts that can be found in the literature, category theory is not another semantic domain in which to formalise the description of components and connectors. Instead, through its universal constructions, it provides the very semantics of *interconnection*, *configuration*, *instantiation* and *composition*, i.e. that which is related to the gross modularisation of complex systems.

1.3.5 Service-Oriented Software Development

However, there is more to category theory than the ability to support such interaction-based views of system behaviour. By relying on "local naming" (as shown by the impossibility of distinguishing between singleton sets), category theory requires all such interactions to be modelled explicitly and outside the participating objects. This is one of the aspects that distinguishes service from object-oriented system development.

Clientship, i.e. the ability to establish client/supplier relations between objects, leads to systems of components that are too tightly coupled and rigid to support the levels of agility that are required to operate in environments that are "business time critical", namely those that make use of Web-services, Business-to-Business (B2B), Peer-to-Peer (P2P), or otherwise operate in what is known as "internet-time". Because interactions in object-oriented approaches are based on *identities*, in the sense that through clientship objects interact by invoking specific methods of specific objects (instances) to get something specific done, the resulting systems are too rigid to support such levels of agility. Any change in the collaborations that an object maintains with other objects needs to be performed at the level of the code that implements that object and, possibly, on the level of the objects with which the new collaborations are established. On the contrary, interactions in a service-oriented approach should be based only on an abstract description of what is required, thus decoupling "what one wants to be done" from "who does it and how". This is precisely the discipline of interconnection that category theory enforces.

Our last claim, in support of the last paragraph of Sect. 1.1, is that category theory is definitely the mathematics of the Internet age (and beyond)!

1.4 What Can Be Found in This Book

This book emerged from tutorials and courses given in the past few years, both in academia and in more industry-oriented fora like OOPSLA, ObjectWorld, TOOLS, ECOOP and FME. I am particularly grateful to the University of Lisbon and the Technical University of Lisbon for the opportunities that they gave me to lecture on much of the material covered in the first two parts, at both the undergraduate and postgraduate levels. These parts of the book were also covered in lectures given at the University of Coimbra, the Institute for Languages and Administration in Lisbon (ISLA) and the Federal University of Rio Grande do Sul (Brazil). Part III was used in a 20-hour course on CommUnity and software architecture that I gave as part of a postgraduate programme of the University of Pisa, as well as at the Summer School on Generic Programming (Oxford 2002). All this experience showed that there was an audience for this particular way of teaching category theory.

The book is structured in three parts, leaving room for different reading/teaching paths to be followed. With respect to most other books in the market, this one uses examples of a different nature, focusing on and giving more emphasis to aspects that are less common in other fields of computing. It also adopts a different pace altogether.

- Part I, i.e. Chaps. 2–5, covers some of the basics of category theory, including the "bare essentials" that are addressed in any book, from graphs to universal constructions and functors. However, a different emphasis and tone are used that are meant to be more appealing and accessible to software engineers. It is hoped that mathematically mature readers may appreciate a different way of exposing and illustrating these familiar concepts and constructions.
- The material included in Part II, Chaps. 6 and 7, is of a more advanced nature, not only because it is more challenging from a mathematical point of view but also because it makes appeal to an additional level of maturity in so far as computing science is concerned, namely the use of multiple formalisms as supporting complementary viewpoints. There, the reader will find material that only a few other books cover in comparable depth, with a strong emphasis on functor-based constructions like fibrations, and ending with a covering of adjunctions that deviates somewhat from the standard coverage. Again, examples are drawn from areas that have normally been confined to papers such as institutions and models of concurrency. The sections in this part should be accessible to anyone with basic knowledge of category theory, but a quick travel through Part I will help the reader become familiar with the notation and the examples that are used in Part II.
- Part III offers the chance of seeing category theory at work in a more ambitious project – giving semantics to CommUnity, a prototype language for architectural modelling. This part can be ignored by readers who are not particularly interested in the applications to software engineering. On the other hand, it can be followed, to a large extent, without the material exposed in Part II. This means that a novice to category theory but who is interested in software engineering, or anyone whose goal is to understand CommUnity or set up a course that aims at teaching (any subset of) the material in Part III, can safely skip Part II. It may be necessary to go back to Part II in order to understand in full the mathematical structures that relate to software architecture, but this may be done once the reader feels more at ease with the mathematics.

Parts I and II use examples taken directly from Meyer's book on Eiffel [86] to illustrate definitions and constructions. It is hoped that readers who are not familiar with the particular notation of Eiffel, but are used to object-oriented modelling, will be able to understand the examples without much effort. The choice for Eiffel instead of a more modern language has to do with two facts. On the one hand, it *was* modern when the book started to be written.... On the other hand, it is one of the object-oriented languages that has solid foundations, which allows it to be used to illustrate many of the mathematical constructions that we cover in the book.

The use of Eiffel allows us to illustrate applications of category theory to the more static aspects of system modelling, namely to what is related with classification (inheritance). Other examples are brought to bear from specification theory that involve logics and algebraic models for system behaviour in order to show how the more dynamic and evolutionary aspects can also be handled. These aspects come together in Part III.

This means that different reading/teaching paths can be established that are based on the examples: one can follow the "Eiffel path" for a lighter approach, perhaps more suitable for "practitioners"; or one can follow the "specification theory path" that will enable the "scientists" to visit more challenging places and reach the end of Part II. But who are the true software engineers if not the people who can combine both? Therefore, the best approach is to read the book from the beginning to the very end!

Part I

Basics

2 Introducing Categories

2.1 Graphs

A distinctive attribute of category theory as a mathematical formalism is that it is essentially graphical. This means that most concepts and properties can be defined, proved and/or reasoned about using diagrams of a formal nature. This diagrammatical nature of category theory is one aspect that makes it so applicable to software engineering. Therefore, it is not surprising that the first definition in this book is that of graphs.

2.1.1 Definition – Graphs

A graph is a tuple $<G_0,G_1,src,trg>$ where:

- G_0 is a collection[1] (of nodes).
- G_1 is a collection (of arrows).
- *src* maps each arrow to a node (the source of the node).
- *trg* maps each arrow to a node (the target of the node). ♦

We usually write $f{:}x \rightarrow y$ to indicate that $src(f)=x$ and $trg(f)=y$. Between two nodes there may exist no arrows, just one in either direction, or several arrows, possibly in both directions.

The attentive reader will have noticed that our very first definition still uses set theory, even if only informally. This may appear confusing, especially after our lengthy discussion of the merits of category theory with respect to set theory. However, we need a meta-language for talking about graphs (and categories, and), which cannot, of course, be the object language itself (i.e. that of category theory). Hence, we use the informal language that is typical of mathematics, which, as also acknowledged in the introduction, is full of set-theoretic concepts.

2.1.2 Examples – Sets and Functions

1. The most "popular" graph in this book (and in any other book we know) is the graph whose nodes are the sets and whose arrows are the total functions.

[1] Questions of "size" arise here because we shall soon be talking about the graph of graphs and constructions of a similar nature. This is why we use the term "collection" instead of "set". See, for instance, [80] for such questions.

2. To illustrate the fact that different graphs may share the same nodes, we introduce what is, perhaps, the second-most popular graph in the book – the graph whose nodes are, again, the sets, but whose arrows are the partial functions, i.e. functions that may be undefined on given elements of the source set. ♦

2.1.3 Example – Class Inheritance Hierarchies

A typical example of the use of graphs in computing is class inheritance hierarchies. These are graphs whose nodes are object classes and for which the existence of an arrow between two nodes (classes) means that the source class inherits from the target class. The following inheritance diagram is taken from [86]. It shows that class *home_business* inherits from both *business* and *residence*, each of which inherits from *house*.

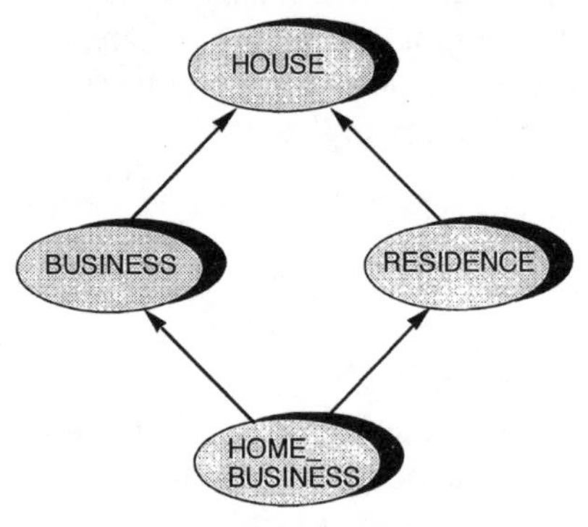

In class inheritance hierarchies, there exists at most one arrow between two nodes: either a class inherits from another or it does not. Many graphs that we encounter in software engineering are like this one. They reflect partial orderings and similar associations. However, even in the case of inheritance, arrows can carry more information. For instance, when one class inherits from another one, some renaming of the features of the original class may be required. Such renamings may be associated with the arrows of a class inheritance diagram.

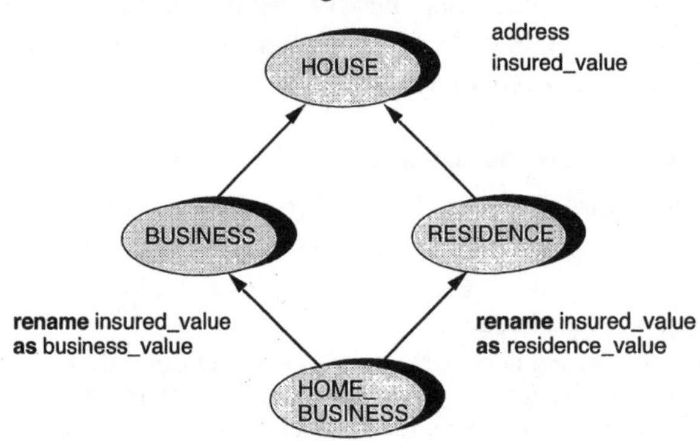

The figure above is taken, once more, from [86]. It depicts the renamings involved in the inheritance hierarchy previously used as an example. Each such enriched class inheritance diagram defines a subgraph of the graph of sets and (total) functions – classes are represented through their sets of features and the renamings through the functions that they induce. Notice that the arrows of the class inheritance hierarchy and the functions that operate the renamings point in opposite directions.

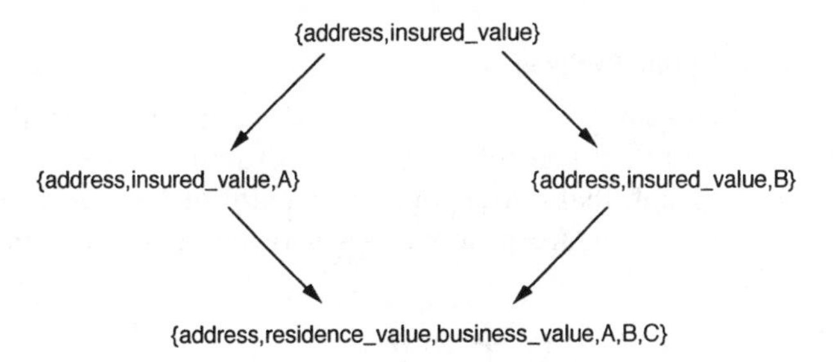

♦

2.1.4 Example – Transition Systems

Another very common example of graphs in computing is transition systems. Every transition system constitutes a graph whose nodes are the states and whose arrows are the transitions. Below we show the transition system that models the behaviour of a vending machine that accept coins and, for each accepted coin, delivers a cake or a cigar, after which it is ready to accept another coin.

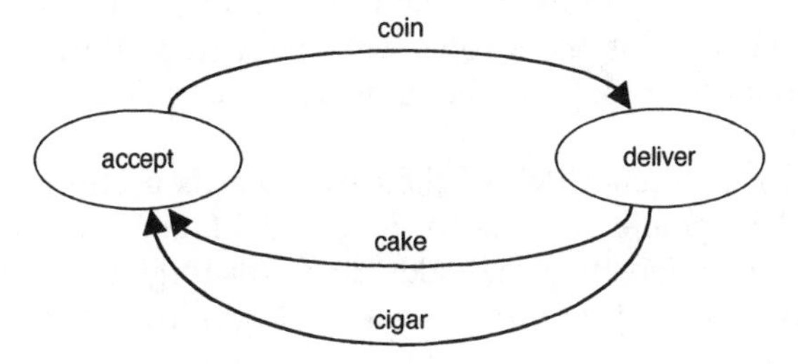

♦

As hinted in the introduction, there are many (and deep) relationships between logic and category theory. The next two examples convey some of the simplest relationships. See [22, 68, 70, 74] for advanced textbooks on this subject.

2.1.5 Example – Logical Entailment

One of the possible views that one can have of a logic is through the notion of a sentence being a consequence of, or derivable from, another sentence. This notion of consequence can be represented by a graph whose nodes are sentences and whose arrows correspond to logical entailment.

The following is an example of three nodes and two arrows of the graph that captures logic entailment in propositional logic:

$$A \wedge B \vdash A \vee B \dashv C \wedge B$$ ♦

We can add detail to entailment and distinguish between different possible ways in which a sentence can be a consequence of another, i.e. by taking arrows to be proofs.

2.1.6 Example – Proof Systems

Every proof system constitutes a graph whose nodes are sentences and whose arrows are proofs. The following is an example of two nodes and two arrows of the graph that corresponds to natural deduction in propositional logic, i.e. the nodes are propositions and the arrows are proofs in natural deduction:

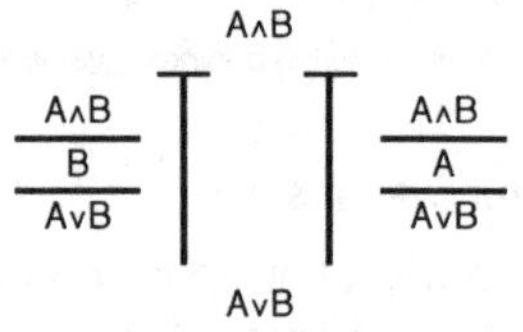

$$\begin{array}{ccc} & A \wedge B & \\ \dfrac{\dfrac{A \wedge B}{B}}{A \vee B} & \Big\downarrow \quad \Big\downarrow & \dfrac{\dfrac{A \wedge B}{A}}{A \vee B} \\ & A \vee B & \end{array}$$

♦

One should be aware of the difference between a graph and the graphical representation that is chosen for its nodes and arrows. The latter is normally chosen according to the traditional notation of the domain of application. This is why we deliberately used turnstiles for arrows above. In the case of the proof system, we could even have omitted the turnstile because the proof itself is the arrow. However, from a graphical point of view, using both the proof and the turnstile seems to make the representation more clear.

Graphs have a "social life" of their own that is useful to know about. For instance, the graphs introduced in Pars. 2.1.5 and 2.1.6 are intuitively related through an operation that adds detail (or forgets, depending on the point of view), namely proofs. Such relationships between graphs are called graph homomorphisms.

2.1.7 Definition – Graph Homomorphisms

A *homomorphism* of graphs $\varphi:G\rightarrow H$ is a pair of maps $\varphi_0:G_0\rightarrow H_0$ and $\varphi_1:G_1\rightarrow H_1$ such that for each arrow $f:x\rightarrow y$ of G we have $\varphi_1(f):\varphi_0(x)\rightarrow\varphi_0(y)$ in H. That is, nodes are mapped to nodes and arrows to arrows but preserving sources and targets. ♦

The figure below illustrates a situation in which the node component of a homomorphism (represented by the dashed arrows) is neither injective (two nodes are collapsed) nor surjective (one of the nodes of the lower graph has no counterpart above). When the node component is not injective, arrows may get mapped to endo-arrows, i.e. arrows with the same source and target, as illustrated. In the example, the arrow component (left implicit) is injective but not surjective.

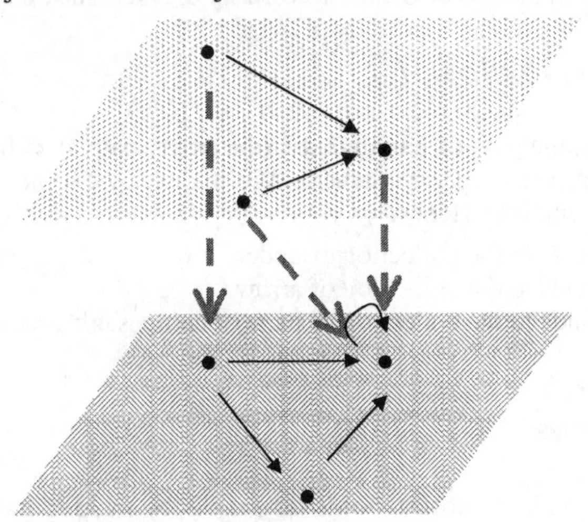

2.1.8 Example – Abstracting Entailment from Proofs

There is a "canonical" homomorphism between proof systems and consequence systems that consists of the identity on nodes and collapses any non-empty set of proof arrows between two nodes into just one consequence arrow. That is, this homomorphism "forgets" the details of the proofs, retaining just the fact that one exists to justify the consequence relation. For instance, when applied to the example in Par. 2.1.6 it would deliver just one arrow from $A \wedge B$ to $A \vee B$ as in Par. 2.1.5. ♦

Notice that the relationship between the inheritance graph used in Par. 2.1.3 and the corresponding graph of feature sets and renamings cannot be captured by a homomorphism because the source and target of arrows is reversed. Absence of a homomorphism in these circumstances seems to be accidental because the direction of the arrows is somewhat arbitrary; we could well have chosen a graph with the same collections of nodes but with the arrows reversed.

2.1.9 Definition – Duality

The dual of a graph $G=<G_0,G_1,src,trg>$ is $G^{op}=<G_0,G_1,trg,src>$, i.e. the graph obtained by reversing the direction of the arrows. A homomorphism from a graph G to the dual H^{op} of a graph H (or from G^{op} to H) is said to be *contravariant* between G and H. ♦

An essential step towards defining the notion of category concerns paths in graphs, i.e. what supports the move from direct to more "global" social relationships.

2.1.10 Definition – Paths in a Graph

Let G be a graph and x, y nodes of G. A *path* from x to y of length $k>0$ is a sequence $f_1...f_k$ of arrows of G (not necessarily distinct) such that:

- $src(f_1)=x$.
- $trg(f_i)=src(f_{i+1})$ for $1 \leq i \leq k-1$.
- $trg(f_k)=y$.

For every x, the path of length 0 at x (the empty path at x) from x to x is, by convention, the empty sequence. We denote by G_k the collection of paths of G of length k. Hence:

- G_0 corresponds to the collection of nodes.
- G_1 corresponds to the collection of arrows.
- G_2 corresponds to the collection of pairs of composable arrows. ♦

2.2 Categories

Categories provide an abstraction over graphs by making paths the basic working elements – what are called *morphisms*; paths provide richer information about "social life" than just one-to-one relationships. For that purpose, categories add to graphs an identity map that "converts" nodes to morphisms (null paths), and a composition law on morphisms that internalises path construction. Morphism composition is required to be associative as for path concatenation. This means that morphisms have no internal structure that can be derived from the order of composition.

2.2.1 Definition – Categories

A category $\boldsymbol{C}$ is a triple $\langle G,;,id \rangle$ where:

- G is a graph, often denoted by ***graph(C)***.
- ; is a map from G_2 into G_1 (called the *composition* law); for every fg in G_2, we denote by $f;g$ the arrow that results from the composition.
- id is a map from G_0 into G_1 (called the *identity* map); for every node x, we denote by id_x its identity arrow.

and, for every f in G_1, fg in G_2, and fgh in G_3:

- $src(f;g)=src(f)$ and $trg(f;g)=trg(g)$.
- $src(id_x)=trg(id_x)=x$.
- $(f;g);h=f;(g;h)$.
- If $f{:}x \rightarrow y$, $id_x;f=f;id_y=f$. ♦

The nodes (rep. arrows) of G are also called the *objects* (rep. *morphisms*) of $\boldsymbol{C}$. The collection of objects of $\boldsymbol{C}$ is denoted by $|\boldsymbol{C}|$. We often use the notation $c{:}\boldsymbol{C}$ to indicate that c is an object of $\boldsymbol{C}$ or a $\boldsymbol{C}$-object. Given $\boldsymbol{C}$-objects x and y, $hom_{\boldsymbol{C}}(x,y)$ denotes the collection of morphisms from x to y. Such sets are also called *hom-sets*.

The properties required are straightforward: the first two establish the types of the composite and identity arrows, respectively; the other two establish associativity of composition and the identity laws.

2.2.2 Example – "The" Category of Sets

In the category ***SET*** objects are sets, morphisms are (total) functions between them, composition is functional composition – i.e. $(f;g)(x)=g(f(x))$ – and the identity map assigns to every set the identity function on that set. Because function composition is associative and the identity map is both a left and right identity for function composition, all the conditions are met.

In mathematics, function composition is usually denoted by the symbol and the order of the arguments reversed, i.e. given $f{:}x\to y$ and $g{:}y\to z$, the composed function $f;g$ is also denoted by $g\ \ f$. This is the "application order": $(g\ \ f)(a)=g(f(a))$ for every $a\in x$. Most textbooks on category theory adopt this alternative notation for the composition law. Ultimately, the choice between one notation and the other is a matter of taste or convenience.[1] Our choice is motivated by the fact that it is closer to the diagrammatic notation (arrow sequencing) and hence supports the diagrammatic forms of reasoning that normally appeal to software engineers. Besides, as evidenced by the definition, the application order derives too much from set membership, i.e. reasoning with it is normally too close to the traditional set-theoretic approach from which we are trying to get away. ♦

2.2.3 Example – Graphs

The category ***GRAPH*** has graphs as objects and its morphisms are the graph homomorphisms. The composition law is defined as follows: for every pair φ and ψ of graph homomorphisms such that $src(\psi)=trg(\varphi)$, $(\varphi;\psi)_0=(\varphi_0;\psi_0)$ and $(\varphi;\psi)_1=(\varphi_1;\psi_1)$. The identity map is defined as follows: for every graph G, $(id_G)_0=id_{G_0}$ and $(id_G)_1=id_{G_1}$.

Proof

By taking graphs as nodes and graph homomorphisms as arrows, we do obtain a graph. The identity and associativity properties are inherited for each component (node and arrow) from the corresponding properties of functions between sets. ♦

2.2.4 Example – Logical Entailment

The category ***LOGI*** has as objects sentences, and morphisms correspond to the existence of a logical entailment. One can think of morphisms as "certificates" of the fact that the source sentence entails the target.

---

[1] In fact, this was one of the major decisions that had to be made before starting writing! Many people are put off reading a book because they are used to the "other" notation. Hence, ultimately, the decision was made taking into account the intended audience (and for "pedagogical" reasons as explained).

Proof

Becase there is at most one morphism between two objects, the identity and composition equations are trivially satisfied. We just have to prove the existence of endomorphisms (reflexivity) and a composition law (transitivity). But (A $\vdash$A) is a tautology, and, from (A $\vdash$B) and (B $\vdash$C), we conclude (A $\vdash$C) (cut rule). ♦

This property show that every preorder defines a category.

2.2.5 Proposition – Preorders

Every preorder $<S,\leq>$, i.e. every set equipped with a reflexive and transitive relation, defines a category $S_{\leq}$ as follows:

- $|S_{\leq}|=S$ and there is a morphism between x and y in $S_{\leq}$ iff $x\leq y$.
- Composition is defined by applying the transitivity law.

Identity morphisms are defined by applying the reflexivity laws. ♦

2.2.6 Example – Inheritance Hierarchies

In Eiffel, the relationship "ancestor" is defined as the "reflexive and transitive closure" of the inheritance hierarchy: class *A* is an ancestor of class *B* iff *A* is *B* itself or *A* is an ancestor of a parent of *B* (i.e. of a class from which *B* inherits) [86]. Given an inheritance graph *G* between classes, e.g. as in Par. 2.1.3, the category ***ancestor(G)*** is generated from the graph by completing it with the arrows that result from reflexivity (identities) and transitivity (compositions).

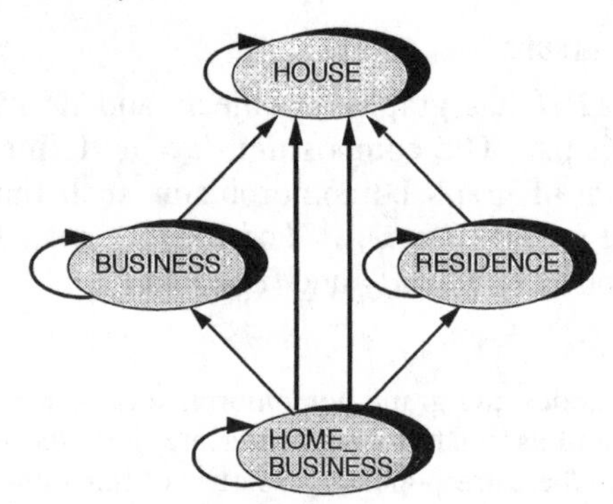

♦

The construction of ***ancestor(G)*** from *G* can be generalised as follows.

2.2.7 Proposition – Category Generated from a Graph

Every graph *G* generates a category ***cat(G)***[1] whose objects are the nodes and whose morphisms are the paths of the graph. Identities are empty paths. Composition is concatenation of paths. ♦

---

[1] The observant reader may have noted that we have already departed from our convention of using only uppercase characters for the names of categories! This is not a random aberration but reflects the fact that generating a category from a

2.2.8 Example – Runs

Every state transition system generates the category of its runs defined to be the paths of the underlying graph. ♦

2.2.9 Example – Proofs

The category ***PROOF*** has sentences as objects, and morphisms are proofs, i.e. a morphism $f:A \rightarrow B$ is a specific proof of B from A. The identity morphisms are the trivial proofs of sentences from themselves (empty proofs). Proof composition corresponds to sequence concatenation – the cut rule. ♦

Category theory supports and encourages forms of diagrammatic reasoning, also called "diagram chasing", a practice that is consistent with the modern culture in computing. Contrary to what often happens in software engineering, the notion of diagram is formal and the reasoning that can be done with diagrams is mathematical.

2.2.10 Definition – Diagrams

Let $\boldsymbol{C}$ be a category and I a graph. A *diagram* in $\boldsymbol{C}$ with *shape I* is a graph homomorphism $\delta:I \rightarrow \boldsymbol{graph(C)}$. Given a node x (resp. arrow u) of I, we normally denote its image $\delta_0(x)$ (resp. $\delta_1(u)$) by δ_x, (resp. δ_u). ♦

The fact that a diagram is being defined as a graph homomorphism and not a graph may surprise some software engineers. This is because there is sometimes confusion between the "form" and the "contents" of a diagram. The homomorphism defines a labelling of the graph I, which establishes the shape or form of the diagram. That is, a diagram in a category can be seen as a graph whose nodes are labelled with objects, and the arrows are labelled with morphisms that respect sources and targets. In particular, the labelling does not need to be injective, meaning that different nodes may be labelled with the same object.

An example can be given in terms of configuration diagrams such as those that are used in Chap. 8. In such diagrams, the graph identifies system components and interconnections between them, and the homomorphism their types. For instance, a configuration of a queue serving two independent printers can be given by the diagram below.

The fact that we are using two printers of the same type, and not only one, results from having two nodes labelled *PRINTER*. The same applies to the cables: each printer is connected to the extension cord by its own cable, but they are of the same type. Therefore, we have two nodes labelled with *i-cable*; these cables connect to the same "printer port", so the two arrows to *PRINTER* are labelled with the same morphism *get_job*. However, the cables connect to different ports of *EXTENSION CORD*, hence the corresponding morphisms are different.

---

graph is a map from one category to another. Such maps, called functors, are introduced in Chap. 5 and are denoted using bold, lowercase characters.

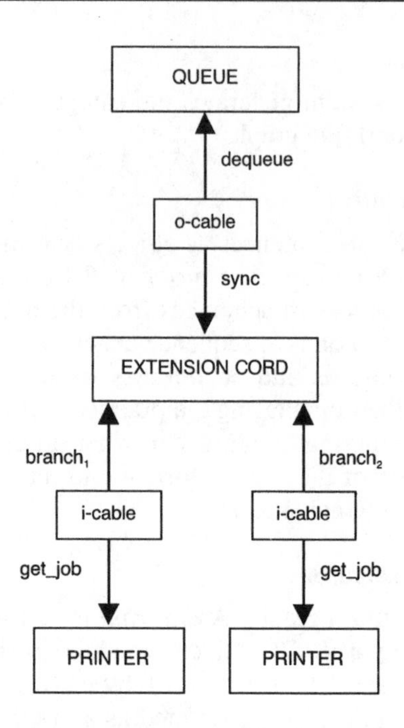

Whereas objects and morphisms provide the main elements of the "vocabulary" that one uses in category theory, diagrams provide the basic "sentences" that one can build to express concepts. In turn, most properties that one can assert in this "language" are expressed in terms of commutative diagrams.

2.2.11 Definition – Commutative Diagrams

Let C be a category and I a graph. A diagram $\delta{:}I\rightarrow\boldsymbol{graph(C)}$ is said to commute iff, for every pair x,y of nodes and every pair of paths $w=u_1...u_m$, $w'=v_1...v_n$ from x to y in graph I,

$$\delta_{u_1};...;\delta_{u_m}=\delta_{v_1};...;\delta_{v_n}$$

holds in $\boldsymbol{C}$. ♦

The property that a diagram commutes establishes a set of equalities between arrows. Hence, diagrams and commutativity provide us with the ability of doing equational reasoning in a visual form, an advantage that has not been fully exploited yet in software engineering. See [49] for a more developed use of these possibilities for mathematical reasoning. To indicate that a diagram commutes, we decorate it with the symbol ◎.

We close this section with an example that is very typical of the applications of category theory to computer science: automata. See [27] for an extended account. This example provides a good illustration of how morphisms "preserve structure" and how diagram-chasing facilitates reasoning about categories.

2.2.12 Example – Automata

A (deterministic) automaton consists of an input set X, a state set S, an output set Y, a transition function $f: X{\times}S \rightarrow S$, an initial state $s_0 \in S$ and an output function $g: S \rightarrow Y$. A notion of morphism of automata must be able to capture this structure by indicating how a given automaton can be simulated by another. It must translate the input, state and output sets from one automaton to the other in such a way that the transition functions, the initial states and the output functions of the two automata "agree".

More concretely, we should require of a morphism $A \rightarrow A'$ that:

1. The initial state of A be mapped to the initial state of A'.
2. If we perform a transition in A and map the resulting state into A', we get the state that is obtained by first translating the input and initial state into A' and then applying the transition function of A'.
3. The translation into A' of the output of a state of A consists of the output in A' of the translation into A' of the original state of A.

Hence, a morphism from $A = <X,S,Y,s_0,f,g>$ to $A' = <X',S',Y',s'_0,f',g'>$ is defined to consist of three functions $<h:X \rightarrow X', i:S \rightarrow S', j:Y \rightarrow Y'>$ such that

1. $i(s_0)=s'_0$.
2. $f;i=h{\times}i;f'$.
3. $g;j=i;g'$.

Notice that the equations in points 2 and 3 can be expressed as requests for certain diagrams to commute. For the equation in point 2,

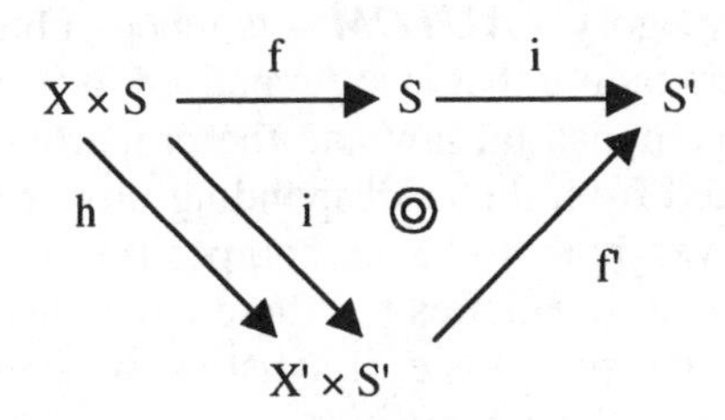

For the equation in point 3,

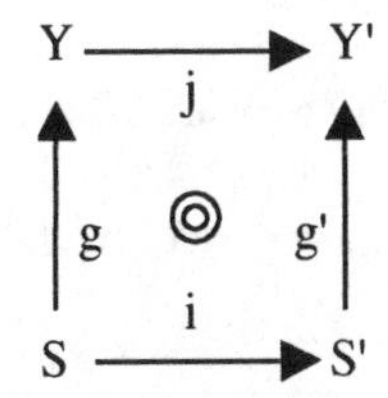

What about the equation in point 1? Can it be expressed as a commutative diagram as well? The answer seems to be less obvious or immediate because the equation is very "set-theoretical": it is an equality between elements (initial states), not functions. However, we have already mentioned in Chap. 1 that an element $s_0 \in S$ can be identified with a morphism $s_0{:}\{\bullet\} \to S$ from an arbitrary singleton set – an observation that we shall formalise in Sect. 4.1, but which can remain at an intuitive level for the purpose of this discussion. Given this, the first equation can be rewritten as $s_0;i=s'_0$ and expressed via:

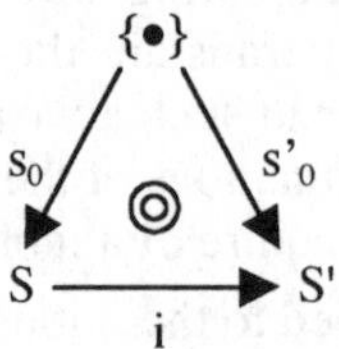

We should make clear that the choice between using diagrams or text should be made according to the intended readership or usage. The point that we want to make here is that it is *possible* to use diagrams to express and reason about properties in a formal way, and that there is some *unified* or *universal* side to this practice. However, we shall abstain from giving recommendations as to when and for whom diagrammatic notations should be used.

The composition law of automata is defined to be that of functions between sets, i.e. composition applies internally to each of the three components of the morphisms. Likewise, the identity for an automaton A is defined to be the triple that consists of the three identity functions over the sets of input, states and outputs.

The proof that a category – ***AUTOM*** – is indeed obtained in this way is very revealing of a procedure that is systematised in Sect. 3.2. Because the associativity of the composition law and the properties of the identity are automatically inherited from the corresponding properties of functions, all that needs to be proved is that (1) the composition of two morphisms is, indeed, a morphism (i.e. it satisfies the three equations above) and (2) the identity is, indeed, a morphism (i.e. it satisfies the same three equations). We prove just the case of the composition

$$<h,i,j>;<h',i',j'>=<h;h',i;i',j;j'>$$

of morphisms

$$<X,S,Y,s_0,f,g> \to <X',S',Y',s'_0,f',g'>$$

and

$$<X',S',Y',s'_0,f',g'> \to <X'',S'',Y'',s''_0,f'',g''>$$

1. $(i;i')(s_0)=i'(i(s_0))=i'(s'_0)=s''_0$.
2. $f;(i;i')$
   $=(f;i);i'$
   $=(h \times i;f');i'$
   $=h \times i;(f';i')$
   $=h \times i;(h' \times i';f'')$

$=(h{\times}i;h'{\times}i');f''$
$=(h;h)'{\times}(i;i');f''$.

3. $g;(j;j')=(g;j);j'=(i;g');j'=i;(g';j')=i;(i';g'')=(i;i');g''$.

This proof is best understood in terms of the following commutative diagram(s):

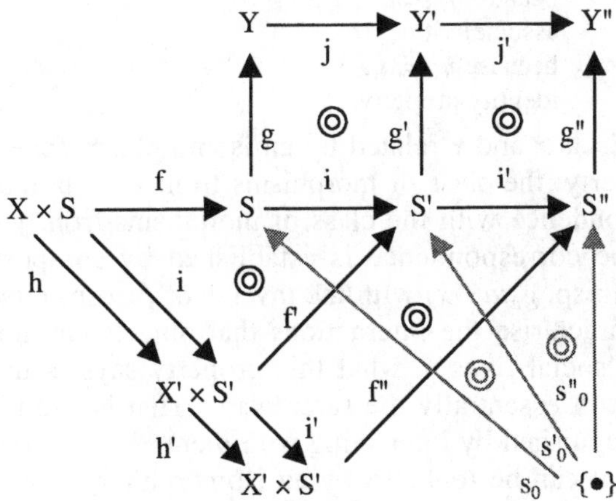

◆

2.3 Distinguished Kinds of Morphisms

There are several classes of morphisms that have special properties worth knowing about because they allow us to recognise situations in which standard results or constructions apply.

2.3.1 Definition/Proposition – Isomorphisms

Let ***C*** be a category and x,y objects of ***C***. A morphism $f{:}x{\to}y$ of ***C*** is an *isomorphism* iff there is a morphism $g{:}y{\to}x$ of ***C*** such that: $f;g=id_x$ and $g;f=id_y$. In these conditions, x and y are said to be isomorphic. ◆

2.3.2 Example

1. In ***SET***, a morphism is an isomorphism iff it is bijective.
2. In ***LOGI***, sentences are isomorphic iff they are logically equivalent. ◆

2.3.3 Exercise

What about isomorphic objects in ***PROOF***? ◆

2.3.4 Definition/Proposition – Inverses

Given an isomorphism $f{:}x{\to}y$ there is a unique morphism $g{:}y{\to}x$ such that: $f;g=id_x$ and $g;f=id_y$. This morphism is called the *inverse* of f.

Proof

The morphism g mentioned in the definition above is unique: given $h{:}y{\rightarrow}x$ in the same circumstances, we have

h

$= h;id_x$	identity property,
$= h;(f;g)$	because $f;g=id_x$,
$= (h;f);g$	associativity,
$= id_y;g$	because $h;f=id_y$,
$= g$	identity property.

◆

Any two objects x and y related by an isomorphism $f{:}x{\rightarrow}y$ have a very important property: the class of morphisms from x (resp. into x) is in one-to-one correspondence with the class of morphisms from y (resp. into y). This one-to-one correspondence is established by composing any morphism $f_a{:}x{\rightarrow}a$ (resp. $g_a{:}a{\rightarrow}x$) with the inverse of f (resp. f itself). Because morphisms characterise the interactions that objects can hold with other objects (their "social lives"), what this property says is that isomorphic objects interact in essentially the same way. That is, isomorphic objects cannot be distinguished by interacting with them. In any context where an object is used, it can be replaced by an isomorphic one by using the isomorphism and its inverse to re-establish the interconnections: any incoming arrow

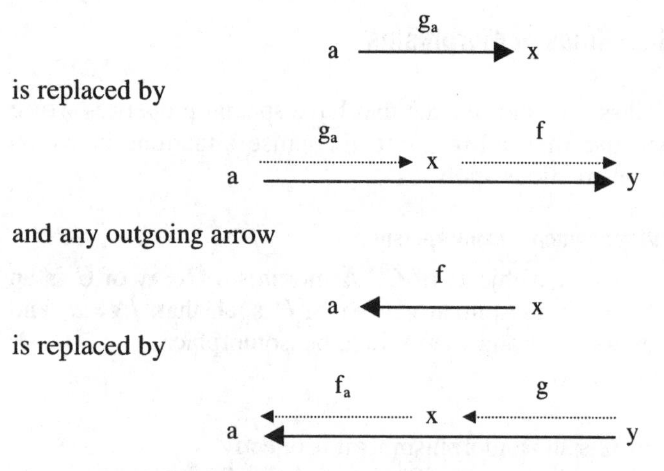

is replaced by

and any outgoing arrow

is replaced by

Hence, it is usual to treat any two isomorphic objects as being "essentially" the same. Again, we should point out that this property holds in so far as the structure ("social life") that is captured by the category is concerned. For a different category over the same objects, revealing other structural aspects, the isomorphism may not carry through. For instance, in object-oriented software development, two objects may be isomorphic in so far as having the same interfaces and exhibiting the same behaviour at their interfaces, but may have completely different implementations.

There are other classes of morphisms that are important to mention, namely those that generalise well-known properties in set theory, like injective and surjective functions. The way this generalisation is made reveals a lot of the way category theory operates and how it relates to and differs from set theory. For instance, the typical characterisation of an injective function $f{:}x \rightarrow y$ in set theory is:

For every $a,b \in x$, $f(a)=f(b)$ *implies* $a=b$.

As expected, this characterisation uses set membership. As motivated already, in category theory we have to replace it by interactions (morphisms). We have already mentioned that elements $a \in x$ of sets can be identified with morphisms $a{:}\{\bullet\} \rightarrow x$ from singleton sets. Hence, the set-theoretic definition amounts to saying:

For any (total) functions $a,b{:}\{\bullet\} \rightarrow x$, $a;f=b;f$ *implies* $a=b$.

$$\{\bullet\} \underset{b}{\overset{a}{\rightrightarrows}} x \xrightarrow{f} y$$

When considered in this way, an injective function can be characterised as not interfering with independent observations that are made on the source. The categorical characterisation consists precisely in generalising this definition to arbitrary morphisms instead of just elements: for every pair of morphisms $g,h{:}z \rightarrow x$, $g;f=h;f$ implies $g=h$.

$$z \underset{h}{\overset{g}{\rightrightarrows}} x \xrightarrow{f} y$$

What is even more interesting is the fact that the generalisation carries through, in a dual way that we will formalise in the next chapter, to the characterisation of surjective functions. Surjectivity is all about interference with the social life of the target of the arrow. If we take a non-surjective function $f{:}A \rightarrow B$ and $b \notin f(A)$, then we can have two functions $g,h{:}B \rightarrow C$ that coincide in $f(A)$, i.e. such that $f;g=f;h$, and yet are different because they disagree on the way they map b.

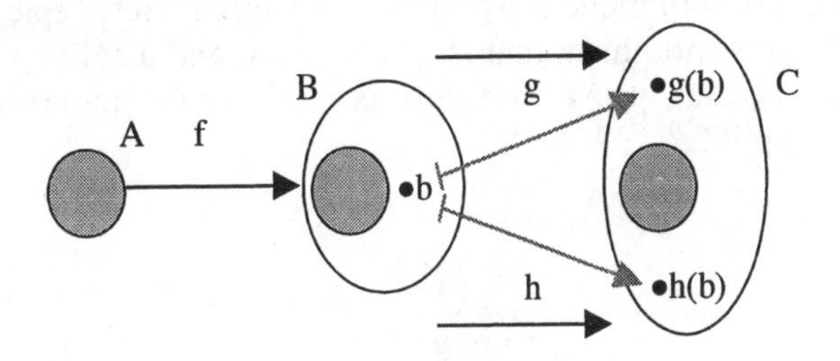

2.3.5 Definition – Monos and Epis

Consider an arbitrary category $\mathbf{C}$ and morphism $f:x\to y$ in $\mathbf{C}$.

1. f is a *monomorphism*, or a mono, or monic, iff, for every pair of morphisms $g,h:z\to x$, $g;f=h;f$ implies $g=h$.
2. f is an *epimorphism*, or an epi, or epic, iff, for every pair of morphisms $g,h:y\to z$, $f;g=f;h$ implies $g=h$. ♦

Monos and epis satisfy many of the properties that we know from injective and surjective functions. They are left here as exercises.

2.3.6 Exercises

Consider an arbitrary category $\mathbf{C}$.

1. Prove that isomorphisms are both epic and monic.
2. Prove that the composition of monos (resp. epis) is also a mono (resp. epi).
3. Prove that if $f;g$ is monic (resp. epic), then so is f (resp. g). ♦

A final word of caution though. The analogy with set theory (well exploited in [4]) cannot be carried too far: the converse of the property referred to in the first exercise does not hold for arbitrary categories! That is, not all morphisms that are both epic and monic are isomorphisms. This is because, to be an isomorphism, an arrow needs to have a left and a right inverse, which, contrary to what happens in set theory, is not guaranteed simply by being monic and epic.

2.3.7 Definition – Split Monos and Epis

Consider an arbitrary category $\mathbf{C}$ and morphism $f:x\to y$ in $\mathbf{C}$.

1. f is a *split* monomorphism, or a split mono, or split monic, iff it admits a right inverse, i.e. iff there is $g:y\to x$ such that $f;g=id_x$.
2. f is a split epimorphism, or a split epi, or split epic, iff it admits a left inverse, i.e. iff there is $g:y\to x$ such that $g;f=id_y$. ♦

2.3.8 Exercises

Consider an arbitrary category $\mathbf{C}$.

1. Prove that every split mono (resp. split epi) is monic (resp. epic).
2. Prove that every morphism that is a split mono and a split epi is an isomorphism. In fact, prove that only one, either the mono or the epi, needs to be "split". ♦

3 Building Categories

There are many ways in which new categories can be built from existing ones. One of the advantages of building new categories by "inheriting" from old ones is that it is easier to prove that the construction yields, indeed, a category. This chapter presents some elementary constructions that illustrate the point, together with examples of categories that relate to software development.

3.1 Some Elementary Operations

There are a number of elementary operations for constructing categories. The first example reflects the fact that the choice for the direction of the morphisms has no essential significance in the sense that, if we reverse the direction of all the morphisms, we obtain a category that has the same structural properties as the original one.

This fact does not mean, however, that the direction of the morphisms should not be carefully chosen when defining a category. It is often the case that there is a "natural" direction for morphisms, either given by the mathematical nature of the morphisms or by their use in practice. If reversed, the direction of the morphisms may make it more difficult or less immediate to understand the intended meaning and the particular constructions one may want to define on objects and morphisms. Hence, if you are going to become a "practising category theorist", be prepared to answer questions like "but aren't the arrows going the wrong way?" which, many times, just means "you do not seem to belong to my club!"...

3.1.1 Definition – Dual of a Category

For any category $\boldsymbol{C}$ we can construct its *dual* or *opposite* $\boldsymbol{C}^{op}$:

- $\boldsymbol{C}^{op}$ has the same objects and arrows as $\boldsymbol{C}$.
- The arrows of $\boldsymbol{C}^{op}$ go in the reverse direction: if $f{:}A{\rightarrow}B$ in $\boldsymbol{C}$ then $f{:}B{\rightarrow}A$ in $\boldsymbol{C}^{op}$.
- Arrow composition is in the reverse direction: $f;g$ in $\boldsymbol{C}^{op}$ is $g;f$ in $\boldsymbol{C}$. ♦

3.1.2 Example

Following [86], the dual of the category ***ancestor(G)*** can be named ***descendant(G)***. ♦

The fact that every category and its opposite have the same structural properties reflects a general *duality principle* that applies to all definitions and results. Every concept in category theory has a dual, which is the concept that is obtained by reversing the direction of the arrows, i.e. the concept that holds in the dual category. Every result in category theory has a dual that holds for the dual concepts. We shall have plenty of occasions to illustrate how the duality principle is applied in practice.

Another elementary construction consists in taking the product of two categories:

3.1.3 Definition/Proposition – Product Category

Given categories $\boldsymbol{C}$ and $\boldsymbol{D}$, their *product* $\boldsymbol{C}\times\boldsymbol{D}$ is such that its objects are the pairs $<c,d>$, where $c{:}\boldsymbol{C}$ and $d{:}\boldsymbol{D}$, and the morphisms $<c,d>\to<c',d'>$ are all the pairs $<f,g>$, where $f{:}c\to c'$ in $\boldsymbol{C}$ and $g{:}d\to d'$ in $\boldsymbol{D}$. Composition of morphisms is defined componentwise, and the identity for an object $<c,d>$ is the pair $<id_C,id_D>$.

Proof

The properties of composition and the identities are trivially inherited from the corresponding properties of the components. ♦

There are also ways of extracting categories from other mathematical structures. For instance, we mentioned in Par. 2.2.5 that every preorder defines a category. A simpler construction views sets as categories.

3.1.4 Definition/Proposition – Discrete Categories

Every set S determines a category whose objects are the elements of S and whose morphisms are just the identities, i.e. for every $s\in S$ $hom(s,s)$ is a singleton (consisting of the identity for s), and $hom(s,s')$ is empty for every $s,s'\in S$ such that $s\neq s'$. Such categories are said to be *discrete*.

Proof

Immediate. Note that, if we take S to be the set of nodes of a graph with no arrows, then this construction is a special case of Par. 2.2.7. ♦

The discrete category defined by a set may have many objects, as many as the elements of the set, but has only the minimum number of morphisms – the identities. The next example illustrates the other extreme: many morphisms, but only one object. These categories correspond to monoids.

3.1.5 Definition/Proposition – Monoids as Categories

A monoid is a triple $<M,*,1>$ where M is a set, $*$ is an associative binary operation on M, and 1 is an identity for $*$, i.e. $m*1=1*m=m$ for every

$m \in M$. Any monoid defines a category that consists of only one object, which we can denote by "•", and whose morphisms $\bullet \to \bullet$ are the elements of M. Composition of morphisms is given by the operation $*$ and the identity morphism is 1.

Proof

The properties of composition and the identity are trivially inherited from the corresponding properties of the monoid. ♦

3.1.6 Exercise

Characterise the duals of the categories defined by monoids. Are these categories their own duals? ♦

3.2 "Adding Structure"

The most typical way of building a new category is, perhaps, by adding structure to the objects of a given category (or a subset thereof). The expression "adding structure" has a broad meaning that the reader will only fully apprehend after building a few categories. The morphisms of the new category are then the morphisms of the old category that "preserve" the additional structure.

The following example is as typical as any other and is used throughout the book for applications related to systems modelling.

3.2.1 Example – Pointed Sets

The category $\boldsymbol{SET}_\perp$ (of pointed sets) is defined as follows. Its objects are the pairs $<A,\perp_A>$, where A is a set and $\perp_A$ is an element of A called the designated element. The morphisms between two pointed sets $<A,\perp_A>$ and $<B,\perp_B>$ are the total functions $f{:}A \to B$ such that $f(\perp_A)=\perp_B$.

The category over which $\boldsymbol{SET}_\perp$ is being built is, obviously, $\boldsymbol{SET}$, the category of sets and total functions. The additional structure that is being added to the objects of the base category, sets, is the designation of an element (not the element itself because it is already in the set). The morphisms of the new category are all the morphisms of the base category that preserve the additional structure, which in this case means preserving the designation, i.e. mapping the designated element of the source to the designated element of the target.

This way of building new categories makes it easier to conduct proofs as illustrated next.

Proof

The attentive reader will have noticed that the definition of $\boldsymbol{SET}_\perp$ omitted two fundamental ingredients: the definition of the composition law and of the identity map. This is because they are, by default, inherited from the base category. Because the laws of composition and identity are satisfied in the base category, they do not need to be checked again for the new category. Hence, all that needs to be checked is that (1) the composition law is closed for $\boldsymbol{SET}_\perp$, i.e. that the composition of two functions that are morphisms in $\boldsymbol{SET}_\perp$ is also a morphism of $\boldsymbol{SET}_\perp$, and (2) that the identities are, indeed, morphisms in $\boldsymbol{SET}_\perp$:

1. Given morphisms $f{:}\langle A,\perp_A\rangle \to \langle B,\perp_B\rangle$ and $g{:}\langle B,\perp_B\rangle \to \langle C,\perp_C\rangle$, we have to check that $(f;g)(\perp_A)=\perp_C$. But $f(\perp_A)=\perp_B$ because f is a morphism of pointed sets. Hence, $(f;g)(\perp_A)=g(f(\perp_A))=g(\perp_B)$. On the other hand, because g is also a morphism of pointed sets, $g(\perp_B)=\perp_C$.
2. Given a pointed set $\langle A,\perp_A\rangle$, we now have to prove that the identity map id_A for A as a set is a morphism of pointed sets, i.e. is such that $id_A(\perp_A)=\perp_A$. But this is true because id_A is precisely the identity function on A. ◆

Notice the similarities between the structure of this proof and the proof outline given for ***AUTOM*** in Par. 2.2.12. This style of proof is typical of categories built from other categories by "adding structure".

The particular use of pointed sets that we have in mind is for modelling the behaviour of objects at their interfaces (methods). A very abstract and general model of object behaviour is one in which the events that can potentially occur during the lifetime of an object consist of all possible subsets of the set of its methods. This means that we work in a model in which every object can potentially handle concurrent method calls. In Chap. 4 we study how we can model restrictions to the degree of concurrency that an object can impose on its methods.

The empty set represents an event in which the object is not involved. This explicit but abstract representation of environment events has been shown to be very important for modelling the behaviour of concurrent and distributed systems [13].

The fact that the events that we are using for modelling object behaviour consist of sets of method calls can be abstracted away to recognise a more general notion of process alphabet as used in concurrency. The application of categorical techniques in concurrency theory is particularly rich and revealing of the power of category theory for systematising constructions and establishing relationships between different models. The application to object behaviour that we will use for illustration purposes throughout the book touches some of these aspects, but the reader interested in a more complete picture of the breadth of the field should consult [21, 24, 97, 106]. Specific applications of these categorical techniques to object-oriented modelling can be found in [21, 24].

In this more general model of process behaviour, a process alphabet is a pointed set. Each element of the set represents an event whose occurrence may be witnessed during the lifetime of the process. The designated element of the set represents an event of the environment, i.e. an event in which the process is not involved. The notion of morphism in this model captures the relationship that exists between systems and components of systems. More precisely, every morphism identifies the way in which the target is embedded, as a component, in the source. The fact that the morphism is a function between the alphabets is somewhat intuitive: for every event *a* occurring in the lifetime of the system, we have to know how the component participates in that event; if the component is not involved in *a*, then the morphism should map *a* to the designated element of the alphabet of the component, i.e. to the element that has been designated to represent the events in which the component does not participate. On the other hand, any event occurring in the environment without the participation of the system cannot involve the component either; hence, every morphism needs to preserve the designated element.

For instance, consider a producer–consumer system. We can assign the events *produce* and *store* to the producer and *consume* and *retrieve* to the consumer. An event of the system during which both *produce* and *consume* take place concurrently is mapped to *produce* by the morphism *prod: system→producer* that identifies the producer as a component of the system. On the other hand, an event of the system during which the consumer executes *retrieve* and the producer remains idle is mapped by *prod* to the designated element of the alphabet of the producer.

Our last example of constructing new categories from base ones is also very typical. It provides so-called *comma categories*. Several different notations can be found in the literature for comma categories. We shall use the same notation as [80]. The particular form of comma categories that we are about to define are also called over/under-cone categories in [22] and (co)slice categories in [49].

3.2.2 Example – Comma Categories

Given a category ***C*** and an object *a*:***C***, we define the category of objects under *a*, denoted $a\downarrow C$, as follows. Its objects are all the pairs <*f*,*x*> where *f* is a morphism of the form *f:a→x* in ***C***. The morphisms between *f:a→x* and *g:a→y* are all the ***C***-morphisms *h:x→y* such that *f;h=g*.

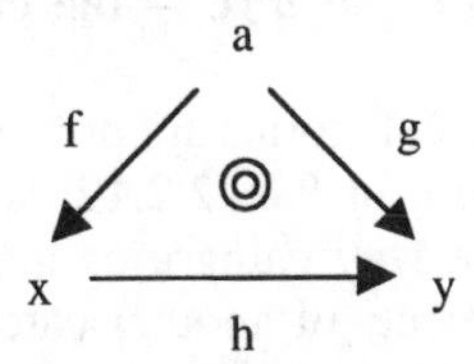

Proof:

In this case, the additional structure that is being added to every object *x*:***C*** is a ***C***-morphism *f:a→x*. Again, we inherit the composition law and the identity map from the base category and prove only the closure properties.

This is also a good example to illustrate how reasoning in category theory is often done at a graphical level. Because objects are now morphisms of the base category, it is more convenient to represent a morphism h between $f{:}a \to x$ and $g{:}a \to y$ by a diagram. As already mentioned, the symbol ◎ indicates that the diagram commutes, i.e. all compositions of morphisms in the diagram along paths that have the same source and target are equal. In the case above, commutativity of the diagram expresses the equation $f;h=g$.

Given composable morphisms i and j

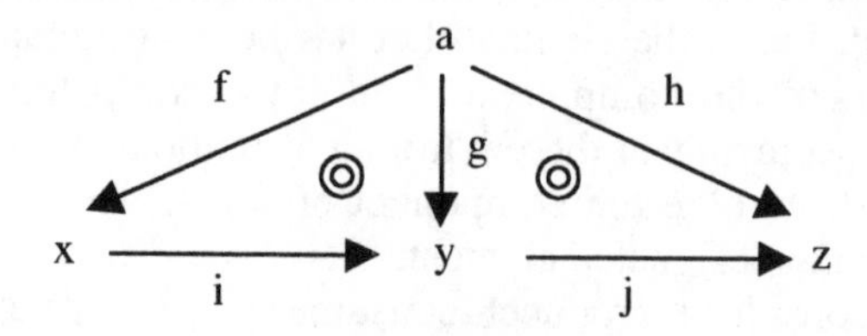

we have to prove that the composition $(i;j)$ in $\boldsymbol{C}$ is indeed a morphism in $a{\downarrow}\boldsymbol{C}$, i.e. $f;(i;j)=h$. There is another way of asserting the property that needs to be proved which makes appeal to the kind of diagrammatic reasoning that we have claimed to be typical of category theory: we have to prove that, for any diagram of the form above, the outer triangle commutes if the inner triangles also commute. This can be done as follows:

1. $f;i=g$ because i is a morphism in $a{\downarrow}\boldsymbol{C}$,
2. $(f;i);j=g;j$ from (1),
3. $g;j=h$ because j is a morphism in $a{\downarrow}\boldsymbol{C}$,
4. $(f;i);j=h$ from (2) and (3),
5. $f;(i;j)=(f;i);j$ from the associativity of the composition law in $\boldsymbol{C}$,
6. $f;(i;j)=h$ from (4) and (5).

The fact that id_x is an identity for $f{:}a \to x$ results from the property $f;id_x=id_x$. ♦

For instance, when applied to a category of the form ***ancestor****(G)*, this construction returns, for every class c, the classes that inherit from c, themselves organised as a hierarchy. When applied to ***PROOF*** it returns, for every sentence, the inferences that can be made from that sentence.

3.2.3 Exercise

Characterise the dual category of $a{\downarrow}\boldsymbol{C}$ – the category of objects over a, usually denoted by $\boldsymbol{C}{\downarrow}a$. ♦

Generalisations of comma categories are presented in Sect. 6.1.

Together with the example in Par. 2.2.12, we hope that we have provided enough insight into a way categories are often defined, that is by "adding structure" to the objects of another category. Because there is no abstract characterisation of this technique, we have relied on the examples to make apparent that there is a systematic procedure for checking the definition of a category built in this way.

3.3 Subcategories

There is another intuitive, and often useful, way of building new categories from old: by forgetting some of the objects and some of the morphisms to create a *subcategory* of the original one. The only proof burden associated with this method is in making sure that we do not through away too much (or too little).

3.3.1 Definition – Subcategories

Given categories $C=<G_C,;_C,id_C>$ and $D=<G_D,;_D,id_D>$, we say that D is a *subcategory* of C iff

- $|D|\subseteq|C|$, i.e. every object of D is an object of C.
- For any objects x and y of D, $hom_D(x,y)\subseteq hom_C(x,y)$, i.e. the morphisms in D are also morphisms in C.
- The composition laws and the identity maps in the two categories agree, i.e. given composable morphisms f and g of D, their composition $f;_Dg$ in D is the same as their composition $f;_Cg$ in C, and for every object x of D, its identity $id_D(x)$ in D is the same as $id_C(x)$ in C. ♦

Notice that, for D to be a subcategory of C, it is not enough to have inclusions between the sets of objects and of morphisms; the additional structure given by the identities and the composition law has to be preserved. Examples of subcategories are given as we go along, most of the time as exercises.

3.3.2 Examples

1. By keeping just the sets that are finite and all the total functions between them, we define a subcategory ***fSET*** of ***SET***.
2. By keeping all sets but just the functions that are injective, we define a subcategory ***INJ*** of ***SET***. This is because identities are injective and the composition of injective functions is still injective.
3. Given any category, we can forget all its morphisms except for the identities and obtain a (discrete) subcategory.
4. We can extend the morphisms of ***SET*** to include all partial functions between sets. Because the identity function is trivially partial and partial functions compose in the same way as total functions, we obtain a category of which ***SET*** is a subcategory. The category of partial functions is called ***PAR***.
5. When discussing pointed sets, we have motivated the notion of alphabet of object behaviour by identifying events with sets of methods. That is, we have worked with pointed sets that consist of powersets, the empty set being the designated element of each powerset. We can show that we define a subcategory of $SET_\perp$ by choosing as morphisms $2^B\rightarrow 2^A$

maps that compute inverse images for functions $A \rightarrow B$, i.e. a morphism $g:2^B \rightarrow 2^A$ is such that, for some $f:A \rightarrow B$, $g(B')=f^{-1}(B')$ for every $B' \subseteq B$. We call this category ***POWER***. ♦

The cases in which we throw away some of the objects but keep all the morphisms between those that remain, like we did for finite sets, deserve a special designation:

3.3.3 Definition – Full Subcategories

Given categories ***C*** and ***D*** we say that ***D*** is a *full* subcategory of ***C*** iff ***D*** is a subcategory of ***C*** and, for any ***D***-objects x and y, $hom_D(x,y)=hom_C(x,y)$. ♦

3.3.4 Exercise

Check which of the subcategories defined in Par. 3.3.2 are full. ♦

We have mentioned several times that, intensionally, a category captures a certain notion of structure, which we called a "social life" for its objects. It is only natural to expect that, when selecting a subcategory, we obtain a different but, nevertheless related, notion of structure. For instance, if ***D*** is a subcategory of ***C***, two objects that are isomorphic in ***C*** are not necessarily isomorphic in ***D***. A trivial illustration of this fact can be obtained by realising that the discrete subcategory of ***C*** that is obtained by forgetting all the morphisms except the identities (see Par. 3.3.2) is such that any object is only isomorphic to itself (it has no social life). Indeed, the more structure a category provides through its morphisms, the more observational power we have over its objects and, hence, the more isomorphisms we are able to establish. The reverse property, however, holds and is left as an exercise.

3.3.5 Exercise

Let ***D*** be a subcategory of ***C***. Show that any two objects that are isomorphic in ***D*** are necessarily isomorphic in ***C***. ♦

The relationship between isomorphisms and subcategories is, in fact, very revealing of the way categories can be used to capture notions of structure. Recall that a subcategory is full iff it is obtained by selecting a subclass of the objects but leaving the morphisms between the selected objects unchanged. Typically, this happens when one wants to select just the objects that satisfy a certain property, e.g. sets that are finite as in the example in Par. 3.3.2. In this case, the converse of the property proved in the exercise above also holds. This is because, by not interfering with the morphisms, we are keeping the structure of the objects intact.

3.3.6 Exercise

Let ***D*** be a full subcategory of ***C***. Show that any two ***D***-objects that are isomorphic in ***C*** are also isomorphic in ***D***. ♦

In the cases where, like finiteness of sets, the discriminating property does not distinguish between isomorphic objects, we obtain a subcategory that is not only full but contains all objects that are isomorphic, i.e. share the same substructure.

3.3.7 Definition – Isomorphism-Closed Full Subcategories

A full subcategory ***D*** of a category ***C*** is said to be *isomorphism-closed* iff every ***C***-object that is isomorphic to a ***D***-object is also a ***D***-object. ♦

So far, we have looked at the formation of a subcategory as the result of a selection of special objects and morphisms from a given category. In certain circumstances, this selection is actually the result of a process of "abstraction", i.e. certain objects are selected because they are "canonical" with respect to a set of objects that share a given property.

As an example, consider automata as defined in Par. 2.2.12. Intuitively, only the states that are reachable, in the sense that they can be reached through the transition function from the initial state and some input sequence, are important for determining the "social life" of automata in terms of their ability to simulate other automata. Consider then the full subcategory ***REACH*** of ***AUTOM*** that consists of all reachable automata. The idea that, for the purposes of simulating other automata, every automaton can be represented by the automaton that is obtained by removing all non-reachable states can be formalised as follows:

- Every automaton A is related to a "canonical" reachable automaton R through a morphism $c{:}R \to A$. More concretely, if $A=<X,S,Y,s_0,f,g>$, then $R=<X,S_R,Y,s_0,f_R,g_R>$ where S_R is the subset of S consisting of the states that are reachable, and f_R (resp. g_R) is the restriction of f (resp. g) to S_R. Notice that R is well-defined because f returns reachable states when applied to reachable states. The morphism c consists of the inclusion of S_R into S and the identities on X and Y. Notice that the inclusion satisfies the three properties of simulations in a trivial way.
- Every inclusion $c{:}R \to A$ as defined above satisfies the following property: given any reachable automata R' and simulation $h{:}R' \to A$, there is a unique morphism of reachable automata $h'{:}R' \to R$ such that $h=h';c$.

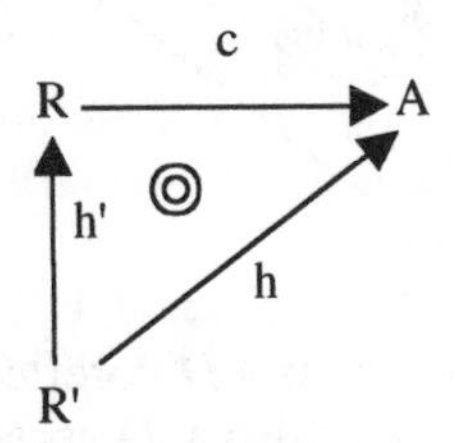

- In other words, R is the reachable automaton that is "closest" to A in the sense that any other reachable automata that A can simulate can also be simulated by R. Indeed, if $R'=<X',S',Y',s'_0,f',g'>$ and $h=<h_X,h_S,h_Y>$, then the equation $h=h';c$ can only be satisfied if $h'_S:S'\to S_R$ coincides with h_S on the whole domain S', i.e. if h_S only returns reachable states. But this is the case because R' is itself reachable. Finally, the equation fully determines the morphism, thus establishing the uniqueness of h'.

This relationship between automata and reachable automata is an instance of what in category theory is called a coreflective subcategory.

3.3.8 Definition – Coreflective Subcategories

Let $\boldsymbol{D}$ be a subcategory of a category $\boldsymbol{C}$.

1. Let c be a $\boldsymbol{C}$-object. A $\boldsymbol{D}$-coreflection for c is a $\boldsymbol{C}$-morphism $i:d\to c$ for some $\boldsymbol{D}$-object d such that, for any $\boldsymbol{C}$-morphism $f:d'\to c$ where d' is a $\boldsymbol{D}$-object, there is a unique $\boldsymbol{D}$-morphism $f':d'\to d$ such that $f=f';i$, i.e. the following diagram commutes:

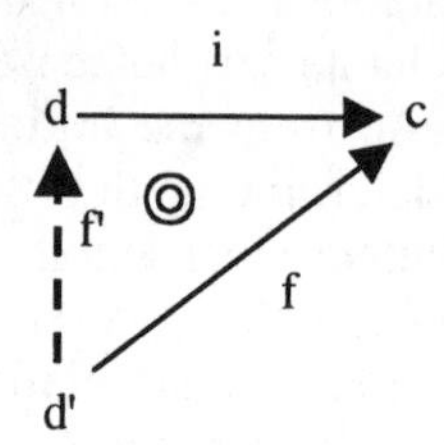

2. We say that $\boldsymbol{D}$ is a *coreflective subcategory* of $\boldsymbol{C}$ iff every $\boldsymbol{C}$-object admits a $\boldsymbol{D}$-coreflection. ♦

Coreflections are essentially unique in the following sense.

3.3.9 Proposition

Let $\boldsymbol{D}$ be a subcategory of a category $\boldsymbol{C}$.

1. Let c be a $\boldsymbol{C}$-object. If $i:d\to c$ and $j:e\to c$ are both $\boldsymbol{D}$-coreflections for c, then there is a $\boldsymbol{D}$-isomorphism $f:e\to d$ such that $j=f;i$, i.e. the following diagram commutes:

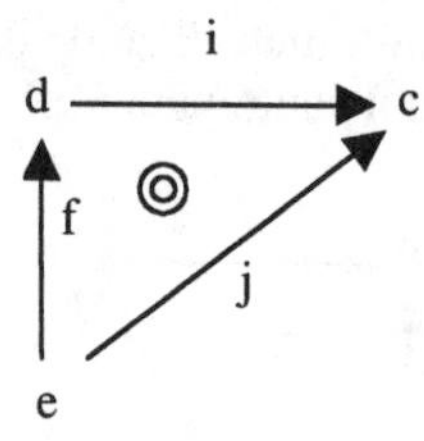

2. Let c be a $\boldsymbol{C}$-object. If $i:d\to c$ is a $\boldsymbol{D}$-coreflection for c, and $f:e\to d$ is a $\boldsymbol{D}$-isomorphism, then $f;i:e\to c$ is also a $\boldsymbol{D}$-coreflection for c.

Proof

1. The existence of f satisfying the equation results directly from the definition. The fact that f is an isomorphism can be proved as follows:
   - The existence of $g{:}d{\to}e$ such that $i{=}g;j$ can be concluded for the same reasons.
   - The composition $g;f$ is such that $(g;f);i{=}g;(f;i){=}g;j{=}i{=}id_d;i$. We can conclude that $g;f{=}id_d$ by applying the uniqueness requirement of the definition of coreflection to id_d.
   - A similar line of reasoning allows us to conclude that $f;g{=}id_e$.
2. This is trivially proved. ◆

Intuitively, a coreflector for an object is a secretary that manages its social life, i.e. through which all communication must go. Indeed, consider two objects c and c' together with their secretaries d and d', respectively. Consider an interaction $h{:}c{\to}c'$.

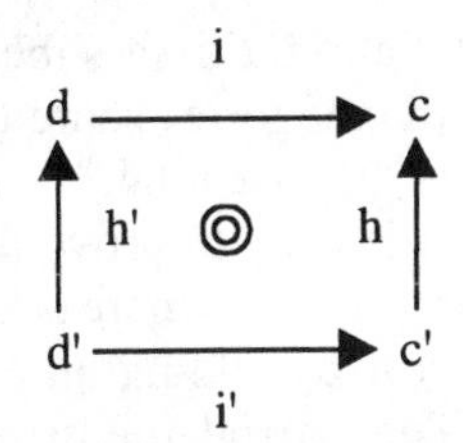

The composition $i';h$ is in the same circumstances as the morphism f in the definition. Hence, we conclude that there is an interaction $h'{:}d'{\to}d$ that factorises (intercepts) h in the sense that $i';h{=}h';i$.

Given that secretaries are also objects of the bigger category, a question arises naturally: Who are their secretaries? Do they talk to each other directly or do they need their own secretaries, and so on? It seems intuitive to expect that all secretaries communicate directly between them, i.e. that the coreflection of a ***D***-object (secretary) is the identity (itself) up to isomorphism. However, this can only be guaranteed if (and only if) ***D*** is a full subcategory.

3.3.10 Proposition

Let ***D*** be a coreflective subcategory of a category ***C***. Then, ***D*** is a full subcategory of ***C*** iff, for every ***D***-object d, id_d is a ***D***-coreflection for d.

Proof

1. Let ***D*** be a full subcategory of ***C*** and d a ***D***-object. Given any ***C***-morphism $f{:}d'{\to}d$ where d' is also a ***D***-object, the equation $f{=}f';id_d$ establishes $f{=}f'$ uniquely. So, the only way for the identity id_d not to be a coreflector for d is that f is not a ***D***-morphism as required by the definition. Naturally, if the category is full, this cannot happen.

2. Consider now an arbitrary ***C***-morphism $f:d' \rightarrow d$ where both d and d' are ***D***-objects. If id_d is a ***D***-coreflection for d then, by definition, there exists a ***D***-morphism $f':d' \rightarrow d$ such that $f=f';id_d$, i.e. $f=f'$. Hence, f is also a ***D***-morphism. ♦

Indeed, if the subcategory is not full, the secretaries, when playing their roles as secretaries, may have more restricted means of interaction and, therefore, may not be able to behave as they would do as "normal" objects.

The social motivation that we have been using is, in fact, somewhat biased towards objects, i.e. it leads us to consider coreflective subcategories that are full like that of reachable automata. However, any categorical property is ultimately determined by the morphisms.

The following example shows that there are coreflective subcategories that are not full.

3.3.11 Example

We mentioned in Par. 3.3.2 that ***SET*** is a subcategory of ***PAR***, the category of partial functions. Both categories share the same objects (sets), but ***SET*** retains only the functions that are total. Hence, ***SET*** is not a full subcategory of ***PAR***. However, it is easy to prove that every set A admits as a coreflector the partial inclusion $A^{\perp} \rightarrow A$ where $A^{\perp}$, also called the "elevation of A", is the set obtained from A by adding an additional element $\perp$ called "bottom" or "undefined". The partial inclusion is the extension of the identity on A that is undefined on $\perp$. Given now an arbitrary partial function $f:B \rightarrow A$ there is a unique elevation of f into a total function $f^{\perp}:B \rightarrow A^{\perp}$ by mapping to $\perp$ all the elements of B on which f is undefined. This is the standard technique that mathematicians use to deal with partiality in the context of total functions. ♦

Hence, in this case, the secretary is also responsible for transforming from the "partial" mode of communication into the "total" mode. This is because, by not being full, the subcategory defines a more specialised mode of interaction. The coreflector operates a transformation from the more general to the more specific mode.

Because category theory distinguishes between incoming and outgoing communication, coreflectors manage, in fact, the incoming communication only. The dual notion, i.e. the notion that is obtained in the dual category, is called reflector and manages the outgoing communication.

3.3.12 Definition – Reflective Subcategories

Let ***D*** be a subcategory of a category ***C***.

1. Let c be a ***C***-object. A ***D***-reflection for c is a ***C***-morphism $o:c \rightarrow d$ for some ***D***-object d such that, for any ***C***-morphism $f:c \rightarrow d'$, where d' is a ***D***-object, there is a unique ***D***-morphism $f':d \rightarrow d'$ such that $f=o;f'$, i.e. the following diagram commutes:

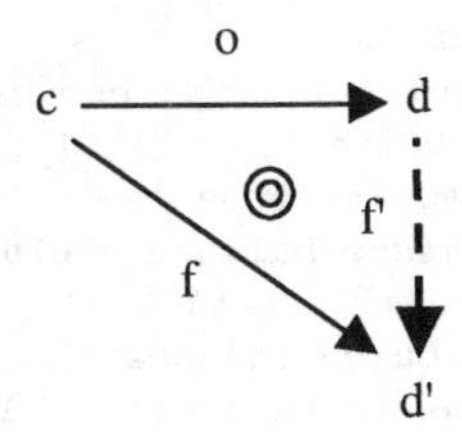

2. We say that ***D*** is a *reflective subcategory* of ***C*** iff every ***C***-object admits a ***D***-reflection. ♦

Being the dual concept of coreflections, reflections satisfy all the dual properties of coreflections. In particular, reflections are essentially "unique" and identities are reflectors iff the subcategory is full. For consistency, in informal discussions, we shall call coreflectors "insecretaries" or "cosecretaries", and reflectors "outsecretaries" or "secretaries". Examples of reflections are given in Par. 3.6.4.

3.4 Eiffel Class Specifications

In order to illustrate some of the ways in which category theory can be used to capture the semantics of software engineering practice, we introduce a "practical" example related to the Eiffel language. This example is used in later chapters as well.

As stated in [86], "Eiffel software texts – classes and their routines – may be equipped with elements of formal specification, called assertions, expressing correctness conditions". We do not detail the language in which these assertions can be written; it can be found in [86].

For our purposes, an Eiffel class specification e consists of a triple $<\Sigma,P,I>$ where:

- Σ, the class signature, is a set (of features). Each feature has its own signature which consists of a pair $<arg,res>$ of sequences of types. Features can be attributes, functions or routines, the sets of which are denoted by $att(\Sigma)$, $fun(\Sigma)$ and $rou(\Sigma)$, respectively.
- P provides, for every $r \in rou(\Sigma)$, a pair $<pre_r,pos_r>$ of Boolean expressions (the pre- and the postcondition of the routine, respectively).
- I is a Boolean expression (the invariant of the class).

The semantics intended for such a specification is the one popularised under the designation "design by contract". Every client of the class when calling a routine r should make sure that the precondition for r holds, in which case the postcondition is assured to hold after the call is completed. The class invariant is a condition that is guaranteed by every creation procedure of the class and maintained by any routine.

For an example of a class specification consider bank accounts:

```
deferred class    account is
attributes        balance: int, vip: boolean
routines          deposit(i:nat)
                       require true
                       ensure balance = old balance + i
                  withdrawal(i:nat)
                       require balance≥i
                       ensure balance = old balance - i
invariant         vip ⊃ balance≥1000
```

A bank account is capable of handling deposits and withdrawals. These update the balance, captured through an attribute, as specified in the *ensure* clauses. The routine that performs the withdrawals requires the client who makes the call to check that the balance is greater than the amount requested. Accounts that are considered to be *vip* have a balance greater than 1000.

The idea is that specifications are used to indicate *what* software components do, or are required to do, rather than *how* they do it. Changes to the implementations of the specified features are allowed as long as they do not violate the specifications.

Specifications have a "social life" that results from the need to adapt features. Feature adaptation [86] typically comes about when the class is inherited. To be consistent with [86], the notion of morphism that can capture such structural aspects of class specifications has to account for the following operations:

- Rename a feature.
- Merge one or more features; this is called the *join mechanism* and applies to features inherited as deferred (i.e. without a chosen implementation).
- Redefine a feature, changing the original signature, preconditions or postconditions.
- Add conditions to the invariant.

There are other circumstances in which features can be changed, like assigning an implementation to a feature that so far was "deferred". Because they relate only to implementations, and do not involve specifications except for discussing the correctness of the changes, i.e. because specifications are not socially active during these operations, we will not consider them in our formalisation.

The redefinition of a feature is subject to several constraints:

- At the level of its signature, the number of arguments and results cannot be changed; changes to their types can be performed subject to a number of conformance rules that, for simplicity, we ignore.
- Functions can be redeclared as attributes but not vice-versa; routines cannot be redeclared as attributes or functions, and attributes and functions cannot be redeclared as routines.

- Preconditions can be weakened but not strengthened.
- Postconditions can be strengthened but not weakened.

The constraints that apply to pre/postconditions have in mind the preservation of the contract: any client of a redefined routine has the right to expect a service that complies with the original specification. Hence, the client cannot be required to test for more conditions before calling the feature. Upon termination, the client must get at least what he would normally get from the original feature.

All these constraints lead to the following definition:

A morphism $F: e=<\Sigma,P,I> \rightarrow e'=<\Sigma',P',I'>$ of Eiffel class specifications consists of a total function between the class features such that

- For every feature $f \in \Sigma$, $arg_{F(f)}= arg_f$ and $res_{F(f)}= res_f$.
- For every attribute $a \in att(\Sigma)$, $F(a) \in att(\Sigma')$.
- For every function $f \in fun(\Sigma)$, $F(f) \in att(\Sigma') \cup fun(\Sigma')$.
- For every routine $r \in rou(\Sigma)$, $F(r) \in rou(\Sigma')$.
- For every $r \in rou(\Sigma)$, $F(pre_r) \vdash pre'_{F(r)}$ and $pos'_{F(r)} \vdash F(pos_r)$.
- $I' \vdash F(I)$.

As an example of a morphism of Eiffel class specifications, consider the following class specification obtained by inheriting from the previous example:

```
deferred class flexible account is
inherit         account
attributes      credit: nat
redefine        withdrawal(i:nat)
                     require else balance+credit ≥ i
invariant       vip ⊃ balance≥10000
```

That is, a flexible account extends the account with an attribute credit whose value can be added to the balance for satisfying a withdrawal request. As a counterpart to the added flexibility, the minimum balance for the account to be considered "vip" is now 10000.

The syntax of Eiffel is already such that some of the constraints are automatically met:

- The redeclaration of routines is such that preconditions are changed with a ***require else*** clause. Its semantics is to add the specified condition to the inherited precondition as a disjunct (hence the weakening).
- The postconditions are changed with a ***ensure then*** clause. Its semantics is to add the specified condition to the inherited postcondition as a conjunct (hence the strengthening).
- The invariant clause is added as a conjunct to the invariant inherited from the parent class (hence the strengthening).

It is easy to prove that class specifications and their morphisms constitute a category ***CLASS_SPEC***. In Chap. 4 we account for other properties of specifications, like the "join semantics rule".

3.5 Temporal Specifications

In this section, we develop a category whose objects are specifications of process behaviour in temporal logic. This category is used throughout the rest of the book in examples. Whereas our purpose with Eiffel class specifications was to show how category theory can apply to "real-life" modelling techniques, with temporal specifications we will try to illustrate, in the simplest way we know, typical properties of specification theory as applied to concurrent systems.

This example will also serve two other purposes. On the one hand, it is used to show how relationships between different domains can be developed once they have been formalised in category theory. This is done by relating temporal specifications with a very simple process model that we started building in Par. 3.2.1. The other purpose is to show that there is a degree of "universality" in the kinds of constructions that are typically used across different domains for system modelling, and that this universality is very easily made evident and formally characterised through the use of category theory. Before proceeding, we should make clear that much of this particular topic was jointly developed with Félix Costa. The essential part of the technical details can be found in [32, 33].

Specifications are often identified with theories (or theory presentations) in a given logic, an idea that has been around for quite a long time [17], although only more recently explored from the point of view of the temporal logic approach to reactive system specification [38]. In fact, process specifications are usually given as theory *presentations* rather than theories. By a theory presentation, we mean a pair consisting of a signature and a set of sentences – the nonlogical axioms of the specification. The signature identifies the vocabulary symbols that are proper to the object being identified and the sentences provide a description of the properties that are being specified about that object.

We shall be modelling the behaviour of concurrent processes at the level of the actions that are provided by their interfaces. Hence, every signature identifies a set of actions in which a process can engage itself. An example of a signature is that of the specification of a vending machine able to accept coins and deliver cakes and cigars – *{coin,cake,cigar}*.

3.5.1 Definition – Signatures

A signature of linear temporal logic is a set, the elements of which are called action symbols. ♦

The language in which the sentences that specify the behaviour of the process are written is that of linear temporal logic. The action symbols provide atomic propositions in the definition of the language associated with a signature.

3.5.2 Definition – Language

The set of *temporal propositions* ***prop***(Σ) for a signature Σ (Σ-propositions, for short) is inductively defined as follows:

- Every action symbol is a temporal proposition.
- ***beg*** is a temporal proposition (denoting the initial state).
- If ϕ is a temporal proposition so is $(\neg\phi)$.
- If ϕ_1 and ϕ_2 are temporal propositions so are $(\phi_1 \supset \phi_2)$, $(\phi_1 \mathbf{U} \phi_2)$. ♦

The temporal operator is **U** (until). Its semantics is defined below; informally, $(\phi_1 \mathbf{U} \phi_2)$ holds if, from tomorrow, ϕ_2 will eventually hold and, until then, but not necessarily then, ϕ_1 will hold. Other temporal operators such as ***X*** (next or tomorrow), ***F*** (eventually) and ***G*** (always) can be defined as abbreviations [67]. We often make use of the operator ***W*** (weak until): $(\phi_1 \mathbf{W} \phi_2)$ holds if, from tomorrow, either ϕ_2 will eventually hold and until then, but not necessarily then, ϕ_1 will hold, or ϕ_2 will forever be false and ϕ_1 true.

3.5.3 Definition – Semantics

The language of linear temporal logic is interpreted over infinite sequences of sets of actions. That is, an interpretation structure for a signature Σ is a sequence $\lambda \in (2^{\Sigma})^{\omega}$. These are canonical Kripke structures for linear, discrete, propositional logic [110]. Each infinite sequence represents a possible behaviour for the process being specified. The sets of actions represent the events that take place during the lifetime of the process. As already explained in Par. 3.2.1, the event that consists of the empty set of actions represents a transition performed by the environment without the participation of the process.

A Σ-proposition ϕ is said to be true for $\lambda \in (2^{\Sigma})^{\omega}$ at state i, which we write $\lambda \models_{\Sigma,i} \phi$ iff:

- If $\phi \in \Sigma$, $\lambda \models_{\Sigma,i} \phi$ iff $\phi \in \lambda(i)$.
- $\lambda \models_{\Sigma,i} \textbf{\textit{beg}}$ iff $i=0$.
- $\lambda \models_{\Sigma,i} (\neg\phi)$ iff it is not the case that $\lambda \models_{\Sigma,i} \phi$.
- $\lambda \models_{\Sigma,i} (\phi_1 \supset \phi_2)$ iff $\lambda \models_{\Sigma,i} \phi_1$ implies $\lambda \models_{\Sigma,i} \phi_2$.
- $\lambda \models_{\Sigma,i} (\phi_1 \mathbf{U} \phi_2)$ iff, for some $j>i$, $\lambda \models_{\Sigma,j} \phi_2$ and $\lambda \models_{\Sigma,k} \phi_1$ for every $i<k<j$.

The proposition ϕ is said to be *true* in λ, written $\lambda \models_{\Sigma} \phi$, if and only if $\lambda \models_{\Sigma,i} \phi$ at every state i. We also write $\lambda \models_{\Sigma} \Phi$ for a collection of propositions Φ meaning that each proposition of Φ is true in λ, and $\Lambda \models_{\Sigma} \phi$ for a collection Λ of sequences meaning that ϕ is true in every $\lambda \in \Lambda$.

Finally, for every set Φ of Σ-propositions and every Σ-proposition ϕ, ϕ is a consequence of Φ, denoted $\Phi \vdash_{\Sigma} \phi$, if and only if ϕ is true in every sequence that makes all the propositions in Φ true. ♦

The corresponding notion of theory is given as usual for the closure system induced by the consequence relation.

3.5.4 Definition – Theories and Presentations

1. Let Σ be a signature. A subset Ψ of $\boldsymbol{prop}(\Sigma)$ is said to be *closed* if and only if, for every $\phi \in \boldsymbol{prop}(\Sigma)$, $\Psi \vdash_{\Sigma} \phi$ implies $\phi \in \Psi$. By $c_{\Sigma}(\Psi)$ we denote the least closed set that contains Ψ.
2. A *temporal theory* is a pair $<\Sigma,\Phi>$ where Σ is a signature and Φ is a closed set of Σ-propositions.
3. A *theory presentation* is a pair $<\Sigma,\Phi>$ where Σ is a signature and Φ is a set of Σ-propositions. The presented theory is $<\Sigma,c_{\Sigma}(\Phi)>$. ♦

That is, a theory consists of a set of sentences that is closed for consequence: it contains all the theorems that can be derived from its sentences. A presentation is not necessarily closed under consequence: it provides a more "economical" way of specifying the intended behaviour of a system. A presentation consists only of a selected set of properties (also called the axioms of the presentation) that are required of the system, leaving the computation of the properties that can be derived from this selected set (its theorems) to the proof theory of the logic. Whereas theories, in general, and in particular for temporal logic, are infinite, presentations can be finite or, at least, recursively enumerable. Hence, specifications are usually identified with presentations, not with theories.

3.5.5 Exercise

Prove that the operators c_{Σ} satisfy the following properties:

- Reflexivity: for every $\Phi \subseteq \boldsymbol{prop}(\Sigma)$, $\Phi \subseteq c_{\Sigma}(\Phi)$.
- Monotonicity: for every $\Phi,\Gamma \subseteq \boldsymbol{prop}(\Sigma)$, $\Phi \subseteq \Gamma$ implies $c_{\Sigma}(\Phi) \subseteq c_{\Sigma}(\Gamma)$.
- Idempotence: for every $\Phi \subseteq \boldsymbol{prop}(\Sigma)$, $c_{\Sigma}(c_{\Sigma}(\Phi)) \subseteq c_{\Sigma}(\Phi)$. ♦

3.5.6 Example – A Vending Machine

As an example, consider the following specification of a vending machine:

```
specification   vending machine is
signature       coin, cake, cigar
axioms          beg ⊃ ((¬cake∧¬cigar) ∧
                     (coin ∨ (¬cake∧¬cigar)Wcoin))
                coin ⊃ (¬coin)W(cake∨cigar)
                (cake∨cigar) ⊃ (¬cake∧¬cigar)Wcoin
                cake ⊃ (¬cigar)
```

This machine is able to accept coins, deliver cakes and deliver cigars. The machine is initialised so as to accept only coins (first axiom). Once it accepts a coin it can deliver either a cake or a cigar (second axiom), but not both (fourth axiom). After delivering a cake or a cigar it is ready to accept more coins (third axiom). ♦

As already mentioned, the model of process behaviour that we adopt reflects an (abstract) synchronous, multi-processor architecture in which, at each transition, several actions may be executed concurrently. We are going to show that this synchronous flavour is captured through the notion of specification morphism (interpretation between theories) as a mathematical model of the relationship between systems and their components.

An interpretation between theories is typically defined as a mapping between their signatures that preserves theorems. Notice, once again, the idea of structure preservation presiding to the definition of morphisms. The structure being preserved in this case is given by the properties of the processes involved as captured through the theorems of the specifications.

3.5.7 Definition – Interpretations Between Theories

1. Let Σ and Σ' be signatures. Every function $f:\Sigma\to\Sigma'$ extends to a *translation map* $\mathbf{prop(f)}:\mathbf{prop}(\Sigma)\to\mathbf{prop}(\Sigma')$ as follows:
   - $\mathbf{prop}(f)(\mathbf{beg})=\mathbf{beg}$.
   - If $a\in\Sigma$ then $\mathbf{prop}(f)(a)=f(a)$.
   - $\mathbf{prop}(f)(\neg\phi)=(\neg\mathbf{prop}(f)(\phi))$.
   - $\mathbf{prop}(f)(\phi_1\supset\phi_2)=(\mathbf{prop}(f)(\phi_1)\supset\mathbf{prop}(f)(\phi_2))$.
   - $\mathbf{prop}(f)(\phi_1 U\phi_2)=(\mathbf{prop}(f)(\phi_1)U\mathbf{prop}(f)(\phi_2))$.
2. An *interpretation* between two theories (or theory morphism) $<\Sigma_1,\Phi_1>$ and $<\Sigma_2,\Phi_2>$ is a map $f:\Sigma_1\to\Sigma_2$ such that $\mathbf{prop}(f)(\Phi_1)\subseteq\Phi_2$.
3. A *morphism* between two presentations $<\Sigma_1,\Phi_1>$ and $<\Sigma_2,\Phi_2>$ is a map $f:\Sigma_1\to\Sigma_2$ such that $\mathbf{prop}(f)(c_{\Sigma_1}(\Phi_1))\subseteq c_{\Sigma_2}(\Phi_2)$. ♦

3.5.8 Proposition

1. Temporal theories and interpretations between theories define a category called $\mathbf{THEO}_{LTL}$.
2. Presentations of temporal theories and their morphisms define a category called $\mathbf{PRES}_{LTL}$, which admits $\mathbf{THEO}_{LTL}$ as a subcategory.

Proof

1. Although we have not made it explicit, the composition law and identity map that we have in mind for theories are the ones inherited from signatures as sets. Hence, this is another example of constructing a category by adding structure: the underlying category is that of sets, and the structure being added is the collection of theorems that constitute the specification. Therefore, the properties of the composition law and identity map are inherited from ***SET***; we only have to prove the closure properties, i.e. that the composition of theory morphisms as functions is still a theory morphism and that the identity function is a theory morphism. The latter is trivial because the translation map induced by the identity on signatures is itself the identity. The former is proved as follows:
   - First of all, given $f:\Sigma\to\Sigma'$ and $g:\Sigma'\to\Sigma''$, we have to prove that $\mathbf{prop}(f;g) = \mathbf{prop}(f);\mathbf{prop}(g)$. That is, the translation map induced by a composite func-

tion is the composition of the translation maps induced by the components. This can be proved by structural induction and is left as an exercise.

- Given theory morphisms $f{:}{<}\Sigma,\Phi{>}\rightarrow{<}\Sigma',\Phi'{>}$ and $g{:}{<}\Sigma',\Phi'{>}\rightarrow{<}\Sigma'',\Phi''{>}$, we have $\mathbf{prop}(f;g)(\Phi)=\mathbf{prop}(g)(\mathbf{prop}(f)(\Phi))$ as proved above. Because f is a theory morphism, $\mathbf{prop}(f)(\Phi)\subseteq\Phi'$, which implies $\mathbf{prop}(f;g)(\Phi)\subseteq\mathbf{prop}(g)(\Phi')$. Because g is a theory morphism, we also have $\mathbf{prop}(g)(\Phi')\subseteq\Phi''$, which implies $\mathbf{prop}(f;g)(\Phi)\subseteq\Phi''$. Hence, $(f;g)$ is a theory morphism.

2. The proof that theory presentations and their morphisms define a category can proceed exactly in the same way as for theories. The fact that $\mathbf{THEO}_{LTL}$ is a subcategory of the resulting category has a very trivial proof:
   - Clearly, every theory is a presentation: it just happens to be closed.
   - Given a theory morphism $f{:}{<}\Sigma,\Phi{>}\rightarrow{<}\Sigma',\Phi'{>}$ we have to prove that it defines a presentation morphism, i.e. that $\mathbf{prop}(f)(c_\Sigma(\Phi))\subseteq c_{\Sigma'}(\Phi')$. Because Φ and Φ' are closed, $c_\Sigma(\Phi)=\Phi$ and $c_{\Sigma'}(\Phi')=\Phi'$. Because f is a theory morphism, $\mathbf{prop}(f)(\Phi)\subseteq\Phi'$.
   - The properties of the composition and the identity are trivially proved. ♦

For simplicity, we overload the notation and use f instead of $\mathbf{prop}(f)$ unless there is a risk of confusion.

Checking that a signature map defines a morphism between two presentations by checking that every theorem of the source is translated to a theorem of the target is not "practical". Typically, one would prefer to check that only the axioms of the presentation are translated into theorems, from which it should follow that the theorems are preserved. Although this is not true for an arbitrary logic, it is a property of the temporal logic that we defined above as proved below. In order to do so, we need to be able to relate the models of the signatures involved.

3.5.9 Definition/Proposition – Reducts

1. Let Σ and Σ' be signatures and $f{:}\Sigma\rightarrow\Sigma'$ a total function. Given an interpretation structure $\lambda'\in(2^{\Sigma'})^\omega$ for Σ', we define its *reduct* $\lambda'|_f$ as the interpretation structure for Σ defined by $\lambda'|_f(i)=f^{-1}(\lambda'(i))$. That is, the reduct of a sequence λ' of Σ'-actions is the sequence that consists, at each point i, of the Σ-actions that are translated through f into $\lambda'(i)$.
2. Let Σ and Σ' be signatures, $f{:}\Sigma\rightarrow\Sigma'$ a total function, $\lambda'\in(2^{\Sigma'})^\omega$ an interpretation structure for Σ', and ϕ a Σ-proposition. Then, for every i, $\lambda'\models_{\Sigma',i}f(\phi)$ iff $\lambda'|_f\models_{\Sigma,i}\phi$ (the *satisfaction condition*).

Proof

The proof of point 2 is by induction on the structure of ϕ. The base case (action symbols) results directly from the definition of reduct: $f(\phi)\in\lambda'(i)$ iff $\phi\in f^{-1}(\lambda'(i))$ $(=\lambda'|_f(i))$. The induction step presents no difficulties. ♦

3.5.10 Proposition – Presentation Lemma

1. Given a morphism $f{:}\Sigma_1\rightarrow\Sigma_2$ and $\Phi\subseteq\mathbf{prop}(\Sigma_1)$, $f(c_{\Sigma_1}(\Phi))\subseteq c_{\Sigma_2}(f(\Phi))$.

2. Given presentations $<\Sigma_1,\Phi_1>$ and $<\Sigma_2,\Phi_2>$, a map $f:\Sigma_1\rightarrow\Sigma_2$ is a morphism between them iff $f(\Phi_1)\subseteq c_{\Sigma_2}(\Phi_2)$. This result is usually called "the presentation lemma".

Proof

1. Let $\phi\in c_{\Sigma_1}(\Phi)$. We have to prove that $f(\phi)\in c_{\Sigma_2}(f(\Phi))$. For that purpose, consider an arbitrary interpretation structure $\lambda'\in(2^{\Sigma})^{\omega}$ in which $f(\Phi)$ is true. From what was just proved, all the propositions in Φ are true in $\lambda'|_f$. But, then, so is ϕ because $\phi\in c_{\Sigma_1}(\Phi)$. Applying the same result, but in the other direction, we know that $f(\phi)$ is true in λ'.
2. All that needs to be proved is that $f(c_{\Sigma_1}(\Phi_1))\subseteq c_{\Sigma_2}(\Phi_2)$ iff $f(\Phi_1)\subseteq c_{\Sigma_2}(\Phi_2)$.
   - The forward implication is an immediate consequence of the reflexivity of closure, which allows us to derive that $\Phi_1\subseteq c_{\Sigma_1}(\Phi_1)$.
   - Consider now the reverse implication
     a. $f(\Phi_1)\subseteq c_{\Sigma_2}(\Phi_2)$ — assumption,
     b. $c_{\Sigma_2}(f(\Phi_1))\subseteq c_{\Sigma_2}(c_{\Sigma_2}(\Phi_2))$ — from a, closure operators being monotone,
     c. $c_{\Sigma_2}(f(\Phi_1))\subseteq c_{\Sigma_2}(\Phi_2)$ — from b, closure operators being idempotent,
     d. $f(c_{\Sigma_1}(\Phi_1))\subseteq c_{\Sigma_2}(\Phi_2)$ — from c and the result proved in 1.

◆

3.5.11 Example – A Regulated Vending Machine

As an example of the use of morphisms for system modelling, consider the following specification:

```
specification    regulated vending machine is
signature        coin, cake, cigar, token
axioms           beg ⊃ (¬cake∧¬cigar∧¬token) ∧
                     (coin ∨ (¬cake∧¬cigar)Wcoin)
                 coin ⊃(¬coin)W(cake∨cigar)
                 coin ⊃(¬cigar)Wtoken
                 (cake∨cigar) ⊃(¬cake∧¬cigar)Wcoin
                 cake ⊃(¬cigar)
                 token ⊃(¬cake∧¬cigar∧¬coin)
```

That is, the machine is now extended to be able to accept tokens, and is regulated in such a way that it will only deliver a cigar if, after having accepted a coin, it receives a token. The last axiom says that coins, cakes and cigars cannot be used as tokens.

This specification results from the previous vending machine through the superposition of some mechanism for controlling the sales of cigars. In Par. 6.1.24, we can see how a regulator can be independently specified and connected to the vending machine in order to achieve the required superposition. That is, the regulated vending machine can be defined as a configuration of which the original vending machine, as well as the regulator, are components.

To finalise, we show how theory morphisms can be used to identify components of systems. For instance, in the example above, we can establish a morphism between the specifications of the vending machine and the regulated vending machine. The signature morphism maps the actions of the vending machine to those of the regulated vending machine that have the same names. To prove that the signature morphism defines an interpretation between the two specifications, we have to check that every axiom of the specification of the vending machine as given in Par. 3.5.6 is translated to a theorem of the regulated vending machine:

- The initialisation condition is preserved: it is strengthened with $\neg token$.
- The second axiom is translated directly into the second axiom.
- The third axiom is translated directly into the fourth axiom.
- The fourth axiom is translated directly into the fifth axiom.

The fact that morphisms identify components of systems is captured by the following result.

3.5.12 Proposition

Let $<\Sigma_1,\Phi_1>$ and $<\Sigma_2,\Phi_2>$ be theories and $f:\Sigma_1\rightarrow\Sigma_2$ a signature morphism. Then, f defines a theory morphism iff, for every model λ' of $<\Sigma_2,\Phi_2>$, $\lambda'|_f$ is a model of $<\Sigma_1,\Phi_1>$.

Proof

1. Assume that f defines a theory morphism and let λ' be a model of $<\Sigma_2,\Phi_2>$. To prove that $\lambda'|_f$ is a model of $<\Sigma_1,\Phi_1>$, let $\phi\in\Phi_1$. Because f is a theory morphism, $f(\phi)\in\Phi_2$ and, hence, $f(\phi)$ is true in λ'. But, by the fundamental property of reducts, ϕ is true in $\lambda'|_f$. Hence, $\lambda'|_f$ is a model of $<\Sigma_1,\Phi_1>$.
2. Assume now that, for every model λ' of $<\Sigma_2,\Phi_2>$, $\lambda'|_f$ is a model of $<\Sigma_1,\Phi_1>$. Let $\phi\in\Phi_1$. We have to prove that $f(\phi)\in\Phi_2$. Let λ' be a model of $<\Sigma_2,\Phi_2>$. Because $\lambda'|_f$ is a model of $<\Sigma_1,\Phi_1>$, ϕ is true in $\lambda'|_f$. By the fundamental property of reducts (opposite direction), $f(\phi)$ is true in λ'. ♦

That is, given a morphism between two temporal specifications, every behaviour of the target is projected back to a model of the source. Because the reduct identifies the actions of the source that occur at each point in the execution of the target, we can say that an interpretation between theories identifies in every behaviour that is according to the target specification, a behaviour that is according to the source specification. Hence, the morphism recognises in the source a component of the target.

Notice that it is not necessary that every behaviour of the source (component) be recovered through the reduct. Indeed, as a component of a larger system, some behaviours of the component may be lost because of interactions with other components. This is the case above. The behaviours in which a coin is followed immediately by a cigar are not part of any model of the regulated vending machine because a token is required after the coin before the cigar can be delivered.

3.6 Closure Systems

In the previous section, we presented several categories related to the specification of systems on the basis of properties stated in the language of temporal logic. We show in the next chapters that these categories satisfy a number of properties that make them useful to modularise specifications. However, these properties are in no way particular to the temporal logic that was chosen to express specifications. They are shared by a number of structures that generalise what, sometimes, are called "closure systems". This section presents the initial step in a sequence of constructions that will end up in the definition of institutions [62], π-institutions [45] and general logics [85]. These are all categorical generalisations of the notions of logic and associated concepts like theories that bring out the structural properties that make them useful for specification.

3.6.1 Definition – Closure Systems

A *closure system* is a pair $<L,c>$ where L is a set and $c:2^L \rightarrow 2^L$ is a total function satisfying the following properties:

- Reflexivity: for every $\Phi \subseteq L$, $\Phi \subseteq c(\Phi)$.
- Monotonicity: for every $\Phi,\Gamma \subseteq L$, $\Phi \subseteq \Gamma$ implies $c(\Phi) \subseteq c(\Gamma)$.
- Idempotence: for every $\Phi \subseteq L$, $c(c(\Phi)) \subseteq c(\Phi)$. ♦

3.6.2 Definition/Proposition – Category of Closure Systems

We define the category ***CLOS*** of closure systems whose morphisms $f:<L,c> \rightarrow <L',c'>$ are the maps $f:L \rightarrow L'$ s.t. $f(c(\Phi)) \subseteq c'(f(\Phi))$ for all $\Phi \subseteq L$.

Proof

Again an example of adding structure to ***SET***.

1. The identity function on sets is trivially a morphism of closure systems.
2. Consider now two morphisms $f:<L,c> \rightarrow <L',c'>$ and $g:<L',c'> \rightarrow <L'',c''>$, and $\Phi \subseteq L$. Because f is a morphism, $f(c(\Phi)) \subseteq c'(f(\Phi))$. This also implies that $g(f(c(\Phi))) \subseteq g(c'(f(\Phi)))$. Because g is a morphism, $g(c'(f(\Phi))) \subseteq c''(g(f(\Phi)))$. Hence, $(f;g)(c(\Phi)) \subseteq c''((f;g)(\Phi)))$. ♦

As an example of closure systems we can give, for every temporal signature Σ, the pair $clos(\Sigma)$ formed by ***prop(Σ)*** as defined in Par. 3.5.2 together with the closure operator defined in Par. 3.5.4. Notice that the result proved in Par. 3.5.10 shows that every signature morphism induces a morphism between the corresponding closure systems.

3.6.3 Definition/Proposition – Theories in Closure Systems

Consider a closure system $<L,c>$.

1. We say that $\Phi \subseteq L$ is *closed* iff $\Phi = c(\Phi)$.

2. We define the category $\boldsymbol{THEO}_{<L,c>}$ whose objects (theories) are the closed subsets of L and morphisms are given by inclusions.
3. We define the category $\boldsymbol{PRES}_{<L,c>}$ whose objects (theory presentations) are the subsets of L and morphisms given by the preorder $\Phi \leq \Gamma$ iff $c(\Phi) \subseteq c(\Gamma)$.
4. We define the category $\boldsymbol{SPRES}_{<L,c>}$ whose objects (strict presentations) are the subsets of L ordered by inclusion.

Proof

These are trivial examples of categories that are given through preordered sets. ♦

We call the objects of $\boldsymbol{SPRES}_{<L,C>}$ "strict presentations" because the morphisms require that the axioms be preserved. The more attentive reader will have noticed that these notions do not coincide exactly with the definitions used in Sect. 3.5 because they do not contemplate changes of language. We generalise these notions in later sections to account for relationships between theories in different languages. The following proposition defines several relationships between these categories. For simplicity, we omit the underlying closure system.

3.6.4 Proposition

1. ***THEO*** is a full subcategory of ***PRES*** (and of ***SPRES***), and ***SPRES*** is a subcategory of ***PRES***.
2. ***THEO*** is a reflective subcategory of ***PRES*** (and of ***SPRES***), but ***SPRES*** is not.
3. ***THEO*** is a coreflective subcategory of ***PRES*** (but not of ***SPRES***) and so is ***SPRES***.

Proof

1. The proof that ***THEO*** is a (full) subcategory of ***SPRES*** and ***PRES*** is trivial. The result that ***SPRES*** is a subcategory of ***PRES*** is an immediate consequence of the monotonicity of the closure operator. This subcategory is not full because theorems may be preserved even if the axioms are not.
2. The first results are about closure as a reflector. We prove that, given an arbitrary presentation Φ, its reflector is its closure $c(\Phi)$. There are two parts in the proofs: (1) the very existence of the reflector as a morphism; (2) its universal property. All the reasoning evolves around the following diagram:

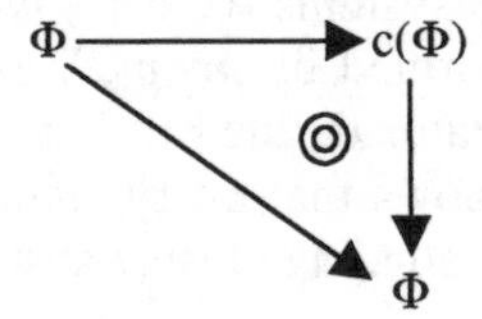

- In the case of ***THEO*** as a subcategory of ***PRES***, part 1 translates to $c(\Phi) \subseteq c(c(\Phi))$, which is a corollary of the reflexivity of the closure relation. Part 2 translates to $c(\Phi) \subseteq c(\Phi')$ *implies* $c(\Phi) \subseteq \Phi'$ because the morphism on the right is in ***THEO***. This holds because, being a theory, $c(\Phi') = \Phi'$.

- In the case of ***THEO*** as a subcategory of ***SPRES***, part 1 translates to the inclusion $\Phi \subseteq c(\Phi)$ because the morphism is now strict. Again, this is a corollary of the reflexivity of the closure relation. Part 2 translates to $\Phi \subseteq \Phi'$ *implies* $c(\Phi) \subseteq \Phi'$ because the morphism on the right is in ***THEO*** and the one on the left is strict. This is a consequence of the monotonicity of the closure operator and the fact that, being a theory, $c(\Phi')=\Phi'$.
- The fact that ***SPRES*** is not a reflective subcategory of ***PRES*** can be inferred from part 2: $c(\Phi) \subseteq c(\Phi')$ *implies* $c(\Phi) \subseteq \Phi'$ does not necessarily hold because, being a presentation, Φ' is not necessarily closed.
- Summarising, on the side of reflections, these results capture the fact that, for the purposes of "outward communication", i.e. for being interpreted, presentations can delegate on theories but not on strict presentations because, whereas the former will use the full theory for the interpretation, the latter will only look at relationships between axioms.

3. The second set of results is about the coreflective properties of closure. The relevant "diagram" now is:

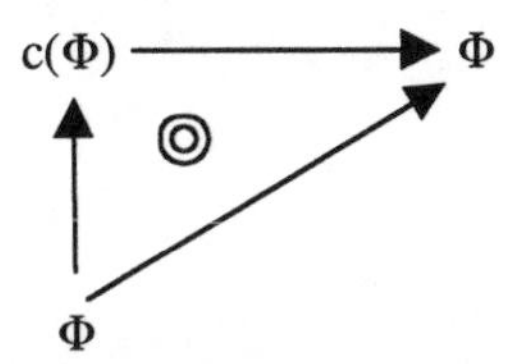

- In the case of ***THEO*** as a subcategory of ***PRES***, part 1 translates to the inclusion $c(c(\Phi)) \subseteq c(\Phi)$, which is a corollary of the idempotence of the closure operator. Part 2 translates to $c(\Phi') \subseteq c(\Phi)$ *implies* $\Phi' \subseteq c(\Phi)$ because the morphism on the left is in ***THEO***. This holds because of reflexivity.
- The fact that ***THEO*** is not a coreflective subcategory of ***SPRES*** can be inferred from part 1: the inclusion $c(\Phi) \subseteq \Phi$ does not necessarily hold because, being a presentation, Φ is not necessarily closed.
- In the case of ***SPRES*** taken as a subcategory of ***PRES***, part 1 translates again to the inclusion $c(c(\Phi)) \subseteq c(\Phi)$. Because the morphism on the left is strict, part 2 translates once again to $c(\Phi') \subseteq c(\Phi)$ *implies* $\Phi' \subseteq c(\Phi)$, which is a consequence of reflexivity. ◆

Summarising, on the side of coreflections, the existence of a coreflection from strict into loose interpretations reflects (pun intended) the fact that in order to interpret another theory presentation, a given theory presentation just needs to consider the axioms of the source. On the other hand, because strict presentations are not expressive enough to interpret theories, they do not admit theories as coreflectors, i.e. they cannot allow theories to handle their in-communication.

Globally, this result shows that theories and theory presentations with the looser notion of morphism define, essentially, the same structure. The same does not happen with strict presentations. Strict presentations can coreflect presentations and be reflected by theories, but not the other way around because they are not expressive enough to capture closure.

4 Universal Constructions

We have already hinted to the fact that category theory provides a totally different approach to the characterisation of a given domain of objects, namely to the fact that objects are characterised by their "social life", or interactions, as captured by morphisms. In this chapter, we take this view one step further by showing how certain objects, or constructions, can be characterised in terms of standard relationships that they exhibit with respect to the rest of the universe (of relevant constructions). It is in this sense that these constructions are usually called "universal". At the same time, we shift our focus from the social life of individual objects to that of groups or societies of interacting objects.

Indeed, in this chapter, we shift some of the emphasis from the manipulation of objects to that of diagrams as models of complex systems. Diagrams can be taken as expressions of what are often called configurations, be it configurations of running systems as networks of simpler components, the way complex programs (text) or specifications are put together from modules, the inheritance structures according to which program modules are organised, etc. This where the term *universal* conjures up general principles such as Goguen's famous dogma [57]: "given a category of widgets, the operation of putting a system of widgets together to form a super-widget corresponds to taking a colimit of the diagram of widgets that shows how to interconnect them".

One of our goals with this book is to make the software community aware that, like in the other domains of science and engineering, there are universal laws or dogmas in our area that can manifest themselves in different ways but be unified under a "categorical roof". For instance, we show that the relationship between composition and conjunction, as debated in [112], is universal in a very precise, mathematical way, which is precisely the one captured by Goguen's dogma.

This chapter is only an entry point to the ways category theory can address the complexity of system development. Subsequent chapters will introduce further constructions, techniques and applications.

4.1 Initial and Terminal Objects

At least at first sight, the first universal constructions that we define are not so much constructions but identifications of distinguished objects. What distinguishes them from the other objects are what are usually called "universal properties", a term that is associated, and used interchangeably in the literature, with "universal construction".

4.1.1 Definition – Initial Objects

An object x of a category $\boldsymbol{C}$ is said to be *initial* iff, for every object y of $\boldsymbol{C}$, there is a unique morphism from x to y. ♦

4.1.2 Proposition

1. Any two initial objects in a given category are isomorphic.
2. Any object isomorphic to an initial object is also initial.

Proof

1. Let x and y be initial objects, and $f{:}x{\to}y$ and $g{:}y{\to}x$ the associated morphisms given by their universal properties. Consider now the morphism $(f;g){:}x{\to}x$. Because x is initial, we know that there is only one morphism with x as source and target. Hence, $(f;g)=id_x$. The same reasoning applied to y returns $(g;f)=id_y$, showing that x and y are isomorphic.
2. Left as an exercise. ♦

Hence, we usually refer to *the* initial object, if one exists, and denote it by 0. The unique morphism from an initial object 0 into an object a is denoted by 0_a.

$$0 \overset{0_a}{\dashrightarrow} a$$

4.1.3 Examples

1. In ***SET***, the initial object is the empty set. This is because the empty set can be mapped to any other set in a unique way: through the empty function. This justifies the notation 0 usually adopted for initial objects.
2. In ***LOGI***, the initial object is $\perp$ (any contradiction). This is because anything can be derived from a contradiction. This is consistent with the use of the notation 0 (as the truth value for false) for initial objects.
3. In ***PAR***, the initial object is also the empty set: all functions from the empty set are, in fact, total; hence it is not surprising that we get the same initial object as in ***SET***.
4. In $\boldsymbol{SET_\perp}$, the initial objects are the singletons $<\{a\},a>$. This is because there is one and only one way of mapping the designated object of a pointed set: to the designated element of the target pointed set. Notice

that, although $SET_{\perp}$ was built over SET, the initial objects of the two categories do not coincide. Indeed, when adding structure to a given category to make another category, the universal constructions may change precisely because the structure has changed. Nevertheless, the spirit of emptiness is still there – $<\{a\},a>$ has no "proper" elements. ♦

4.1.4 Definition – Terminal Objects

An object is terminal in a category C iff it is initial in C^{op}. That is, x is terminal in C iff, for every object y of C, there is a unique morphism from y to x. ♦

Once again, any two terminal objects are isomorphic and, therefore, we usually refer to *the* terminal object of a category, if one exists. A terminal object is usually denoted by 1 and the unique morphism from an object a is denoted by 1_a.

$$1 \xleftarrow{\;1_a\;} a$$

4.1.5 Examples

1. In SET, the terminal objects are the singletons. This is because there is one, and only one way of mapping any given set to a singleton: by mapping all the elements of the source set (even if there is none...) to the element of the singleton. Any set with more than one element is clearly not terminal because it provides a choice for the target image of any element of the source set, i.e. it does not satisfy the uniqueness criterion. The empty set is not terminal because it only admits total functions from itself, i.e. it does not satisfy the existence criterion.

   Notice that all singletons are indeed isomorphic in SET. This is consistent with what we said before about the way category theory handles sets as objects: because we are not allowed to look into a set to see what elements it has, there is no way we can distinguish two singletons in terms of their structural properties. This also justifies the use of the notation 1 for arbitrary terminal objects.

   On the other hand, any singleton can be used to identify the different elements of any non-empty set A by noticing that each element $a \in A$ defines, in a unique way, a function $a{:}1 \rightarrow A$. Indeed, following the idea that morphisms characterise the "social life" of the objects of a category, the social relationships that a singleton set can establish with an arbitrary non-empty set characterise precisely the elements of that set.

   The reader may wish to return to Pars. 2.2.12 and 2.3.5 to see examples that make use of this analogy, the notation $\{\bullet\}$ being used for "the" terminal set. This idea can be generalised to any terminal object 1_C in an arbitrary category C as a mechanism for identifying what, in category

theory, are called points [75], constants [12] or global elements [22] of an object x: morphisms of the form $1 \rightarrow x$.
2. In ***LOGI***, the terminal object is T (any tautology). This is because any tautology can be derived from any other formula. Again, this is also consistent with the use of the notation 1 (the truth value for true) for terminal objects.
3. In ***PAR***, the terminal object is also the empty set: there is always one and only one way of mapping any set to the empty one – through the partial function that is undefined in all the elements of the source! Hence, in ***PAR***, the initial and the terminal objects coincide. Objects of an arbitrary category that are both initial and terminal are sometimes called *null* [80] or *zero* [1] objects.
   Notice that, in spite of being a subcategory of ***PAR***, ***SET*** has different terminal objects. On the one hand, the undefined function not being total, the empty set cannot play the role of terminal object in ***SET***. On the other hand, because sets in ***PAR*** have a richer social life, i.e. more morphisms, the singletons in ***PAR*** relate differently to other sets than they do in ***SET***. In particular, besides the constant (total) function, they also admit the (partial) undefined one. Indeed, categories cannot be expected to share the same kind of initial/terminal objects with their subcategories, unless there is some special structural property that justifies so (see exercise below).
4. In $\boldsymbol{SET_\perp}$, the terminal objects are also the singletons $<\{a\},a>$. Hence, in $\boldsymbol{SET_\perp}$, the initial and the terminal objects also coincide. ♦

4.1.6 Exercises

1. Let $\boldsymbol{C}$ be an arbitrary category and $a{:}\boldsymbol{C}$. Show that id_a is initial in $a \downarrow \boldsymbol{C}$. Further show that if a is terminal in $\boldsymbol{C}$, id_a is terminal in $a \downarrow \boldsymbol{C}$.
2. Show that the category $\boldsymbol{SET_\perp}$ of pointed sets as introduced in Par. 3.2.1 "corresponds" to the comma category $1 \downarrow \boldsymbol{SET}$ as defined in Par. 3.2.2. This correspondence is what is Sect. 5.1 is called an isomorphism of categories. What can be said about the category $\boldsymbol{SET} \downarrow 1$ as defined in Par. 3.2.3?
3. Show that, in any preorder (see Par. 2.2.5), initial (resp. terminal) objects coincide with the minimum (resp. maximum), if they exist.
4. Show that, if $\boldsymbol{D}$ is a full subcategory of $\boldsymbol{C}$, any initial (resp. terminal) object of $\boldsymbol{C}$ that is an object of $\boldsymbol{D}$ is also initial (resp. terminal) in $\boldsymbol{D}$. ♦

4.1.7 Example – Eiffel's Inheritance Structure

The inheritance structure of Eiffel has an initial object – the class *NONE* that inherits from every other class, and a terminal object – the class *GENERAL* from which every other class inherits.

The following diagram is, once again, copied from [86].

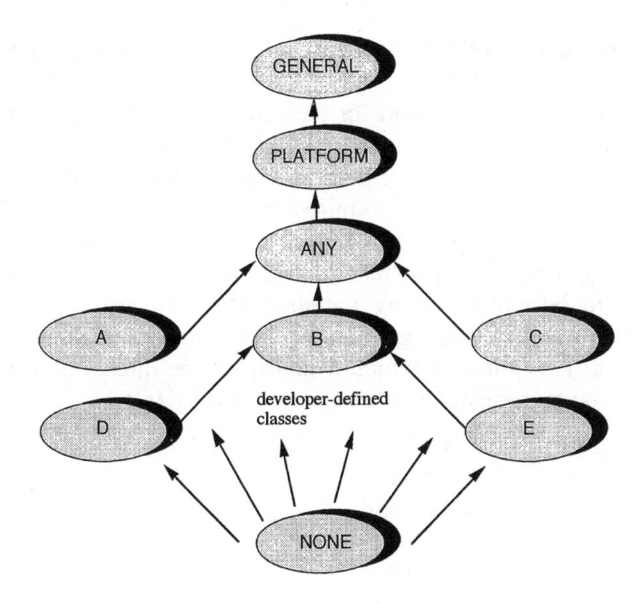

♦

4.2 Sums and Products

Intuitively, the universal constructions that are the theme of this chapter concern the possibility of finding objects that are able to capture the social lives of whole collections of objects and morphisms showing how the objects relate to one another. This is because, for instance, we are interested in the study of properties of whole systems, e.g. emergent behaviour, rather than isolated objects. The obvious questions that we have to answer are: What exactly is meant by a collection? What is the social life of such a collection of interacting objects?

Leaving the first question out of the discussion for a while, and relying on an intuitive level of understanding for the time being, we start with a simple example. Not the simplest, though: this would be for collections consisting of only one object, in which case only isomorphic objects would have a social life that is able to capture the one of the given object. The simplest non-trivial example is that of a collection of two objects with no interactions between them, for instance two processes running in parallel with no communication between them, or two software modules with no dependencies between them.

Consider first the characterisation of the out-going communication for such a collection x, y of objects. We say that, as a collection, x and y interact with other objects v via morphisms $f_x{:}x \rightarrow v, f_y{:}y \rightarrow v$. For instance, to say how a software module v uses x and y collectively, we have to say how it uses each of them in particular. This is because there are no dependencies between x and y, otherwise we would have to express the fact that these dependencies are respected.

An object that is able to stand for the relationships that a collection x, y of objects has towards its environment is called their sum and is denoted by $x+y$. In fact, the sum is more than an object. We need to make explicit how x and y relate to it. This requires morphisms (injections) i_x:$x \rightarrow x+y$ and i_y:$y \rightarrow x+y$. The ability for this object and the connecting morphisms to characterise the relationships that the collection has towards its environment can then be expressed by the property that any such interaction f_x:$x \rightarrow v$, f_y:$y \rightarrow v$ can be performed via $x+y$ in the sense that there is a unique morphism k:$x+y \rightarrow v$ through which f_x and f_y can be intercepted, i.e. $i_x;k=f_x$ and $i_y;k=f_y$. The morphism k is often represented by $[f_x,f_y]$.

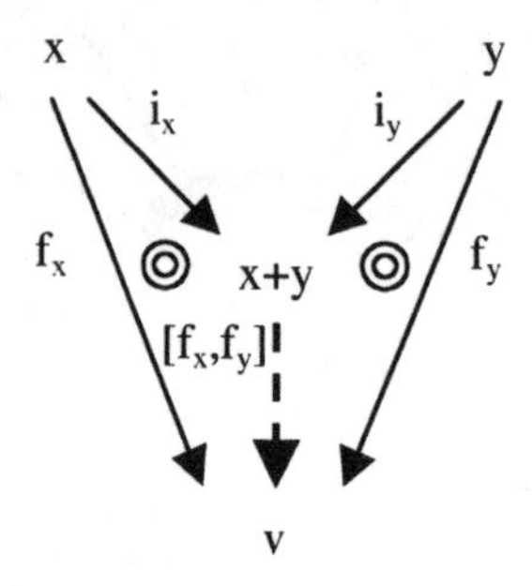

4.2.1 Definition – Sums

Let C be a category and x,y objects of C. An object z is a *sum* (or *coproduct*) of x and y with injections i_x:$x \rightarrow z$ and i_y:$y \rightarrow z$ iff for any object v and pair of morphisms f_x:$x \rightarrow v$, f_y:$y \rightarrow v$ of C there is a unique morphism k:$z \rightarrow v$ in C such that $i_x;k=f_x$ and $i_y;k=f_y$. ♦

In order to express, in diagrams, that the objects and morphisms involved are related by a universal construction, we use the symbols ⟳ and ⟲ according to rules that we will make explicit in each case. For instance, in the case of sums,

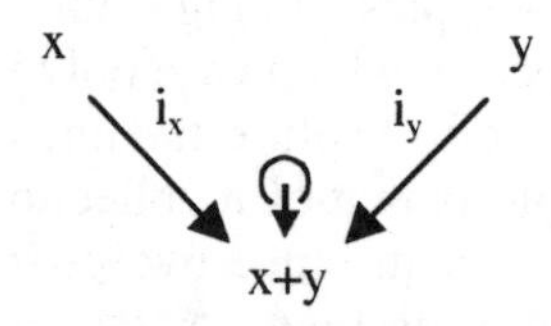

The exact identity of the sum is of no importance, just the way in which it relates to the other objects. Hence, the following property holds.

4.2.2 Proposition

Let C be a category and x,y objects of C. If a sum of x and y exists, it is unique up to isomorphism and is denoted by $x+y$.

Proof

Let z be a sum of x and y with injections $i_x:x\to z$ and $i_y:y\to z$, and w another sum of x and y with injections $j_x:x\to w$ and $j_y:y\to w$. By the universal property of z, we can conclude that there is a morphism $k:z\to w$ such that $i_x;k=j_x$ and $i_y;k=j_y$. Applying the same reasoning to w, we conclude that there is a morphism $l:w\to z$ such that $j_x;l=i_x$ and $j_y;l=i_y$. Consider now the morphism $(k;l):z\to z$. It satisfies:

$i_x;(k;l)$
$= (i_x;k);l$ associativity of the composition law,
$= j_x;l$ universal property of i_x,
$= i_x$ universal property of j_x.

Similarly, we can prove that $i_y;(k;l)=i_y$. The universal property of z, i_x and i_y implies that there is only one morphism $z\to z$ satisfying this property. This implies that $(k;l)=id_z$. Using the universal property of w, j_x and j_y we can prove in exactly the same way that $(l;k)=id_w$. Therefore k is an isomorphism. ♦

4.2.3 Example – Logical Disjunction

1. In ***LOGI***, sums correspond to disjunctions.
2. In ***PROOF***, sums also correspond to disjunctions. Indeed, the sum of A and B is a sentence characterised by morphisms that capture the introduction and elimination rules for the disjunction as a logical operator:

$$i_A:A\to A\vee B \qquad\qquad i_B:B\to A\vee B$$

$$\frac{f_A : A\to C, f_B : B\to C}{[f_A, f_B] : A\vee B\to C}$$

Notice that the conditions required of these three morphisms (derivations) correspond to typical properties of normalisation in proof-theory. ♦

Other well-known properties of disjunction can be captured in such a "universal" way.

4.2.4 Exercise

Let $\boldsymbol{C}$ be a category and 0 an initial object of $\boldsymbol{C}$. Show that, for every object x, the sum $x+0$ exists and i_x is an isomorphism, i.e. $x+0$ "is" x. ♦

4.2.5 Example – Disjoint Union of Sets

In the category ***SET***, the disjoint union $x\oplus y$ (with corresponding injections) is the sum of x and y.

Proof

1. *Existence*: consider an arbitrary object v and pair of morphisms $f_x:x\to v$, $f_y:y\to v$. Define $k:x\oplus y\to v$ as follows: given $A\in x\oplus y$, let $k(A)=f_x(a)$ if $A=i_x(a)$ with $a\in x$ and $k(A)=f_y(a)$ if $A=i_y(a)$ with $a\in y$. This is a proper definition of a total function because, on the one hand, every element of $x\oplus y$ is either in the image of x

through i_x or the image of y through i_y and, on the other hand, these two images are disjoint (which removes any conflict of choice between which case to apply). The conditions $i_x;k=f_x$ and $i_y;k=f_y$ are satisfied by construction.

2. *Uniqueness*: given any other total function $k':x \oplus y \to v$, the conditions $i_x;k'=f_x$ and $i_y;k'=f_y$ define k' completely (and equal to k). ♦

In order to illustrate the construction and show why the union of sets is not (necessarily) their sum, consider the following example where f_x and f_y are set inclusions. The injections are such that $i_x(a)=a_x$ and $i_y(a)=a_y$. By construction, $[f_x,f_y]$ is such that $[f_x,f_y](a_x)=[f_x,f_y](a_y)=a$. The union $\{a,b,c\}$ does not provide a sum through the corresponding inclusions because there is no total function $l:\{a,b,c\} \to \{a_x,a_y,b,c\}$ that satisfies the commutativity conditions. Indeed, these require $l(a)=a_x$ because $a_x=i_x(a)$ and $l(a)=a_y$ because $a_y=i_y(a)$.

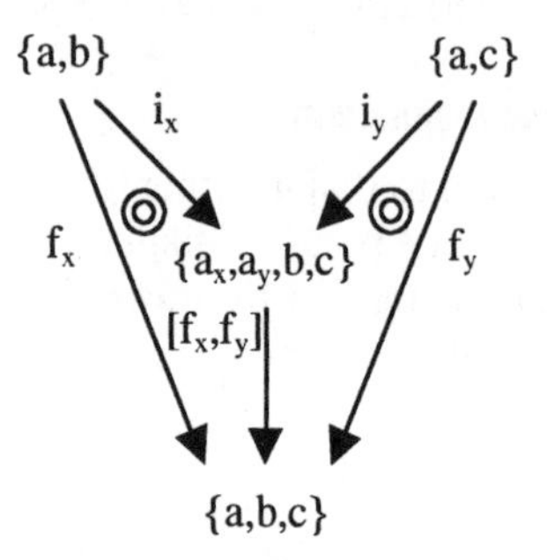

Intuitively, this happens because the union is an operation that "looks" inside the sets to which it is being applied in order not to repeat the elements that they have in common. That is, it is an operation that relies on an interaction between the sets to which it applies, whereas the sum is defined over objects without any relationship between them.

This construction illustrates how, in category theory, all interactions need to be made explicit and external to the entities involved. That is, we cannot rely on implicit relationships such as the use of the same names in the definition of different objects. This may sound too "bureaucratic", but we are far from suggesting that category theory should be used directly, that is "naked", as a language for specification, modelling or even programming. We view category theory as a mathematical framework that supports those activities. Hence, the need for an explicit and external representation of all interactions is, in our opinion, a bonus because it enforces directly fundamental properties of the methodology that is being supported, service-oriented in the most general case, but also component or object-oriented as amply demonstrated in the literature.

In the next section, we address constructions that involve interactions. In the rest of this section, we look at the dual construction of sums: products. We do so extensionally, i.e. by providing the definition of the construction directly. The reason for doing so in spite of knowing already that it is enough to reverse the direction of the arrows, is that, from a "pragmatic" point of view, one is "naturally" biased by the direction of the arrows and, in certain contexts, tradition is that one uses products instead of sums. Our experience in using and teaching category theory is that arguments and misunderstandings are too often caused by being presented with the duals of constructions that otherwise would be very familiar! Hence, it is important that we know these more basic constructions explicitly in both forms rather than having to translate back and forth every time between a category and its dual.

4.2.6 Definition – Products

The dual notion of sum is *product*. That is, letting x, y be objects of a category $\boldsymbol{C}$, a $\boldsymbol{C}$-object z is a product of x and y with projections $\pi_x{:}z{\to}x$ and $\pi_y{:}z{\to}y$ iff for any object v and pair of morphisms $f_x{:}v{\to}x, f_y{:}v{\to}y$ of $\boldsymbol{C}$ there is a unique morphism $k{:}v{\to}z$ in $\boldsymbol{C}$ (often denoted by $<f_x,f_y>$) such that $k;i_x{=}f_x$ and $k;i_y{=}f_y$. ♦

Products handle the relationships from the environment towards collections of two unrelated objects. It is easy to see that, in ***SET***, the product of two sets is given, up to isomorphism, by their Cartesian product. In ***LOGI***, products capture conjunction, which is consistent with its traditional status as the "dual" of disjunction.

4.2.7 Example – Parallel Composition Without Interactions

In the category $\boldsymbol{SET}_\perp$ of pointed sets, products are constructed in the same way as in ***SET***. The Cartesian product of two pointed sets includes all pairs of "proper elements", the pairs of which one and only one of the elements is a designated one, and the pair of designated elements. We claim that, together with the projections, this Cartesian product is still a product of the pointed sets. On the one hand, it is trivial to prove that if we elect the pair of designated elements as the designated element of the product, we obtain morphisms of pointed sets through the original ***SET***-projections. On the other hand, the commutativity requirements ensure that all the universal functions $<f_x,f_y>$ are morphisms of pointed sets, i.e. preserve designated elements. In other parts of the book, we make use of this style of argument in the proof of universal properties for certain categories with structure.

As an example, consider the use of pointed sets for modelling the alphabets of concurrent processes, as suggested in Par. 3.2.1. Products of alphabets give us all the possible events in which both processes participate ($a|b$, as a representation of the pair $<a,b>$, in the case below), plus all possible events in which only one process participates ($\perp_A|b$ and $a|\perp_B$ in the case below), and the "silent" environment event in which none of the processes participates ($\perp_A|\perp_B$ in the case below).

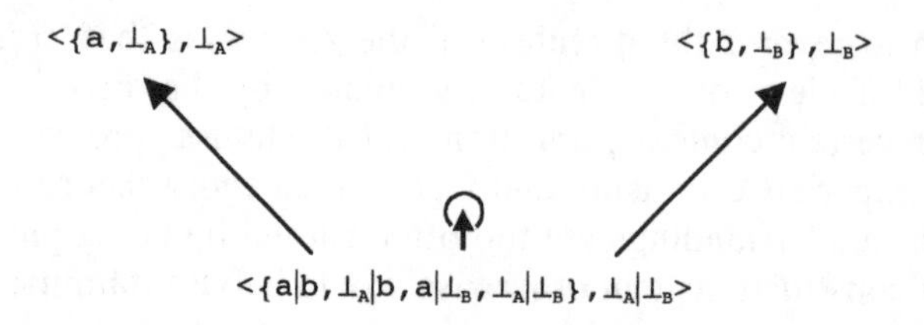

In order to simplify the notation, both textual and graphical, we normally hide the designated element as in

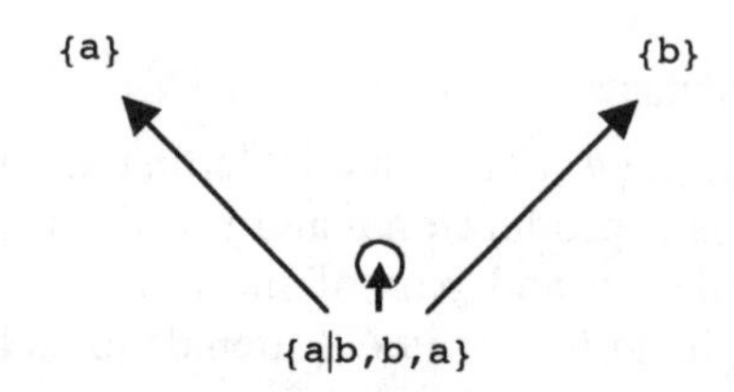

Notice in particular that the events in which only one process participates are represented by the event of that process.

In the specific case in which we identify events with synchronisation sets of method execution, the product provides for joint executions as disjoint unions of synchronisation sets. To be more precise, when we work in ***POWER*** (see Par. 3.3.2), we can take as a representation of a pair $<a,b>$ the set $a \oplus b$. This is because, as sets, the Cartesian product $2^A \times 2^B$ is isomorphic to $2^{A \oplus B}$, and the isomorphism is given precisely by the map that associates pairs $<a,b>$ with sets $a \oplus b$. This is to show that, because universal constructions are only unique up to isomorphism, we can choose from the isomorphism class the representation that suits us best. In this case, it justifies a uniform treatment of concurrent process alphabets as synchronisation sets.

In summary, through products, we obtain the alphabet of the process that is the interleaving of the given ones. ♦

4.2.8 Exercise

Characterise products and sums of partial functions (i.e. in ***PAR***). Show in particular that sums work like in ***SET***, i.e. they compute the disjoint union, but that products, besides the pairs that result from the Cartesian product, include the disjoint union of the two sets as well. That is, products in ***PAR*** are very "similar" to those in $\boldsymbol{SET_\perp}$. Why do you think this is so? What about sums in $\boldsymbol{SET_\perp}$? How do they compare? ♦

4.2.9 Example – Inheritance Without Name Clashes

In Eiffel, products capture "minimal" inheritance without name clashes, a typical example being the following hierarchy, copied again from [86]:

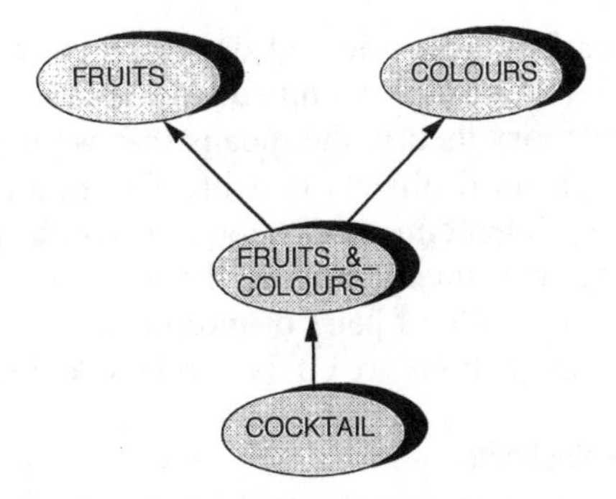

As argued in Par. 2.1.3, when the sets of features of the classes in an inheritance graph are considered, the direction of the corresponding arrows is reversed. As a result, the universal construction on the underlying features is a sum:

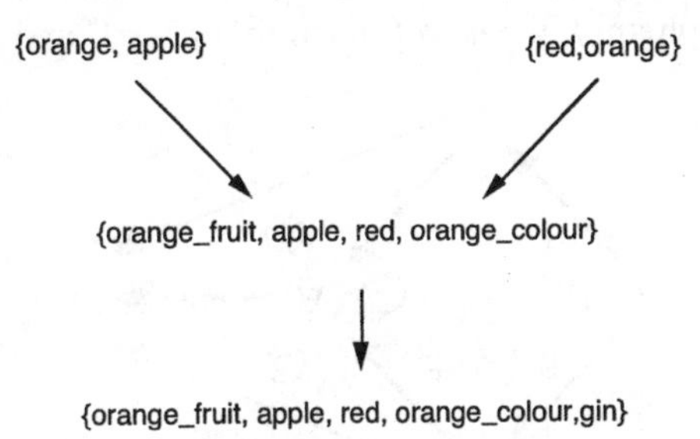

The "automatic renaming" associated with this categorical construction is very useful because it makes sure that no confusion arises from the inadvertent use of the same names for different "things" in different contexts. This is why the injections/projections are an integral part of the concept of sum/product: they keep a record of the renamings that take place, i.e. of "who is who" or, better, "who comes from where".

However, in many situations we want to make "joins", i.e. identify things that are meant to be the same but were included in different contexts. For instance, how to require that two processes synchronise on given events? How to require that features declared in different object classes be identified when performing multiple inheritance? This is the purpose of the universal construction that we illustrate next.

4.3 Pushouts and Pullbacks

A distinguishing factor of category theory, and one that makes it so suitable for addressing architectural concerns and service-oriented development in software engineering, is that such forms of interaction are *exogenous*, i.e. they have to be established *outside* the objects involved. For instance, interactions in object-oriented development are endogenous because feature calling (clientship) is embedded explicitly in the code of the caller (client). In category theory, the means that we have for establishing interactions is through third objects that handle communication between the ones that are being interconnected via given morphisms. In the case of sums, we are dealing with interactions at the level of the sources, i.e. we are interested in the social life of pairs of morphisms $f:x \to y$, $g:x \to z$. In the case of products, the interactions are on the target side, i.e. $f:y \to x$, $g:z \to x$.

4.3.1 Definition – Pushouts

Let $f:x \to y$ and $g:x \to z$ be morphisms of a category $\boldsymbol{C}$. A pushout of f and g consists of two morphisms $f':y \to w$ and $g':z \to w$ such that:

- $f;f'=g;g'$.
- For any other morphisms $f'':y \to v$ and $g'':z \to v$ such that $f;f''=g;g''$, there is a unique morphism $k:w \to v$ in $\boldsymbol{C}$ such that $f';k=f''$ and $g';k=g''$.

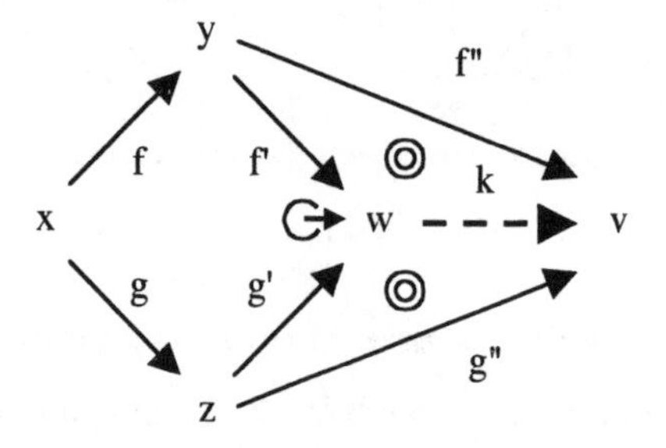

Notice the use of the symbol ↻ as in the case of sums. ♦

4.3.2 Example – Amalgamated Sums of Sets

In $\boldsymbol{SET}$, pushouts perform what are usually called amalgamated sums, i.e. they identify (join) elements as indicated by the "middle object" and corresponding morphisms. For instance, in the example below, the morphisms indicate that elements b and d are to be identified. Because nothing is said about c, the categorical default applies: the two occurrences are to be distinguished because the fact that the same name was used is accidental.

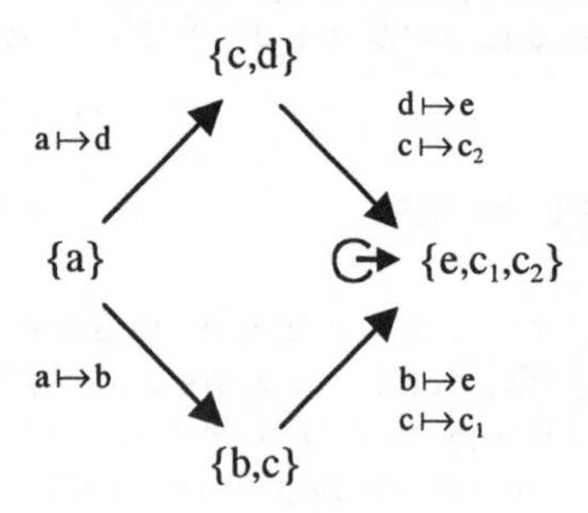

The proof that what we have built is actually a pushout can be outlined as follows. On the one hand, the commutativity requirement is clearly satisfied. On the other hand, for the universal property, consider given functions $f:\{c,d\}\rightarrow A$ and $g:\{b,c\}\rightarrow A$ satisfying the commutativity requirement: $f(d)=g(b)$. Any function $k:\{e,c_1,c_2\}\rightarrow A$ that satisfies the commutativity requirements of the universal property must be such that $k(e)=f(d)$, $k(e)=g(b)$, $k(c_1)=g(c)$ and $k(c_2)=f(c)$. These requirements leave no other choice for defining k, hence uniqueness is ensured. Because these requirements are consistent, which is due to the fact that $f(d)=g(b)$, existence is also ensured. ♦

More interesting than this proof is the mechanism through which pushouts can be systematically constructed in ***SET***. This is revealed once we address the process of computing pushouts in more general terms.

4.3.3 Example – Multiple Inheritance in Eiffel

Pushouts are the universal construction that allows us to join (merge) features during multiple inheritance. For instance, in [86], the construction of *HOME_BUSINESS* merges the home and business addresses:

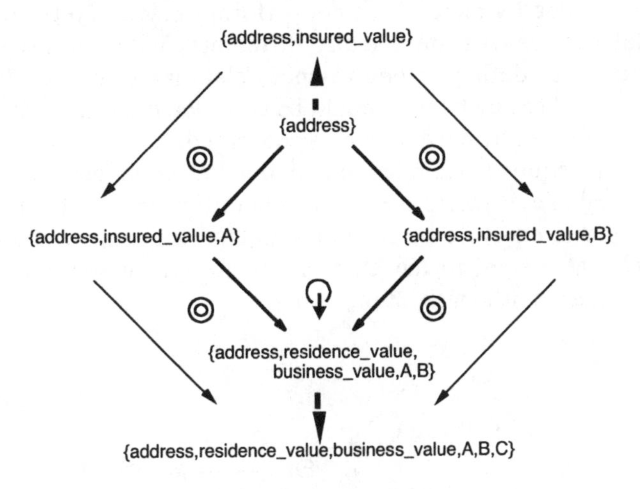

♦

Let us now return to the definition of pushout and explain how the interactions operate. The definition consists of two main requirements: a commutativity condition and a universal property. The universal property is just an instance of a "ritual" that we explain in the next section. For the time being, notice how similar it is to the corresponding property of sums, and try to see where it lies in the definition of initial object.

The commutativity requirement is the one responsible for the "amalgamation" or, more generally, for encapsulating the interactions in place. In the case of ***SET***, the amalgamation takes place as a quotient defined over the sum of the objects for an equivalence relation that is defined by the middle object and the connecting morphisms. If we take two morphisms $f:x \rightarrow y$, $g:x \rightarrow z$ and we can find a sum for y and z, we obtain a square like before except that it does not necessarily commute.

For instance, consider the following diagram:

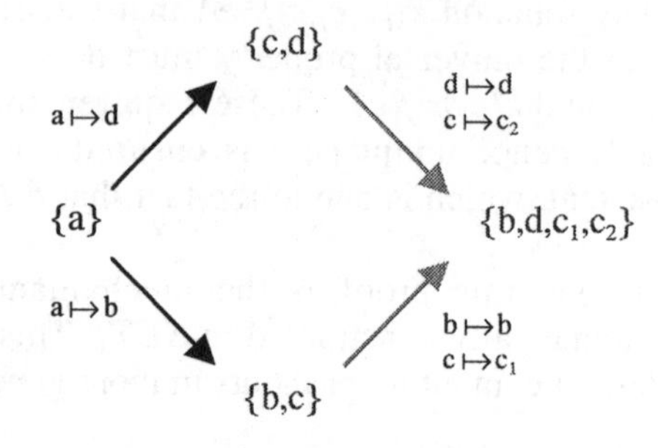

This diagram does not commute because, if we pick a in the middle object and follow all the paths from it, we do not arrive at the same element: one, via b, terminates in c_1 and the other, via d, in c_2.

In order to enforce commutativity, we have to "equalise" the ends of these paths. Category theory does not perform miracles (yet), so it cannot make equal entities that are actually different. What we usually do in mathematics is to define an equivalence relation expressing that, albeit being different, these endpoints should be considered the same as far as the system (as expressed by the diagram) is concerned.

In ***SET***, this equivalence relation is defined as being generated from all pairs $f(i_y(a)) \cong g(i_z(a))$, where $a \in x$. We claim that the quotient set of the sum $y \oplus z$ by this equivalence relation together with the functions $[i_y(\_)]$ and $[i_z(\_)]$ that assign, to the elements of the given sets y and z, their equivalence classes, is a pushout of f and g.

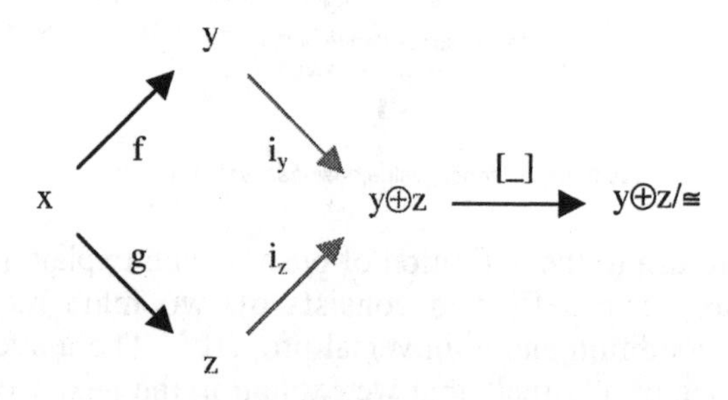

The proof of this result is not that difficult because the whole construction was made to ensure commutativity, i.e. $[i_y(f(a))]=[i_z(g(a))]$ for every $a \in x$, which is precisely the set of pairs $i_y(f(a)) \cong i_z(g(a))$ that generate the equivalence relation.

The universal property can now be derived from the universal properties of the sum (in grey in the figures) and quotients:

1. Consider any other two morphisms $f'':y \to v$ and $g'':z \to v$ such that $f;f''=g;g''$.

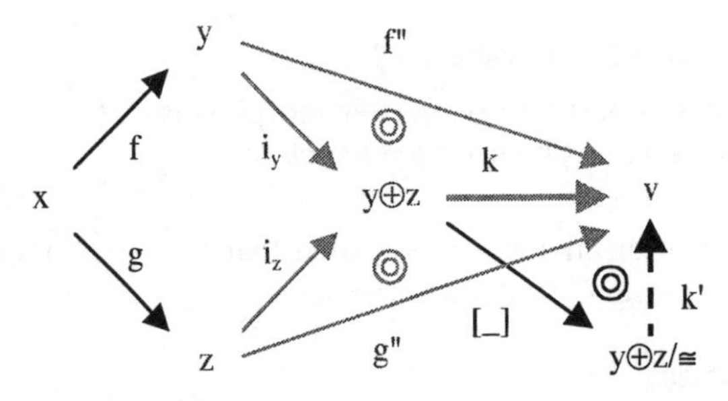

2. We know from the universal property of the sum that there is a unique morphism $k:y \oplus z \to v$ such that $i_y;k=f''$ and $i_z;k=g''$.
3. The morphism k also equalises the pairs $<i_y(f(a)),i_z(g(a))>$:

   $k(i_y(f(a)))$
   $= f''(f(a))$ from 2,
   $= g''(g(a))$ from 1,
   $= k(i_z(g(a)))$ from 2.
4. From the properties of quotients, we know that there is a unique way in which this map can be factorised through $[\_]$ i.e. there is a unique $k':y \oplus z/\cong \to v$ such that $k=[\_];k'$. Using associativity of morphism composition, this provides us with the required unique morphism satisfying $(i_y;[\_]);k'=f''$ and $(i_z;[\_]);k'=g''$.

How can we generalise the quotient construction to categories in general? Taking the diagram above to be over an arbitrary category ***C***, the purpose of the quotient is to make the following diagram commute:

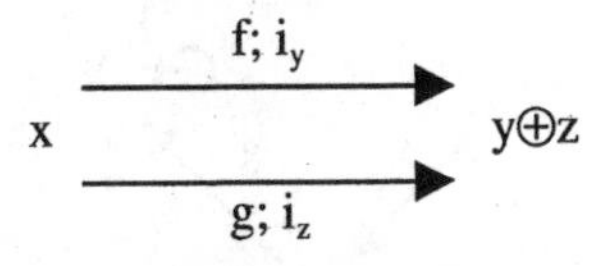

The idea is to do so via a third morphism $e:y \oplus z \to v$ whose purpose is to replace the initial equality $f;i_y=g;i_z$ by $(f;i_y);e=(g;i_z);e$. However, there are many ways of doing so. One of them is to choose v to be a terminal object (if one exists), but this is clearly too intrusive on the given morphisms because it over-equalises them. We prefer to do it in a minimal way, which is what the universal property of quotients provides: for any other morphism $e':y \oplus z \to w$ that also equalises the morphisms, i.e. such that $(f;i_y);e'=(g;i_z);e'$, there should be a unique $k:v \to w$ such that $e;k=e'$.

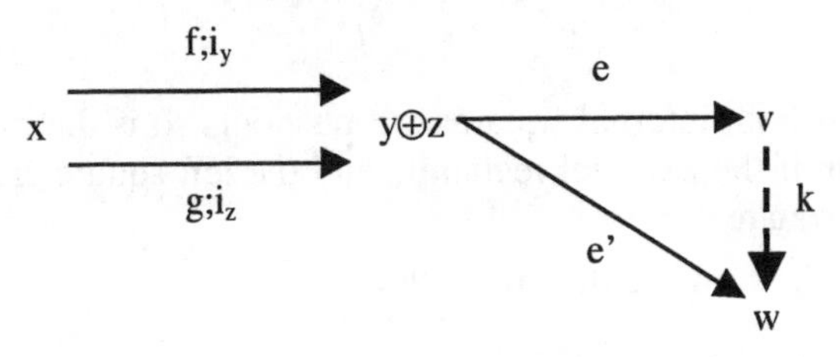

This is precisely another instance of a universal construction.

4.3.4 Definition – Coequalisers

Let C be a category and $f{:}x{\to}y$, $g{:}x{\to}y$ morphisms of C. A *coequaliser* of f and g consists of a morphisms $e{:}y{\to}z$ such that

- $f;e=g;e$.
- For any other morphisms $e'{:}y{\to}v$ such that $f;e'=g;e'$, there is a unique morphism $k{:}z{\to}v$ in C such that $e;k=e'$. ♦

4.3.5 Exercises

1. Prove that pushouts can be obtained from sums and coequalisers.
2. Prove that coequalisers are a particular case of pushouts.
3. Prove that, if initial objects exist, sums can be obtained from pushouts. ♦

The following exercises concern typical properties of pushouts that are quite useful. Their proofs also reveal a lot about the structures involved.

4.3.6 Exercises

1. Prove that the universal arrow e in a coequaliser is epic (thus generalising the fact that a quotient map is surjective).
2. In a pushout diagram as below, prove that if g is epic, so is f'.

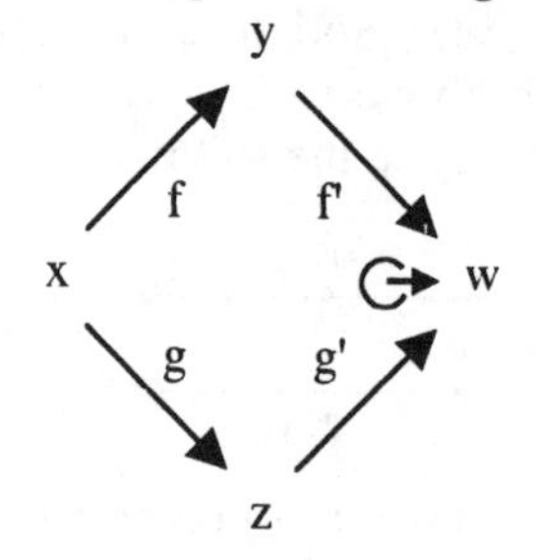

3. Consider a commutative diagram

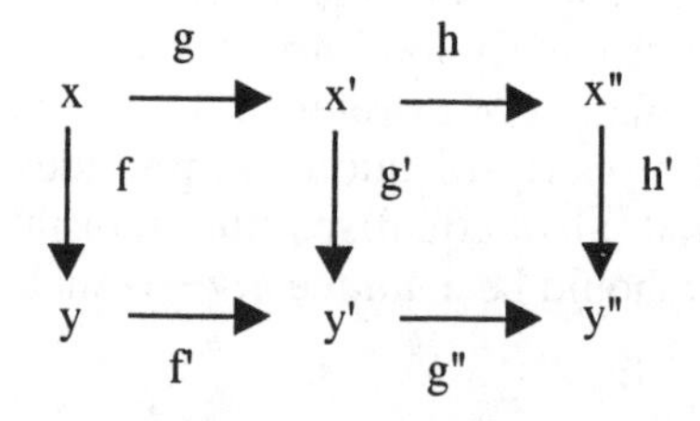

Prove that if both internal squares are pushouts so is the external rectangle, and that if the external rectangle and the left square are pushouts, so is the right square. ♦

Let us consider now the dual of pushouts.

4.3.7 Definition – Pullbacks

Let $f{:}y{\to}x$, $g{:}z{\to}x$ be morphisms of a category $\boldsymbol{C}$. A *pullback* (or fibred product) of f and g consists of morphisms $f'{:}w{\to}y$ and $g'{:}w{\to}z$ such that

- $f';f=g';g$.
- For any other morphisms $f''{:}v{\to}y$, $g''{:}v{\to}z$ such that $f'';f=g'';g$, there is a unique $\boldsymbol{C}$-morphism $k{:}v{\to}w$ such that $k;f'=f''$ and $k;g'=g''$. ♦

Pullbacks can also be explained from products and a construction that equalises arrows, except that this time the equalising needs to be made on the source and not the target side of the arrows. Not surprisingly, this universal construction is named equaliser, the dual of coequalisers.

Whereas, in set-theoretic terms, we saw that equalising on the target side corresponds to taking a quotient to group entities that should be "equal" in the same equivalence relation, equalising on the source side is even conceptually easier: it is enough to restrict the domain by throwing away the elements over which the two functions disagree. That is, we equalise through the inclusion $\{a{\in}y{:}\, f(a)=g(a)\}{\subseteq}y$.

$$\{a{\in}y{:}\ f(a)=g(a)\} \xrightarrow{\ m\ } y \ \overset{f}{\underset{g}{\rightrightarrows}}\ x$$

Hence, computing a fibred product consists of computing a product followed by a "purge" of the pairs that violate the commutativity requirement. We illustrate this procedure with process alphabets.

4.3.8 Example – Parallel Composition With Interactions

We have already argued that in the category $\boldsymbol{SET_\bot}$ of pointed sets, products are constructed in the same way as in $\boldsymbol{SET}$ through Cartesian products. When the pointed sets capture process alphabets, we saw that this construction captures parallel composition without synchronisation in the sense that all the pairs of events are generated in an interleaving semantics. Fibred products allow us to compute parallel composition with synchronisation constraints.

For instance, consider two process alphabets $\langle\{produce,store,\bot_P\},\bot_P\rangle$ and $\langle\{consume,retrieve,\bot_C\},\bot_C\rangle$. The alphabet of the parallel composition of the two processes when required to synchronise in the *store* and *retrieve* events can be obtained through the pullback of:

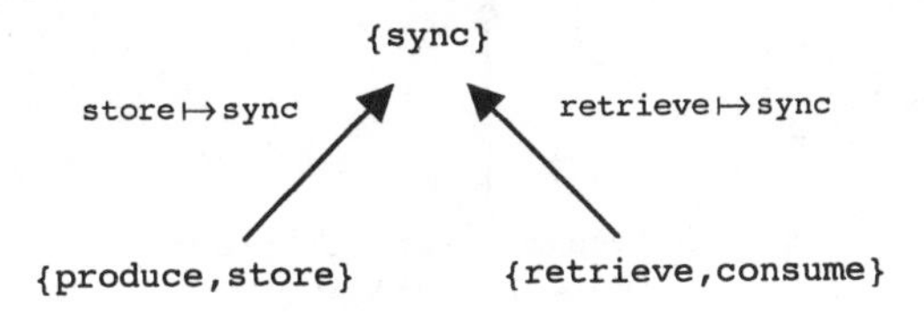

Notice that, as mentioned in Par. 4.2.7, and for simplicity, we have omitted the designated elements from the representations of the pointed sets and will only show the proper events. The middle object models the alphabet of the "cable" that is being used to interconnect the two processes. The maps mean that *store* and *retrieve* are being synchronised, with *sync* providing the place of their "rendez vous".

The product of the two alphabets gives us:

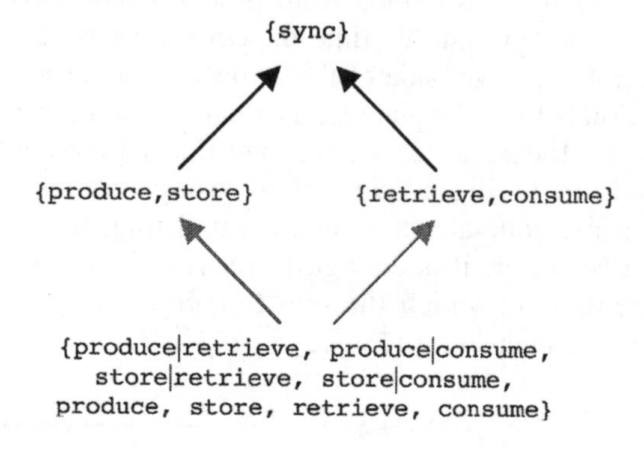

Recall that the events of the given processes are recovered as a result of synchronisations with the designated event, thus capturing system events in which only one of the two processes participates.

Notice that the diagram does not commute. For instance, the system event *store|consume* is mapped on the left to *sync*, but on the right to the designated event of the cable. In order to make the diagram commute, we throw away all events on which the maps to the channel do not agree:

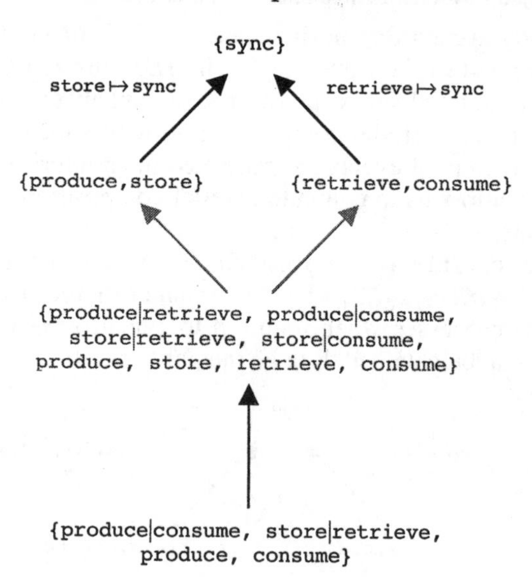

Basically, the events that remain are all possible combinations of executing *produce* and *consume* because there is no synchronisation constraint on them, plus the synchronisation that is explicitly required. ♦

4.3.9 Exercise

Follow up on Par. 4.2.8 by characterising fibred products and amalgamated sums of partial functions (i.e. in ***PAR***) and relating them to ***SET*** and $\boldsymbol{SET}_\perp$. ♦

4.3.10 Exercises

1. Define explicitly the notion of equaliser and prove that pullbacks can be obtained from products and equalisers as suggested.
2. Prove that equalisers are a particular case of pullbacks.
3. Prove that, if terminal objects exist, products can be obtained from pullbacks.
4. Prove that the equalising arrow m is a mono. ♦

4.4 Limits and Colimits

It should be clear by now that the notion of collective behaviour that we wish to capture through universal constructions takes diagrams as the expression of the collection of objects and interactions that constitute what we could call a system. In fact, we tend to use diagrams to deal with complex entities for which the objects of the category provide components and the morphisms the means for interconnecting them. Hence, for instance, a typical use of diagrams is for defining configurations. The universal constructions that we address in this chapter allow us to define the semantics of such complex entities by internalising the configuration and collapsing the structure into an object that captures the collective behaviour.

An aspect of these universal constructions that is important to keep in mind is the fact that they deliver more than an object: this object comes together with morphisms that relate it to the objects out of which it was constructed. It is through these morphisms that we can understand how properties of the system (complex object) emerge from the properties of its components and the interconnections between them. Hence, the constructions are better understood in terms of structures that consist of objects together with configurations to which they relate, what are called (co)cones.

4.4.1 Definition – Cocones

Let $\delta{:}I{\rightarrow}\boldsymbol{C}$ be a diagram in a category $\boldsymbol{C}$. A *cocone* with base δ is an object z of $\boldsymbol{C}$ together with a family $\{p_a{:}\delta_a{\rightarrow}z\}_{a\in I_0}$ of morphisms of $\boldsymbol{C}$, usually denoted by $p{:}\delta{\rightarrow}z$. The object z is said to be the vertex of the cocone, and, for each $a \in I_0$, the morphism p_a is said to be the edge of the cocone at point a. A cocone p with base $\delta:I \rightarrow C$ and vertex z is said to be commutative iff for every arrow $s:a \rightarrow b$ of graph I, $\delta_s;p_b=p_a$. ♦

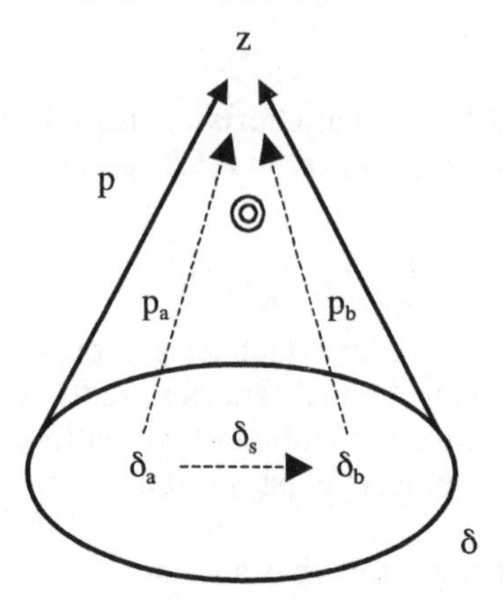

We tend to consider the family p of morphisms as identifying how the source is represented in, or is a component of the target, that is the object z. The commutativity property is important because it ensures that the interconnections that are expressed in the base through the morphisms are also represented in z. Hence, z is an object that is able to represent the base objects and their interactions. However, it may not do so in a "minimal" way. If a minimal representation exists, we call it a colimit of the diagram.

4.4.2 Definition – Colimits

Let $\delta:I \rightarrow C$ be a diagram in a category C. A colimit of δ is a commutative cocone $p:\delta \rightarrow z$ such that, for every other commutative cocone $p':\delta \rightarrow z'$, there is a unique morphism $f:z \rightarrow z'$ such that $p;f=p'$, i.e. $p_a;f=p'_a$ for every edge. Colimit cocones are decorated with ↻.

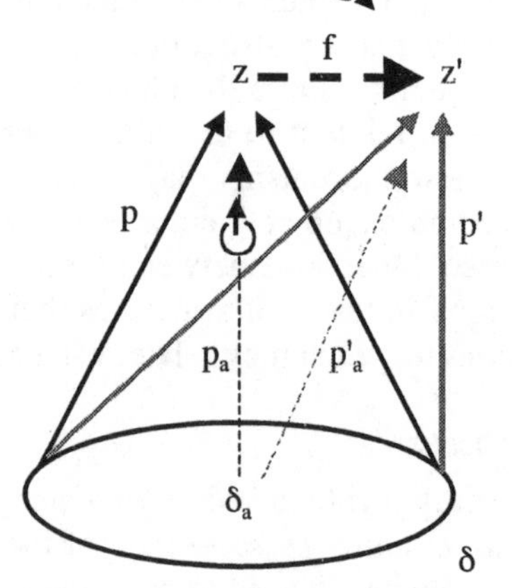

♦

Notice how all the ingredients used in the universal constructions studied in the previous sections are present in this definition. On the one hand, there is a commutativity requirement. On the other hand, a universal property ensures minimality.

4.4.3 Exercise

Show that initial objects, sums, coequalisers and pushouts are instances of colimits by identifying the shape of the base diagrams and checking that the properties required are equivalent. ♦

Cocones over a given base can be organised in a category of their own.

4.4.4 Definition/Proposition

Let $\delta:I\to C$ be a diagram in a category C. A category ***CO_CONE(δ)*** is defined whose objects are the commutative cocones with base δ, and the morphisms f between cocones $p:\delta\to z$ and $q:\delta\to w$ are the morphisms $f:z\to w$ such that $p;f=q$, i.e. $p_a;f=q_a$ for every edge. ♦

4.4.5 Exercises

1. Prove the previous result.
2. Prove that the colimits of δ are the initial objects of ***CO_CONE(δ)***.
3. Conclude that colimits are unique up to isomorphism. ♦

4.4.6 Definition – Cocompleteness

A category is (finitely) cocomplete if all (finite) diagrams have colimits. ♦

There are several results on the finite cocompleteness of categories. A commonly used one is given below.

4.4.7 Proposition

A category C is finitely cocomplete iff it has initial objects and pushouts of all pairs of morphisms with common source. ♦

Using this result and the examples that we studied in the previous sections, we can conclude, for instance, that both ***SET*** and ***LOGI*** are finitely cocomplete. These notions are also straightforward generalisations of well-known constructions over ordered sets, namely least upper bounds and greatest lower bounds. A more "interesting" example can be given over Eiffel class specifications.

4.4.8 Example – Eiffel's "Join Semantics" Rule

The category ***CLASS_SPEC*** is finitely cocomplete.

Proof

Once more, the proof is much more interesting than the result. Using the previous result, we show that ***CLASS_SPEC*** admits initial objects and pushouts.

1. It is easy to prove that the class specification that has no features is initial.
2. On the other hand, pushouts work as follows:
   - Given a diagram

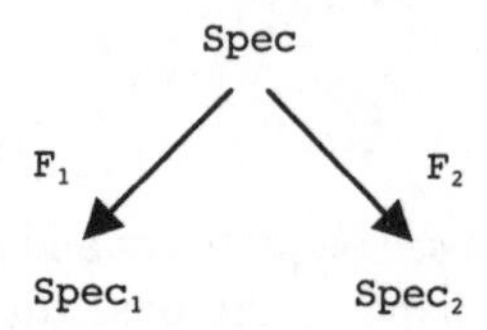

   we first compute the pushout for the underlying diagram of signatures (sets of attributes, functions and routines). This pushout returns, for each feature

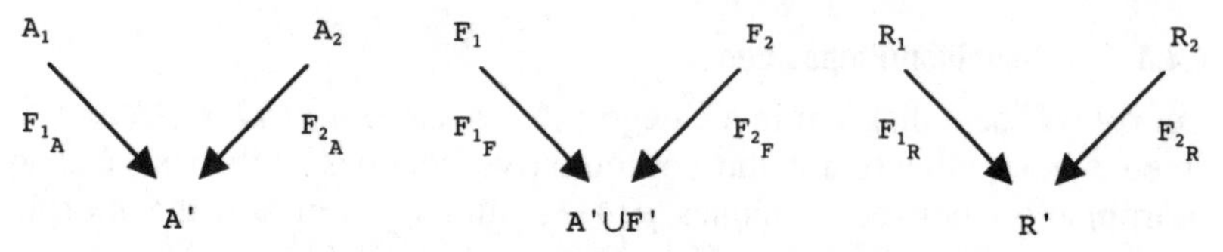

   Recall that, according to the definition of morphism given in Sect. 3.4, functions can be mapped to attributes (but not the other way around)! Hence, equivalence classes can mix together attributes and functions. The pushout classifies an equivalence class as an attribute iff it contains at least one attribute, and as a function iff it only contains functions. Routines are not mixed together with functions or attributes.
   - For every routine $r' \in R'$, we compute its pre/postconditions as follows:
   If $r'=F_{1_R}(r_1)=F_{2_R}(r_2)$ then
   $$pre_{r'} = F_1(pre_{r_1}) \vee F_2(pre_{r_2}) \text{ and } post_{r'} = F_1(post_{r_1}) \wedge F_2(post_{r_2}).$$
   If $r'=F_{1_R}(r_1) \notin F_{2_R}(R_2)$ then $pre_{r'} = F_1(pre_{r_1})$, and $post_{r'} = F_1(post_{r_1})$.
   If $r'=F_{2_R}(r_2) \notin F_{1_R}(R_1)$ then $pre_{r'} = F_2(pre_{r_2})$, and $post_{r'} = F_2(post_{r_2})$.
   - The new invariant is $I' = F_1(I_1) \wedge F_2(I_2)$.

The proof that the maps F_1 and F_2 that result from these constructions are indeed morphisms of class specifications is trivial. The commutativity property is inherited from the pushouts in ***SET*** that determine the new attributes and routines. The universal property is left as an exercise. It reflects the universal properties of conjunction and disjunction that we already identified in the previous sections. ♦

Note how this construction matches the *Join Semantics rule* and inheritance of invariants via concatenation of parent invariants [86].

4.4.9 Exercise

Work out in detail the way pushouts operate on attributes and functions of class specifications. ♦

4.4.10 Definition – Cones and Limits

The dual notion of cocone is *cone*, and the dual notion of colimit is *limit*. Limit cones are decorated with . ♦

Hence, limits generalise terminal objects, products, equalisers and pullbacks. From the examples studied in the previous sections, we can also conclude that both ***SET*** and $\boldsymbol{SET}_\perp$ are (finitely) complete.

4.4.11 Example – Parallel Composition with Interactions

In order to illustrate the calculation of limits, consider process alphabets once again. In the previous section, we showed how we can synchronise a consumer and a producer on the *store/retrieve* events. This interconnection is, however, too tight because it ties the consumer completely to that producer and does not allow it to consume from other producers. Hence, the situation that we would like to model now is the one in which the consumer retrieves from a producer but leaves open the possibility of retrieving from other producers as well. This form of interconnection cannot be achieved simply through a channel as before.

What we have to do is to make explicit a communication protocol that can be placed in between the producer and the consumer. The alphabet of this protocol needs to account for the synchronisation between the consumer and the producer, which we model through an event *sync*, and the open communication between the consumer and other possible producers, which we model through an event *open.*

Omitting, as before, the designated events, this configuration is given by the following diagram:

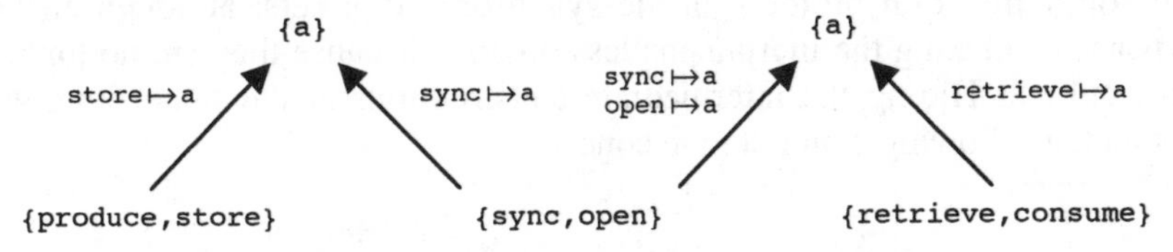

The dual of Par. 4.4.7 can be used to compute the limit of this diagram by computing successive pullbacks.

We start with the obvious pullbacks over each of the interconnections:

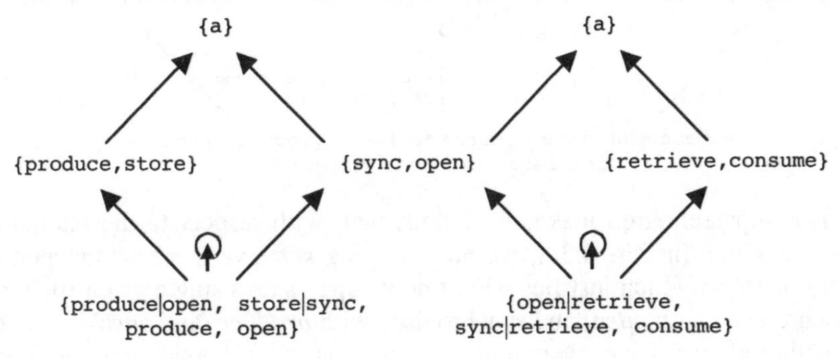

We finalise with the pullback of the new arrows, which provides for the global interaction:

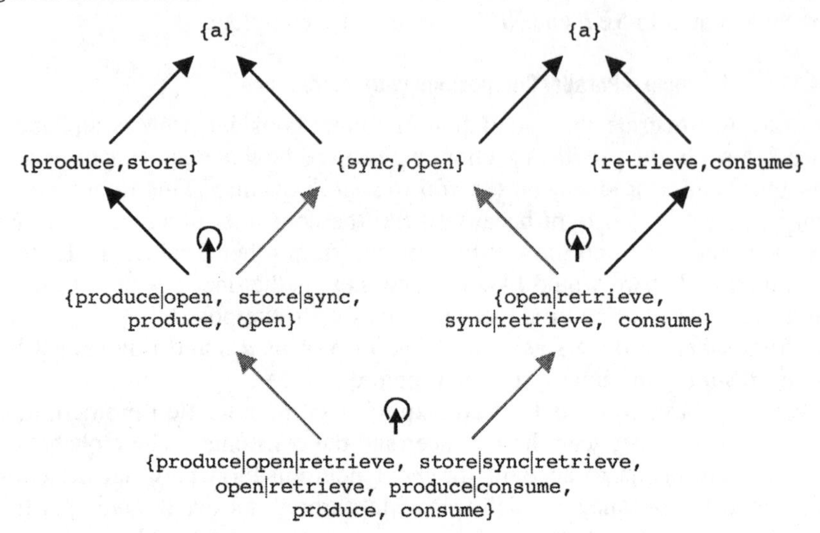

Using the privilege of choosing among the isomorphic representations of the limit the one that best reflects our intuition or purpose, we can simplify the notation. For instance, we can omit the reference to the actions of the communication protocol in the synchronisation sets, although at the expense of making the morphisms less obvious because they are no longer projections. Hiding the intermediate constructions at the same time, we obtain the following commutative cone:

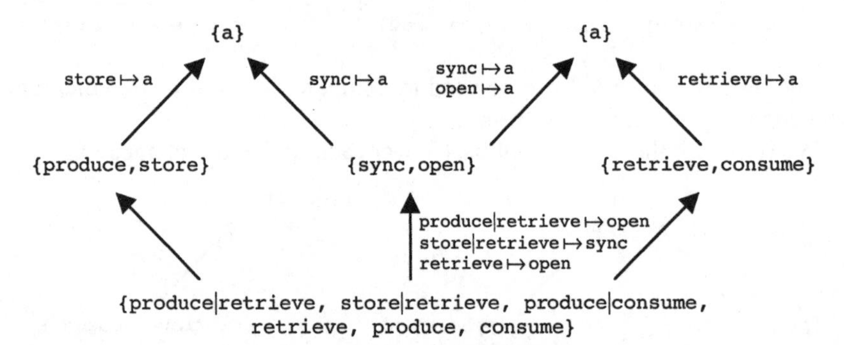

This representation makes it obvious that, with respect to the interconnection given in Par. 4.3.8, we are allowing *retrieve* to occur independently of *store*, which justifies why it now appears as a single event (though synchronised with *open*) and synchronised with *produce* (and *open*). ♦

Although we have systematically motivated universal constructions such as these in terms of parallel composition of processes, with and without interaction, we have remained at the level of the composition of process alphabets, i.e. we have not taken into account their behaviours. There are two reasons for this. On the one hand, all the complexity of interaction resides on alphabets. On the other hand, the way behaviours can be brought into the picture can be used to illustrate another categorical structure and construction. Hence, we defer the completion of this topic to Sect. 6.3 (Pars. 6.3.2 and 6.3.7). For similar reasons, we defer the illustration of the application of universal constructions to configurations of specifications to Sect. 6.1 (Par. 6.1.24) so that we can use it to illustrate another class of categorical structures and constructions.

5 Functors

5.1 The Social Life of Categories

It should come as no surprise that we are also interested in the social life of categories and, therefore, need to define a corresponding notion of morphism. Indeed, we have said more than once that, when defining a category, we are capturing only a particular view or structure of the underlying objects. There may well be more than one such notion of structure for the same collection of objects. For instance, in the case of object-oriented development, we have already mentioned the structure that relates to the way objects can be interconnected into systems, and the view that captures refinement. There may also be relationships that we wish to study between two different domains, e.g. Eiffel specifications and processes. Hence, we need a way of reasoning about relationships between categories.

Because categories are structured objects, more precisely, graphs with the additional structure given by the identity arrows and the composition law, the definition of morphisms between categories (functors) should now look "standard".

5.1.1 Definition – Functors

Let $\boldsymbol{C}$ and $\boldsymbol{D}$ be categories. A *functor* $\varphi{:}\boldsymbol{C}{\rightarrow}\boldsymbol{D}$ is a graph homomorphism (see Par. 2.1.7) from $\boldsymbol{graph(C)}$ into $\boldsymbol{graph(D)}$ such that:

- $\varphi_1(f;g) = \varphi_1(f);\varphi_1(g)$ for each path fg in $\boldsymbol{graph(C)}$.
- $\varphi_1(id_x) = id_{\varphi_0(x)}$ for each node x in $\boldsymbol{graph(C)}$. ♦

5.1.2 Examples

1. For any category $\boldsymbol{C}$, the identity functor $\boldsymbol{id_C}{:}\boldsymbol{C}{\rightarrow}\boldsymbol{C}$ consists of the identity map on objects and the identity map on morphisms.
2. For any given functor $\varphi{:}\boldsymbol{C}{\rightarrow}\boldsymbol{D}$, its dual $\varphi^{op}{:}\boldsymbol{D}^{op}{\rightarrow}\boldsymbol{C}^{op}$ is defined by the graph homomorphism that operates the same mappings on nodes and arrows.
3. We define a functor $\boldsymbol{nodes}{:}\boldsymbol{GRAPH}{\rightarrow}\boldsymbol{SET}$ as follows:
   - $\boldsymbol{nodes}(G_0,G_1,src,trg)=G_0$.
   - $\boldsymbol{nodes}(\varphi{:}G{\rightarrow}H)=\varphi_0{:}G_0{\rightarrow}H_0$

4. We define a functor ***PROOF→LOGI*** by mapping sentences to themselves and every proof between two sentences to a logical implication between them. This is an example of a *forgetful functor*: the details of the proof are forgotten; only the fact that there exists a proof remains.
5. Another example of a forgetful functor is $\boldsymbol{sign}$:$\boldsymbol{PRES}_{LTL}\boldsymbol{\rightarrow SET}$ mapping presentations and their morphisms to the underlying signatures and signature morphisms. A similar forgetful functor is obtained for $\boldsymbol{THEO}_{LTL}$.
6. We define a functor $\boldsymbol{ANCESTOR \rightarrow SET}^{op}$ by mapping classes to the set of their features and inheritance relationships to the functions that rename the features. Notice that the target category is the dual of ***SET***. This is because, as illustrated in Par. 2.1.3, the edges of the inheritance graph and the corresponding graph of features have opposite directions. Functors of the form $\varphi:\boldsymbol{C \rightarrow D}^{op}$ (or $\varphi:\boldsymbol{C}^{op}\boldsymbol{\rightarrow D}$) are sometimes called *contravariant* between ***C*** and ***D***. A functor $\varphi:\boldsymbol{C \rightarrow D}$ can also be called *covariant* between ***C*** and ***D***. ♦

5.1.3 Example – Relating Programs and Specifications

In computing, functors are often used to express relationships between different levels of abstraction, namely by providing a way of abstracting properties from representations. In this context, the abstractions are sometimes called "specifications", and the representations "programs". For instance, although we have worked with specifications of Eiffel classes and not with the programs that implement the methods of such classes, it is clear that such specifications are at a lower level of abstraction than the temporal specifications that we defined in Sect. 3.5. For instance, the temporal specifications can be used to describe general properties of the possible behaviours of a system whereas Eiffel specifications are concerned with more specific operational aspects of the execution of actions, like their effects on the attributes.

The notion of a program satisfying a specification is typically formalised through a (satisfaction) relation. When specifications consist of sets of logical formulas such as those defined in Sect. 3.5, the collection of specifications that are satisfied by a given program is ordered by inclusion and has a maximum: the union of all the sets of properties satisfied by the program. In this case, we can assign to every program P a "canonical" specification ***spec(P)*** – its strongest specification.

We have also seen that morphisms between models of system behaviour can be used to capture notions of simulation or refinement. In the traditional stepwise refinement method initiated by Dijkstra [23], refinement is formalised as a relationship between two units of design: starting with a specification of the intended behaviour of the system, one progresses by adding detail (superposing activities, computation mechanisms, etc) until a design is reached that satisfies certain criteria. These criteria, the notions of "specification" and "unit of design", as well as the notion of "progress" (refinement step) itself, vary according to the formalism that is being used to support development.

When ***spec*** is a functor, this means that the refinement relationship defined on programs is captured by the corresponding notion of refinement at the level of specifications. When specifications are given as theory presentations, this means that program refinement is property-preserving, which is what one usually expects from a refinement relation. Hence, there is a natural way in which the satisfaction relation between programs and specifications can be expected to be functorial.

As an example, consider the following mapping ***spec*** from ***CLASS_SPEC*** to $\boldsymbol{PRES_{FOLTL}}$, where by $\boldsymbol{PRES_{FOLTL}}$ we are denoting the category of theory presentations for the first-order extension of the temporal logic presented in Par. 3.5.4.[1]

- The image of every class signature is itself. That is, the features of the class are taken as symbols of the vocabulary of the logic.
- For every routine r, its specification (pre,pos) is mapped to the temporal formula

$$(r \wedge pre \wedge att \supset Xpos*)$$

By att we denote the conjunction

$$\bigwedge_{a \in att(\Sigma)} (a = x_a)$$

where the variables x_a are all new, and by $pos*$ we denote the formula that is obtained from the expression pos by replacing every occurrence of every expression *(**old** a)*, where a is an attribute, by the variable x_a.
- The invariant I is mapped to itself.

For instance, the specification of the bank account given in Par. 3.5.6 is mapped to the following set of temporal formulas:

```
deposit(i)∧balance=x_balance ⊃ X(balance=x_balance+i)
withdrawal(i)∧balance≥i∧balance=x_balance ⊃ X(balance=x_balance−i)
vip ⊃ balance≥1000
```

To show that we obtain, indeed, a functor, we have to show in particular that morphisms of class specifications are property preserving. Consider a morphism $F{:}e{=}\langle\Sigma,P,I\rangle \rightarrow e'{=}\langle\Sigma',P',I'\rangle$ of ***CLASS_SPEC***.

- Given any routine r of Σ, its image $F(r)$ is such that its specification (pre',pos') is mapped to: $(F(r) \wedge pre' \wedge att' \supset Xpos'*)$. We know that

1. $F(pre) \vdash pre'$	because F is a morphism,
2. $pos' \vdash F(pos)$	because F is a morphism,
3. $att' \supset Xpos'* \vdash F(att) \supset XF(pos**)$	from 2.

[1] For simplicity, and because the properties of the categories for the propositional and first-order versions of temporal logic in which we are interested are the same, we omit the extension. The reader interested in the logic itself can consult [67] as well as [39] for its use in a categorical framework.

from which we can conclude that

$(F(r) \wedge pre' \wedge att' \supset Xpos'*) \vdash F(r \wedge pre \wedge att \supset Xpos*)$

- By definition of morphism, we also have $I' \vdash F(I)$.

Hence, every axiom of ***spec***(Σ) is translated through F to a theorem of ***spec***(Σ'). ♦

It should come as no surprise that even categories can be organised in a category.

5.1.4 Definition/Proposition – The Category of Categories

1. Let $\varphi{:}\boldsymbol{C}\rightarrow\boldsymbol{D}$ and $\psi{:}\boldsymbol{D}\rightarrow\boldsymbol{E}$ be functors. By $\varphi;\psi$ we denote the functor defined by $(\varphi;\psi)_0=\varphi_0;\psi_0$ and $(\varphi;\psi)_1=\varphi_1;\psi_1$. This law of composition is associative and admits identities as defined in Par. 5.1.2, point 1.
2. We can thus define the category ***CAT*** whose objects are the categories and whose morphisms are the functors.

Proof

The proofs are trivial. However, bear in mind that, in defining ***CAT***, there are problems of "size" as noted at the beginning of Sect. 2.1. Again, see a more "mathematically oriented" book such as [80] for matters of size. ♦

Having a notion of morphism between categories, we can apply to categories and functors all the machinery that we have so far presented for manipulating objects and their morphisms. For instance, the notion of product that we defined in Sect. 3.1 corresponds to a universal construction in the category ***CAT***.

5.1.5 Definition/Proposition – Product of Categories

Given categories $\boldsymbol{C}$ and $\boldsymbol{D}$, we define functors $\pi_C{:}\boldsymbol{C}\times\boldsymbol{D}\rightarrow\boldsymbol{C}$ and $\pi_D{:}\boldsymbol{C}\times\boldsymbol{D}\rightarrow\boldsymbol{D}$, called the projections of the product, by mapping objects and morphisms of $\boldsymbol{C}\times\boldsymbol{D}$ to their components in $\boldsymbol{C}$ and $\boldsymbol{D}$, respectively. These functors satisfy the universal property of products in the following sense: given any category $\boldsymbol{E}$, and functors $\varphi_C{:}\boldsymbol{E}\rightarrow\boldsymbol{C}$ and $\varphi_D{:}\boldsymbol{E}\rightarrow\boldsymbol{D}$, there is a unique functor $\psi{:}\boldsymbol{E}\rightarrow\boldsymbol{C}\times\boldsymbol{D}$ such that $\varphi_C=\psi;\pi_C$ and $\varphi_D=\psi;\pi_D$. We normally denote ψ by $<\varphi_C,\varphi_D>$. ♦

We can also define a product over functors:

5.1.6 Definition/Proposition – Product of Functors

The product $\varphi_1\times\varphi_2{:}\boldsymbol{C}_1\times\boldsymbol{D}_1\rightarrow\boldsymbol{C}_2\times\boldsymbol{D}_2$ of two functors $\varphi_1{:}\boldsymbol{C}_1\rightarrow\boldsymbol{D}_1$ and $\varphi_2{:}\boldsymbol{C}_2\rightarrow\boldsymbol{D}_2$ is defined by $(\varphi_1\times\varphi_2)<c_1,c_2>=<\varphi_1(c_1),\varphi_2(c_2)>$ on objects and $(\varphi_1\times\varphi_2)<f_1,f_2>=<\varphi_1(f_1),\varphi_2(f_2)>$ on morphisms. ♦

Functors come in all "shapes and sizes", and as morphisms between categories, they provide the means for characterising the structural properties of categories in the way they relate to other categories. Hence, it is useful to study the properties of functors and the way they allow us to reveal the structure of categories. We start with some elementary properties that result from the functional nature of functors as mappings between the sets of nodes and arrows.

5.1.7 Definition

Let φ:$\boldsymbol{C}$→$\boldsymbol{D}$ be a functor.

1. φ is an *isomorphism* iff there is a functor ψ:$\boldsymbol{D}$→$\boldsymbol{C}$ such that $\varphi;\psi=id_C$ and $\psi;\varphi=id_D$. In this case, $\boldsymbol{C}$ and $\boldsymbol{D}$ are said to be *isomorphic* and ψ to be the *inverse* of φ.
2. φ is an *embedding* iff φ_1 is injective, i.e. φ is injective on morphisms.
3. φ is *faithful* iff all the restrictions $\varphi_1{:}hom_C(x,y)\to hom_D(\varphi_0(x),\varphi_0(y))$ are injective.
4. φ is *full* iff all the restrictions $\varphi_1{:}hom_C(x,y)\to hom_D(\varphi_0(x),\varphi_0(y))$ are surjective. ◆

5.1.8 Exercise

Show that faithful functors are not necessarily embeddings: this is the case only when they are also injective on objects. ◆

5.1.9 Proposition

Let φ:$\boldsymbol{C}$→$\boldsymbol{D}$ and ψ:$\boldsymbol{D}$→$\boldsymbol{E}$ be functors.

1. If φ and ψ are isomorphisms (resp. embeddings, faithful, full), so is $\varphi;\psi$.
2. If $\varphi;\psi$ is an embedding (resp. faithful), so is φ.
3. If $\varphi;\psi$ is full, so is ψ. ◆

5.1.10 Example

Every subcategory $\boldsymbol{D}$ of a category $\boldsymbol{C}$ defines an inclusion functor $\iota_{D,C}$:$\boldsymbol{D}$→$\boldsymbol{C}$. This functor is an embedding, and is full iff $\boldsymbol{D}$ is a full subcategory of $\boldsymbol{C}$. ◆

Embeddings are, intuitively, the best approximations to subcategories. In fact, in a world in which concepts are normally taken up to isomorphism, it seems intuitive not to make any difference between embeddings and inclusions of subcategories because all that is at stake are the identities of the objects involved, not their properties or structure. This intuition is supported by the following result.

5.1.11 Proposition

A functor ψ:$\boldsymbol{D}$→$\boldsymbol{C}$ is a (full) embedding iff there is a (full) subcategory $\boldsymbol{C}'$ of $\boldsymbol{C}$ and an isomorphism ϕ:$\boldsymbol{D}$→$\boldsymbol{C}'$ such that $\psi=\phi;\iota_{C',C}$. ◆

The relationships that we studied in Sect. 3.3 between subcategories and isomorphisms extend to functors.

5.1.12 Proposition

Let φ:$\boldsymbol{C}$→$\boldsymbol{D}$ be a functor.

1. φ *preserves isomorphisms*, i.e. if f:x→y is a $\boldsymbol{C}$-isomorphism, then $\varphi(f)$ is also an isomorphism.
2. If φ is faithful and full, then it *reflects isomorphisms*, i.e. if $\varphi(f)$ for f:x→y is an isomorphism, then f is itself an isomorphism. ♦

The notions of preservation and reflection are extended in Par. 5.2.1 to universal constructions. We end this section with the definition of a class of functors that arise from (co)reflective subcategories.

5.1.13 Definition/Proposition – Reflector

Let $\boldsymbol{D}$ be a reflective subcategory of a category $\boldsymbol{C}$. We define a functor ρ:$\boldsymbol{C}$→$\boldsymbol{D}$ as follows:

- Every $\boldsymbol{C}$-object c has a $\boldsymbol{D}$-reflection arrow η_c:c→d. We define $\rho(c)=d$.
- Consider now a morphism h:c→c'. The composition $h;\eta_{c'}$ is such that the definition of $\boldsymbol{D}$-reflection arrow for c guarantees the existence and uniqueness of a morphism h':$\rho(c)$→$\rho(c')$ such that $h;\eta_{c'}=\eta_c;h'$. We define $\rho(h)=h'$.

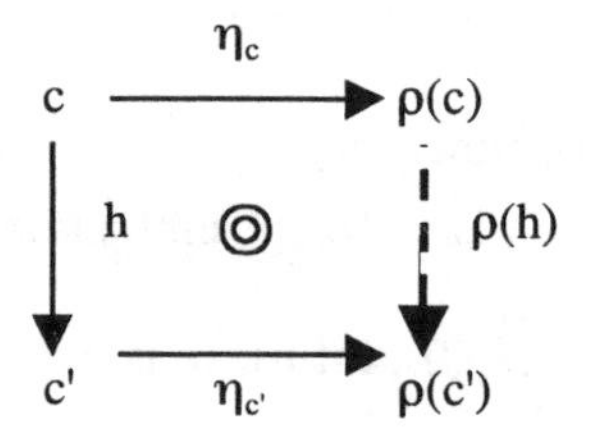

This functor is called a *reflector for* $\boldsymbol{C}$, more precisely *for the inclusion functor* $\iota_{D,C}$:$\boldsymbol{D}$→$\boldsymbol{C}$.

Proof

We have to prove that a functor is indeed defined.

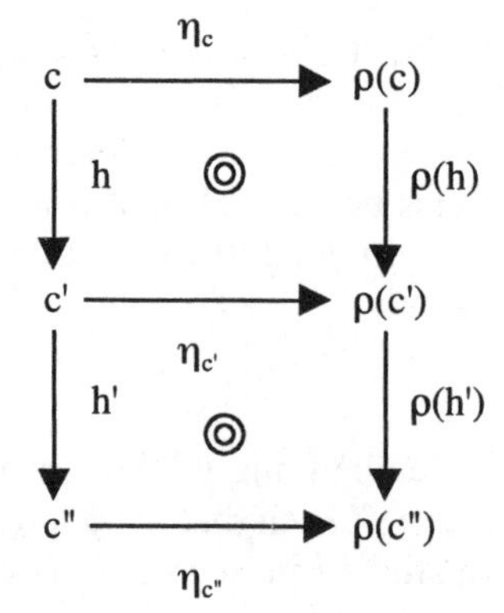

1. The fact that a graph homomorphism is defined (i.e. sources and targets of arrows are respected) is trivially checked.
2. By definition, $\rho(id_c)$ is the unique D-morphism $h':\rho(c)\to\rho(c)$ satisfying $id_c;\eta_c=\eta_c;h'$. Because $id_{\rho(c)}$ also satisfies that equation, we get $id_{\rho(c)}=\rho(id_c)$.
3. Consider now the composition law. The composition $\rho(h);\rho(h')$ is such that $(h;h');\eta_{c''}=\eta_c;(\rho(h);\rho(h'))$. This equality is obtained from the equations satisfied individually by $\rho(h)$ and $\rho(h')$. Hence, by definition, $\rho(h);\rho(h')$ is $\rho(h;h')$. ♦

By duality, we obtain the notion of *coreflector*. Notice that we had already made use of the idea of coreflector in Sect. 3.3 to explain the intuitions behind coreflective subcategories. When applied to the categories defined in Par. 3.5.4 for temporal specifications, we obtain several functors that perform the closure of sets of axioms, both as reflectors (from ***PRES*** and from ***SPRES*** to ***THEO***) and coreflectors (from ***PRES*** both to ***SPRES*** and ***THEO***). Notice in particular that the functor from ***PRES*** to ***THEO*** is both a reflector and a coreflector.

This proposition also shows that the relationship between automata and reachable automata is functorial. The map that eliminates non reachable states as defined after Par. 3.3.7 is a coreflector. We generalise these kinds of functors in Sect. 7.2. They are particular classes of functors that generate free or canonical structures, or approximate other kinds of structures.

5.2 Universal Constructions Versus Functors

As seen in Chap. 4, one of the motivations for the study of universal constructions like (co)limits is the ability to capture the collective behaviour of systems of interconnected components. In the previous section, functors can be seen to provide the means to map and relate different levels of abstraction in system development or different aspects of system description. One of the obvious questions that arises concerns the ability of functors to relate such universal constructions when applied in the categories related by the functors.

For an example of what we mean, let us discuss what we consider to be one of the most profound recent contributions that have been made in the area of programming languages and models: the separation between "computation" and "coordination" [51]. The best introduction we know to this topic can be found in [3]. In a nutshell, this whole area of research evolves around the ability to separate what in systems are the structures responsible for the computations that are performed and the mechanisms that are made available for coordinating the interactions that may be required between them. A decade of research has shown that languages that support this separation of concerns can improve our ability to handle complex systems. Its importance for architectures has led to very close synergies between the two areas and a great deal of cross-fertilisation goes on.

One of the challenges of this area has been to provide a mathematical characterisation of this separation that is independent of the particular languages that were developed for this paradigm. Such a characterisation is important, on the one hand, to establish a systematic study of its properties and relationships to other paradigms and, on the other hand, to extend and support the new paradigm with tools. This is what we started to do in [35] through the use of category theory.

The basic idea of our approach is to model this separation by a forgetful functor ***int*:*SYS*→*INT***, where the category ***SYS*** stands for systems, i.e. for whatever representations (models, behaviours, specifications, etc.) we are using for addressing systems. The category ***INT*** intends to capture the mechanisms that, in these representations, are responsible for the coordination aspects. We refer to the objects of ***INT*** as interfaces following the idea that interconnections should be established only on the basis of what systems make available for interaction with other systems (e.g. communication channels), not at the level of the computations that they perform.

An example that will help us clarify more precisely what we have in mind can be given in terms of the linear temporal specifications introduced in Sect. 3.5. Therein, we motivate the fact that we model the behaviour of concurrent processes at the level of the actions that are provided by their public interfaces in the sense that every signature identifies a set of actions in which a process can engage itself. The axioms of a specification provide an abstraction of the computations (traces) that are performed locally through the properties that they satisfy. The idea is, then, to take signatures as interfaces and characterise the ability of linear temporal logic specifications to separate computation and coordination in terms of properties of the functor $\boldsymbol{PRES_{LTL}}\rightarrow\boldsymbol{SET}$ that maps presentations and their morphisms to the underlying signatures and signature morphisms.

Which properties should we require of ***int*** to capture the proposed separation? Basically, we have to capture the fact that any interconnection of systems is established via their interfaces.

For instance, an important property is that ***int*** should be faithful. This means that morphisms of programs should not induce more relationships between programs than those that can be captured through their underlying interfaces. That is, by taking into consideration the computational part, we should not get additional observational power over the external behaviour of systems.

Another important property concerns the way colimits are computed. We have already mentioned that the colimit of a diagram expressing how a system is configured in terms of simpler components and interconnections between them, returns a model of the global behaviour of the system and the morphisms that relate the components to the global system. If only the interfaces matter for establishing the required interconnections, then the colimit of the diagram of systems should be obtainable from the colimit of the underlying diagram of interfaces. In other words, if we interconnect system components through a diagram, then any colimit of the underlying diagram of interfaces should be able to be lifted to a colimit of the original diagram of system components.

This property can be stated more precisely as follows: given any diagram $\boldsymbol{dia}$:$\boldsymbol{I}\rightarrow\boldsymbol{SYS}$ and colimit $(\boldsymbol{int}(S_i)\rightarrow C)_{i:I}$ of $(\boldsymbol{dia};\boldsymbol{int})$ there exists a colimit $(S_i\rightarrow S)_{i:I}$ of $\boldsymbol{dia}$ such that $\boldsymbol{int}(S_i\rightarrow S)=(\boldsymbol{int}(S_i)\rightarrow C)$. When a functor satisfies a property like this one we say that it lifts colimits.

This means that when we interconnect system components, any colimit of the underlying diagram of interfaces establishes an interface for which a computational part exists that captures the joint behaviour of the interconnected components. Notice that this property does not tell us how to construct the lift; it just ensures that it exists. The operational aspects of lifts are discussed in Chap. 6. This property is really about (non)interference between computation and coordination: on the one hand, the computations assigned to the components cannot interfere with the viability (in the sense of the existence of a colimit) of the underlying configuration of interfaces; on the other hand, the computations assigned to the components cannot interfere in the calculation of the interface of the resulting system.

A kind of inverse property is also quite intuitive: that every interconnection of system components be an interconnection of the underlying interfaces. In particular, that computations do not make viable a configuration of system components whose underlying configuration of interfaces is not. This property is verified when $\boldsymbol{int}$ preserves colimits, i.e. when given any diagram $\boldsymbol{dia}$:$\boldsymbol{I}\rightarrow\boldsymbol{SYS}$ and colimit $(S_i\rightarrow S)_{i:I}$ of $\boldsymbol{dia}$, $(\boldsymbol{int}(S_i)\rightarrow\boldsymbol{int}(S))_{i:I}$ is a colimit of $(\boldsymbol{dia};\boldsymbol{int})$. These two properties together imply that any colimit in $\boldsymbol{SYS}$ can be computed by first translating the diagram to $\boldsymbol{INT}$, then computing the colimit in $\boldsymbol{INT}$, and finally lifting the result back to $\boldsymbol{SYS}$.

So: does the (forgetful) functor $\boldsymbol{sign}$:$\boldsymbol{PRES}_{\boldsymbol{LTL}}\rightarrow\boldsymbol{SET}$ satisfy these properties? The answer is "yes", as shown in Chap. 6. This is because this functor is an instance of a class that captures many useful structures in computing and, hence, is worth studying on its own, which includes the properties that we have just discussed. We shall return to the issue of separating "computation" and "coordination" in several other places in the book, including examples. Summarising, it is useful to classify functors vis-à-vis the way they relate universal constructions in the source and target categories.

5.2.1 Definition

A functor φ:$\boldsymbol{C}\rightarrow\boldsymbol{D}$

1. *preserves*
   - a colimit p:$\delta\rightarrow c$ of δ:$\boldsymbol{I}\rightarrow\boldsymbol{C}$ iff the cocone $\{\varphi(p_a)$: $\varphi(\delta_a)\rightarrow\varphi(c)\}_{a\in I_0}$, denoted by $\varphi(p)$:$\delta;\varphi\rightarrow\varphi(c)$, is a colimit of $\delta;\varphi$.

- colimits of shape $\boldsymbol{I}$ iff it preserves the colimits of all diagrams $\delta{:}\boldsymbol{I}\to\boldsymbol{C}$.
- colimits iff it preverses the colimits of any diagram in $\boldsymbol{C}$.

2. *lifts* colimits iff for any diagram $\delta{:}\boldsymbol{I}\to\boldsymbol{C}$ and colimit $p'{:}\ \delta;\varphi\to d$ of $\delta;\varphi$, there is a cocone $p{:}\delta\to c$ that is a colimit of δ and $p'=\varphi(p)$. The lift is *unique* (or φ is said to lift colimits uniquely) when there is a unique cocone $p{:}\delta\to c$ satisfying the two properties.
3. *reflects* colimits iff for any diagram $\delta{:}\boldsymbol{I}\to\boldsymbol{C}$ and cocone $p{:}\delta\to c$, if $\varphi(p)$ is a colimit *of* $\delta;\varphi$, then p is a colimit of δ.
4. *creates* colimits iff for any diagram $\delta{:}\boldsymbol{I}\to\boldsymbol{C}$ and colimit $p'{:}\ \delta;\varphi\to d$, there is a unique cocone $p{:}\delta\to c$ such that $p'=\varphi(p)$ and, moreover, p is a colimit of δ. ♦

Points 2, 3 and 4 of this definition extend to specific classes of colimits (sums, coequalisers, etc.) as formulated for preservation. The dual notions have the obvious designations. The following exercise can help in understanding the difference between these notions.

5.2.2 Exercises

Prove that

1. If a cocone $p{:}\delta\to c$ is commutative, so is its image by $\varphi(p)$ as defined in point 1.
2. Every functor that creates colimits also reflects them.
3. A functor creates colimits iff it reflects and lifts colimits uniquely. ♦

Note that the requirement on reflecting colimits is essential in case 2. For instance, it is easy to prove that the (forgetful) functor $\boldsymbol{THEO}_{LTL}\to\boldsymbol{SET}$ that maps theories and their morphisms to the underlying signatures and signature morphisms lifts colimits uniquely, but it is easy to see that it does not create them: the colimit will choose the minimal theory among the whole class of theories that have the given signature and give rise to a cocone. This class is only singular when some inconsistency arises from the interconnections and/or the base theories. We can also relate these properties to Par. 4.4.7.

5.2.3 Exercise

Prove that if $\boldsymbol{C}$ is finitely cocomplete then $\varphi{:}\boldsymbol{C}\to\boldsymbol{D}$ preserves finite colimits iff it preserves initial objects and pushouts. ♦

More results and related notions can be found, for instance, in [1, 22].

Part II

Advanced Topics

6 Functor-Based Constructions

Functors provide us not only with the ability to investigate relationships between categories, as shown in Chap. 5 and continued in the first section of this chapter, but also to build new categories based on such relationships. In this chapter we present some of the functor-based constructions that we have found useful in our day-to-day work.

6.1 Functor-Distinguished Kinds of Categories

In this section we show how the properties of certain categories can be derived from the properties of functors that relate them to other categories. That is, we investigate functors as a means of revealing structural properties of categories and of their objects.

We start with the notion of concrete category, which appeared in [80, p. 26] and was extensively explored in [1]. This is also a notion that we have found to be extremely useful in computing because it is often the case that, when building new categories over old ones by adding some structure, we do not interefere with the structure of the original category. This construction by what amounts to a conservative extension can be captured by a faithful functor $\boldsymbol{u}:\boldsymbol{D}\rightarrow\boldsymbol{C}$, where $\boldsymbol{C}$ is the old category and $\boldsymbol{D}$ is the new one. Typically, the functor *forgets* the structure that is being added to the objects of $\boldsymbol{C}$ to produce objects of $\boldsymbol{D}$.

6.1.1 Definition – Concrete Categories

A *concrete* category over a category $\boldsymbol{C}$ is a pair $\langle\boldsymbol{D},\upsilon\rangle$ where $\upsilon:\boldsymbol{D}\rightarrow\boldsymbol{C}$ is a faithful functor. ♦

Notice that we are using the terminology introduced in [1]. The notion of concrete category introduced in [80] is a particularisation of the one above to the case where the category $\boldsymbol{C}$ is $\boldsymbol{SET}$. Concrete categories over $\boldsymbol{SET}$ are called *constructs* in [1]. The category $\boldsymbol{C}$ is sometimes called the *base* category of $\langle\boldsymbol{D},\upsilon\rangle$, and υ is called the *forgetful* or *underlying* functor.

Because the underlying functor is faithful, i.e. injective on hom-sets, for each pair of $\boldsymbol{D}$-objects $\langle x,y\rangle$, $hom_{\boldsymbol{D}}(x,y)$ is usually regarded as a subset of $hom_{\boldsymbol{C}}(\upsilon(x),\upsilon(y))$. Following [1], we shall often use the expression "$f:\upsilon(x)\rightarrow\upsilon(y)$ *is a* $\boldsymbol{D}$*-morphism*" to mean that there exists a (necessarily) unique $\boldsymbol{D}$-morphism $x \rightarrow y$ whose image by υ is f. We have already come across a few concrete categories.

6.1.2 Examples

1. The category ***CLOS*** of closure systems defined in Sect. 3.6 is concrete over ***SET***: the forgetful functor maps closure systems to the underlying sets (languages).
2. The categories $\boldsymbol{PRES}_{LTL}$, $\boldsymbol{SPRES}_{LTL}$ and $\boldsymbol{THEO}_{LTL}$ defined in Par. 3.5.4 are all concrete over the category ***SET*** of sets. The underlying functors, which we name $\boldsymbol{sign}_{LTL}$, "forget" the sets of axioms/theorems, mapping theories and their presentations to the corresponding signatures.
3. Another example is the category ***AUTOM*** of automata as defined in Par. 2.2.12. This category is concrete over the product $\boldsymbol{SET} \times \boldsymbol{SET} \times \boldsymbol{SET}$. The underlying functor forgets the input, output and transition functions and projects every automaton to its sets of inputs, states and outputs. ♦

A typical situation in which concrete categories arise is one in which the underlying functor represents some sort of classification or typing mechanism that is strong enough to extend to the morphisms and, hence, to the structure defined over the objects by the morphisms. In such circumstances, one is usually interested in studying the way all the objects that share the same classification or type relate to one another.

6.1.3 Definition – Fibres

Given a concrete category $\langle \boldsymbol{D}, \upsilon \rangle$ over $\boldsymbol{C}$ and a $\boldsymbol{C}$-object c, the *fibre* of c is the preorder that consists of all the objects d of $\boldsymbol{D}$ that are mapped to c, i.e. such that $\upsilon(d)=c$, ordered by $d_1 \leq_c d_2$ iff $id_c : \upsilon(d_1) \rightarrow \upsilon(d_2)$ is a $\boldsymbol{D}$-morphism. ♦

The structure of the fibres reveals a lot about the properties of the concrete category. A detailed discussion can be found in [1]. We limit our study to three cases that are used in the book.

6.1.4 Definition

A concrete category $\langle \boldsymbol{D}, \upsilon \rangle$ over $\boldsymbol{C}$ is called:

1. *Amnestic* provided that its fibres are partially ordered, i.e. $d_1 \leq_c d_2$ and $d_2 \leq_c d_1$ implies $d_1 = d_2$ for all $\boldsymbol{C}$-objects c and objects d_1, d_2 in the fibre of c.
2. *Fibre-complete* if its fibres are complete lattices (i.e. admit arbitrary meets and joints).
3. *Fibre-discrete* if its fibres are ordered by equality. ♦

Amnestic concrete categories are such that the morphisms of $\boldsymbol{D}$ do not introduce additional properties over the structures that are being superposed on $\boldsymbol{C}$. For instance, $\boldsymbol{SPRES}_{LTL}$ and $\boldsymbol{THEO}_{LTL}$ are amnestic because their morphisms do not introduce any structural properties over the sets of axioms/theorems. However, $\boldsymbol{PRES}_{LTL}$ is not amnestic because its morphisms capture the notion of closure. Two presentations over the same temporal signature can be isomorphic without being equal. On the other hand, strictly isomorphic presentations and isomorphic theories over the same temporal signature are necessarily equal: the former because the consequences of the axioms are not taken into consideration, and the latter because they are already closed under consequence. In other words, morphisms of theories just take into account the elements of a set (the theorems), whereas morphisms of presentations have to compute the closure of a set.

Concrete categories that are fibre-complete superpose over the objects of $\boldsymbol{C}$ information that the morphisms organise in a "convenient" way, allowing for operations to be performed internally within each fibre. We shall see some examples further on. Hence, being fibre-complete is a step further than amnesticity in the degree of interaction that exists between the existing and the superposed structure. Fibre-discrete categories present still a step further: they are such that the extension that $\boldsymbol{D}$ makes over the objects of $\boldsymbol{C}$ is inessential, i.e. it has no intrinsic structure or meaning – it acts just like a comment.

Because concrete categories have added structure. we should provide a notion of functor that reflects that structure.

6.1.5 Definition – Concrete Functors

A *concrete functor* φ between two concrete categories $\langle \boldsymbol{D}_1, \upsilon_1 \rangle$ and $\langle \boldsymbol{D}_2, \upsilon_2 \rangle$ over the same underlying category $\boldsymbol{C}$ is a functor $\varphi: \boldsymbol{D}_1 \rightarrow \boldsymbol{D}_2$ such that $\upsilon_1 = \varphi; \upsilon_2$.

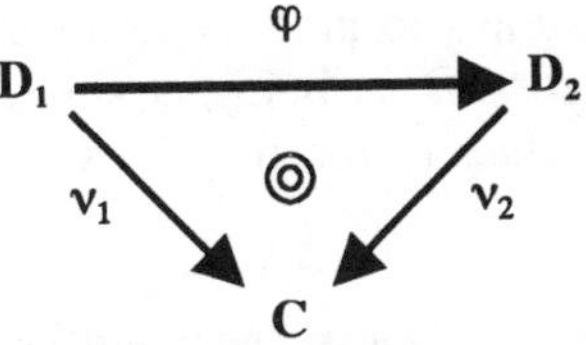

♦

Because the morphisms of a concrete category are imported from its underlying category, concrete functors cannot act on them and, hence, are fully determined by their values on objects.

6.1.6 Proposition

Given concrete functors φ and ψ between two concrete categories $\langle \boldsymbol{D}_1, \upsilon_1 \rangle$ and $\langle \boldsymbol{D}_2, \upsilon_2 \rangle$, $\varphi = \psi$ if, for every $\boldsymbol{D}_1$-object d, $\varphi(d) = \psi(d)$.

Proof

Let $f: d \rightarrow d'$ be a morphism of $\boldsymbol{D}_1$. Given that both functors agree on objects, both $\varphi(f)$ and $\psi(f)$ have the same source and target. Because both φ and ψ are concrete, we have $\upsilon_1(f)=\upsilon_2(\varphi(f))=\upsilon_2(\psi(f))$. Finally, because υ_2 is faithful, we can conclude that $\varphi(f)=\psi(f)$. Hence, both functors agree on morphisms as well. ◆

6.1.7 Remark – Concrete Subcategories

Given that subcategories determine embeddings (see Par. 5.1.10), and that the composition of faithful functors is also faithful (see Par. 5.1.9), every subcategory $\boldsymbol{D}'$ of a category $\boldsymbol{D}$ that is concrete over $\boldsymbol{C}$ with underlying functor υ can be regarded also as a concrete category over $\boldsymbol{C}$ whose underlying functor is the composition $\iota_{D',D};\upsilon$. We then say that $<\boldsymbol{D}',\iota_{D',D};\upsilon>$ is a concrete subcategory of $<\boldsymbol{D},\upsilon>$. ◆

Consider now the notions of reflective and coreflective subcategories that we discussed in Sect. 3.3. Intuitively, for the (co)reflection to be "concrete", i.e. to be consistent with the classification that the underlying functor provides, we would like to remain within the same fibre. That is, we would like that the (co)reflection arrows be identities.

6.1.8 Definition – Concretely (Co)Reflective Subcategories

A concrete subcategory $<\boldsymbol{D}_1,\upsilon_1>$ of $<\boldsymbol{D}_2,\upsilon_2>$ is *concretely (co)reflective* iff, for each object of $\boldsymbol{D}_2$, there is a (co)reflection arrow that is mapped by υ_2 to the identity. ◆

For instance, although ***REACH*** is a coreflective subcategory of ***AUTOM***, it is not concretely coreflective when both categories are considered as being concrete over ***SET×SET×SET***. This is because the reachable automata are not necessarily in the same fibre as the automata from which they are computed – their state space may have been reduced.

"Concreteness" also extends to universal constructions in the sense that, once we consider a category $\boldsymbol{D}$ as being concrete through the underlying functor $\upsilon:\boldsymbol{D}\rightarrow\boldsymbol{C}$, we can classify the way universal constructions in $\boldsymbol{D}$ relate to $\boldsymbol{C}$ through υ. ◆

6.1.9 Definition – Concrete Universal Constructions

Consider a concrete category $<\boldsymbol{D},\upsilon>$ over $\boldsymbol{C}$ and a diagram $\delta:\boldsymbol{I}\rightarrow\boldsymbol{D}$. A (co)limit of δ is said to be a *concrete (co)limit* in $<\boldsymbol{D},\upsilon>$ iff it is preserved by υ. ◆

That is, universal constructions are concrete when they map to the underlying category. In some categories, all universal constructions are concrete. We give an example in Par. 6.3.6. In other concrete categories, some universal constructions may be concrete and others not so.

6.1.10 Exercise

Work out examples of limits and colimits in ***AUTOM*** that are not concrete over ***SET×SET×SET***. ◆

The construction of concrete (co)limits, when they exist, can be systematised through a process that consists in projecting the (co)cones to the base category where a (co)limit is computed and then lifted back to the original category. This process can be extended to more general notions of fibre as discussed further on.

The notion of fibre is not exclusive to concrete categories. It can be generalised to arbitrary functors as follows.

6.1.11 Definition – Fibres

Consider a functor $\varphi{:}\boldsymbol{D}\to\boldsymbol{C}$.

1. Given a $\boldsymbol{C}$-object c, the *fibre of* c, which we denote by $\boldsymbol{D}(c)$, is the subcategory of $\boldsymbol{D}$ that consists of all the objects d that are mapped to c, i.e. such that $\varphi(d)=c$, together with the $\boldsymbol{D}$-morphisms $f{:}d_1\to d_2$ such that $\varphi(f)=id_c$.
2. The functor φ is said to be *amnestic* if, in its fibres, no two distinct objects are isomorphic. That is, if an isomorphism $f{:}d_1\to d_2$ in $\boldsymbol{D}$ is such that $\varphi(f)=id_c$ for some object c of $\boldsymbol{C}$, then f is itself an identity. ♦

The discussion around concrete categories focused basically on the structure of fibres. Another interesting aspect worth discussing is the relationships that morphisms on the underlying category induce over the fibres corresponding to their sources and targets.

For instance, consider again the category $\boldsymbol{THEO}_{LTL}$ of temporal theories defined in Par. 3.5.4 and the forgetful functor $\boldsymbol{sign}_{LTL}$ that maps theories and their morphisms to their signature components. Consider an arbitrary signature Σ. A signature morphism $f{:}\Sigma\to\Sigma'$ provides a translation from the symbols of one signature to symbols of the other. We have seen in Par. 3.5.7 that such a translation extends to the temporal language associated with Σ. Can we further extend this translation to the fibres of Σ, i.e. can we define a mechanism that translates theories with signature Σ to theories over Σ'? What about the reverse? Can we define a mechanism that translates theories with signature Σ' to theories over Σ?

To answer these questions, we have first to provide a reasonable definition of "translation" (and inverse translation) between fibres induced by a morphism. Consider a functor $\varphi{:}\boldsymbol{D}\to\boldsymbol{C}$, a $\boldsymbol{C}$-morphism $f{:}c\to c'$, and a $\boldsymbol{D}$-object d in the fibre of c, i.e. $\varphi(d)=c$. By the "image" of d under f we mean some object $f(d)$ in the fibre of c' that is "closest" to d. By "closest" we mean the following. On the one hand, f can be lifted to a morphism between d and $f(d)$, i.e. there must exist some morphism $g{:}d\to f(d)$ such that $\varphi(g)=f$. On the other hand, for any other object d'' and morphism $g'{:}d\to d''$ that leaves d'' at a "distance" $f'{:}c'\to\varphi(d'')$ of f, i.e. such that $\varphi(g')=f;f'$, this distance can be covered in $\boldsymbol{D}$ in a unique way by a morphism $h{:}f(d)\to d''$ such that $\varphi(h)=f'$ and $g'=g;h$. By inverse translation, we mean the dual.

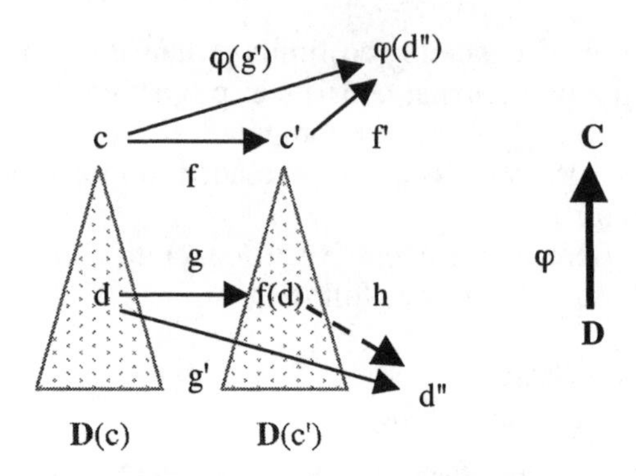

The following definitions are adapted from [12], a good source for the study of fibre-related matters.

6.1.12 Definition – (Co)Cartesian Morphisms

Let φ:***D***→***C*** be a functor and *f*:*c*→*c*' a ***C***-morphism.

1. Let *d*:***D***(*c*), i.e. a ***D***-object such that $\varphi(d)=c$. A ***D***-morphism *g*:*d*→*d*' is *co-Cartesian* for *f* and *d* iff:
   - $\varphi(g)=f$.
   - For every *g*':*d*→*d*" and *f*':*c*'→φ(*d*") such that $\varphi(g')=f;f'$, there is a unique morphism *h*:*d*'→*d*" such that $\varphi(h)=f'$ and *g*'=*g*;*h*.
2. Let *d*':***D***(*c*'), i.e. a ***D***-object such that $\varphi(d')=c'$. A ***D***-morphism *g*:*d*→*d*' is *Cartesian* for *f* and *d*' iff:
   - $\varphi(g)=f$.
   - For every *g*':*d*"→*d*' and *f*':φ(*d*")→*c* such that $\varphi(g')=f';f$, there is a unique morphism *h*:*d*"→*d* such that $\varphi(h)=f'$ and *g*'=*h*;*g*.

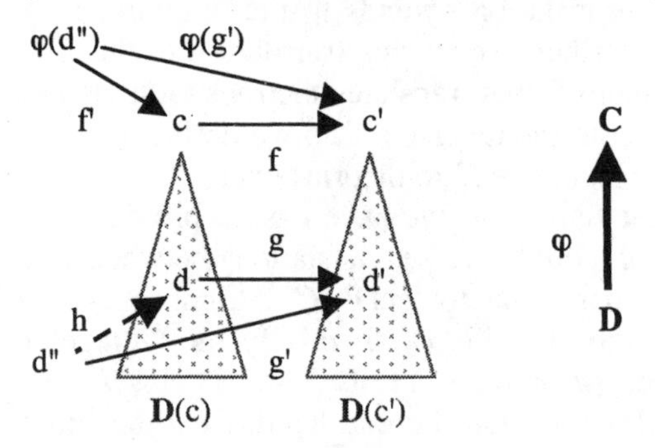

♦

6.1.13 Definition – (Co)Fibrations

Let φ:***D***→***C*** be a functor.

1. We say that φ is a *fibration* if, for every ***C***-morphism *f*:*c*→*c*' and ***D***-object *d*' in the fibre of *c*', there is a Cartesian morphism for *f* and *d*'.

2. We say that φ is a *cofibration* if, for every $\boldsymbol{C}$-morphism $f{:}c \rightarrow c'$ and $\boldsymbol{D}$-object d in the fibre of c, there is a co-Cartesian morphism for f and d. ♦

6.1.14 Example – Specifications as (Co)Fibrations

The forgetful functors that define $\boldsymbol{PRES}_{LTL}$, $\boldsymbol{SPRES}_{LTL}$ and $\boldsymbol{THEO}_{LTL}$ as concrete categories over the category $\boldsymbol{SET}$ are all fibrations and cofibrations at the same time, but with different (co)Cartesian morphisms among themselves. Given a signature morphism $f{:}\Sigma \rightarrow \Sigma'$:

1. In $\boldsymbol{THEO}_{LTL}$, i.e. in the case of theories, $f{:}{<}\Sigma,f^{-1}(\Phi'){>} \rightarrow {<}\Sigma',\Phi'{>}$ is a Cartesian morphism for ${<}\Sigma',\Phi'{>}$, and $f{:}{<}\Sigma,\Phi{>} \rightarrow {<}\Sigma',c(f(\Phi)){>}$ is a co-Cartesian morphism for ${<}\Sigma,\Phi{>}$.
2. In $\boldsymbol{PRES}_{LTL}$, i.e. in the case of presentations, a Cartesian morphism for ${<}\Sigma',\Phi'{>}$ is $f{:}{<}\Sigma,f^{-1}(c(\Phi')){>} \rightarrow {<}\Sigma',\Phi'{>}$, and a co-Cartesian morphism for ${<}\Sigma,\Phi{>}$ is $f{:}{<}\Sigma,\Phi{>} \rightarrow {<}\Sigma',f(\Phi){>}$.
3. In $\boldsymbol{SPRES}_{LTL}$, i.e. in the case of strict presentations, a Cartesian morphism for ${<}\Sigma',\Phi'{>}$ is $f{:}{<}\Sigma,f^{-1}(\Phi'){>} \rightarrow {<}\Sigma',\Phi'{>}$, and a co-Cartesian morphism for ${<}\Sigma,\Phi{>}$ is $f{:}{<}\Sigma,\Phi{>} \rightarrow {<}\Sigma',f(\Phi){>}$.

The differences that we can witness between these three cases have to do with both the properties of the closure operator and the morphisms that characterise the categories. For instance, even if Φ is closed for consequence, $f(\Phi)$ is not necessarily so. Hence, the co-Cartesian morphisms for theories have to return the closure of the image set. This is not necessary for presentations because it is already implicit in the morphisms, nor is it necessary for strict presentations because they do not involve the consequence operator at all. The Cartesian morphisms, on the contrary, do not need to compute the closure of the inverse image set. This is because, if Φ is closed, so is $f^{-1}(\Phi)$. However, presentations need to compute the closure explicitly because they only do it implicitly to the target, not the source. ♦

These examples illustrate the fact that there may be more than one co-Cartesian morphism for a given signature morphism and theory presentation. This is because the set of axioms of the specification can be determined only up to logical equivalence. In the case of theories, the fact that the set of sentences must be closed ensures that the lifts are unique.

Hence, although it is possible to generalise the translation induced by signature morphisms to presentations, there may be more than one way of doing so. This is the general case of any (co)fibration. Amnestic concrete (co)fibrations, however, guarantee uniqueness of the lifting and, hence, of the choice for (co)Cartesian morphisms.

6.1.15 Definition – Cleavages, Cloven Fibrations

Let $\varphi{:}\boldsymbol{D} \rightarrow \boldsymbol{C}$ be a functor. A choice of a Cartesian morphism for every $\boldsymbol{C}$-morphism $f{:}c \rightarrow \varphi(d')$ and $\boldsymbol{D}$-object d' is called a *cleavage*. A fibration equipped with a cleavage is called *cloven*.

We shall often denote by $\varphi_{f,d'}$ the Cartesian morphism selected for f and d' by the cleavage. Given $g':d''\rightarrow d'$ and $f':\varphi(d'')\rightarrow c$ such that $\varphi(g')=f';f$, we know that there is a unique morphism $h:d''\rightarrow d$ such that $\varphi(h)=f'$ and $g'=h;\varphi_{f,d'}$. We denote h by $g'/_{f'}\varphi_{f,d'}$. When f' is id_c, we omit the subscript.

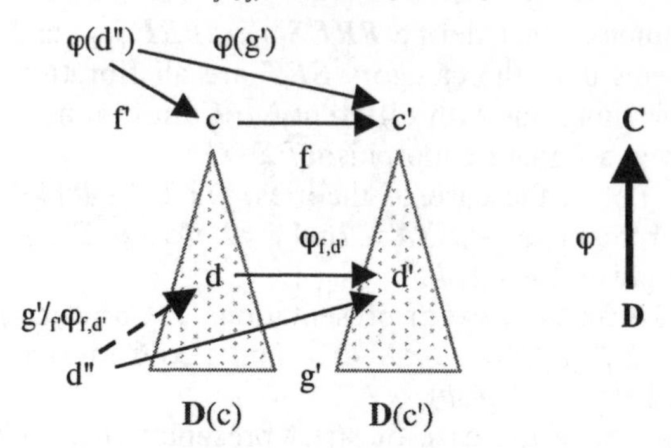

◆

The dual notion is called cocleavage, and a cofibration equipped with a cocleavage is also said to be cloven. Summarising, a cloven (co)fibration φ:***D***→***C*** provides us with a way of lifting morphisms to translations between the objects of their fibres in the sense that every ***C***-morphism $f:c\rightarrow c'$ defines a map from the objects of ***D***(c') to the objects of ***D***(c) in the case of a fibration, and from the objects of ***D***(c) to the objects of ***D***(c') in the case of a cofibration. Can these translations be generalised to morphisms of the fibres? Can we generalise this mapping to a functor between the fibres?

6.1.16 Proposition

Let φ:***D***→***C*** be a cloven fibration and $f:c\rightarrow c'$ a ***C***-morphism.

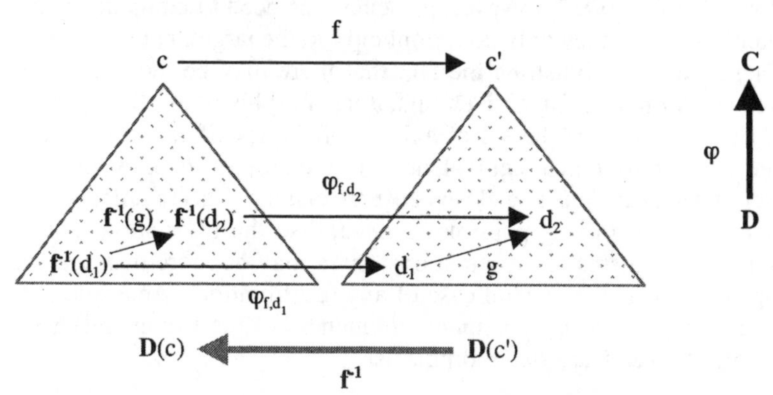

1. The morphism f defines a functor f^{-1}:***D***$(c')\rightarrow$***D***(c) as follows:
   - Given d':***D***(c'), $f^{-1}(d')$ is the source of the Cartesian morphism $\varphi_{f,d'}:d\rightarrow d'$ that the cleavage associates with the fibration.

- Given $g:d_1 \rightarrow d_2$ in $\boldsymbol{D}(c')$, $f^{-1}(g)$ is the morphism $f^{-1}(d_1) \rightarrow f^{-1}(d_2)$ that results from the universal property of the Cartesian morphism $\varphi_{f,d_2}:f^{-1}(d_2) \rightarrow d_2$ when applied to $\varphi_{f,d_1};g$ and $id_{c'}$. That is, $f^{-1}(g)$ is the morphism that we denote by $(\varphi_{f,d_1};g)/\varphi_{f,d_2}$. Notice that this morphism is the only one that satisfies $\varphi_{f,d_1};g=f^{-1}(g);\varphi_{f,d_2}$ and $\varphi(f^{-1}(g))=id_{c'}$.

2. The morphism f defines a functor $f:\boldsymbol{D}(c) \rightarrow \boldsymbol{D}(c')$ in the dual way, i.e. by working on the target side of the co-Cartesian morphism.

Proof

1. There are three properties to prove. The reader is invited to fill in the details:
   - The functor is well defined in the sense that the image objects and morphisms exist and are of the right types.
   - Because $id_{f^{-1}(d)}$ satisfies the properties of the universal property that characterises $f^{-1}(id_d)$, the uniqueness associated with that universal property guarantees that the cleavage has no other choice.
   - The same line of reasoning applies to conclude that $f^{-1}(g_1;g_2)=f^{-1}(g_1);f^{-1}(g_2)$.
2. By duality. ◆

This definition raises immediate questions: What if $f=id_c$? Are f^{-1} and f the identity functor? What if $f=f_1;f_2$? Are f^{-1} and f the compositions $f^{-1}_1;f^{-1}_2$ and $f_1;f_2$, respectively? The answer is twofold: they can be equal, but they do not have to. Consider the first part of the answer, this time instantiated for cofibrations.

6.1.17 Proposition

Let $\varphi:\boldsymbol{D} \rightarrow \boldsymbol{C}$ be a functor.

1. Given a $\boldsymbol{C}$-object c and an object d in the fibre of c, the identity id_d is both a Cartesian and a co-Cartesian morphism for id_c and d.
2. Given $\boldsymbol{C}$-morphisms $f_1:c_1 \rightarrow c_2$ and $f_2:c_2 \rightarrow c_3$, an object d in the fibre of c_1, and co-Cartesian morphisms $g_1:d \rightarrow f_1(d)$ and $g_2:f_1(d) \rightarrow f_2(f_1(d))$, the composition $g_1;g_2$ provides a co-Cartesian morphism for $f_1;f_2$ and d.

Proof

1. Left as an exercise.
2. There are two properties to prove:
   - The composite translation lifts the composition in $\boldsymbol{C}$:
     $\varphi(g_1;g_2)=\varphi(g_1);\varphi(g_2)=f_1;f_2$.
   - The lift has the required couniversal property:
     Let $g':d \rightarrow d''$ and $f':c_3 \rightarrow \varphi(d'')$ be such that $\varphi(g')=f_1;f_2;f'$. The couniversal property of g_1 applied to g' and $(f_2;f')$ gives a unique $h_1:f_1(d) \rightarrow d''$ such that $g_1;h_1=g'$ and $\varphi(h_1)=f_2;f'$. The couniversal property of g_2 to h_1 and f' implies the existence and uniqueness of $h_2:f_2(f_1(d)) \rightarrow d''$ such that $g_2;h_2=h_1$ and $\varphi(h_2)=f'$. We can now prove the properties required of h_2. On the one hand, $\varphi(h_2)=f'$. On the other hand, if we take $h'_2:f_2(f_1(d)) \rightarrow d''$ such that $g_1;g_2;h'_2=g'$ and $\varphi(h'_2)=f'$, we derive $g_2;h'_2=h_1$ because $\varphi(g_2;h'_2)=\varphi(g_2);\varphi(h'_2)=f_2;f'$, $g_1;(g_2;h'_2)=g'$, and h_1 is the unique morphism satisfying these two properties.

Then, we conclude that $h'_2=h_2$ because h_1 is the unique morphism satisfying $g_2;h_2=h_1$ and $\varphi(h_2)=f'$.

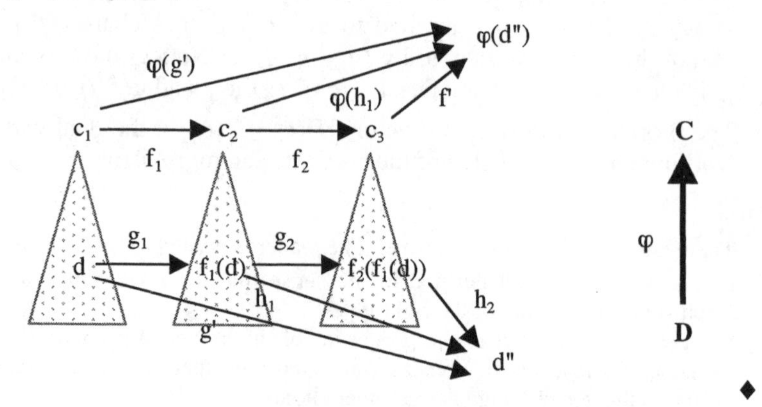

♦

The fact that cleavages do not need to choose (co)Cartesian morphisms satisfying these properties leads to the following definition.

6.1.18 Definition – Split Fibrations

Let φ:***D***→***C*** be a cloven fibration. If, for every ***C***-object c, id_c^{-1} is $id_{C(c)}$ and, for every decomposition $f=f_1;f_2$, f^{-1} is the composition $f^{-1}_2;f^{-1}_1$, then the fibration is said to be *split*. ♦

6.1.19 Remark

More alert readers may have noticed that the universal property of the (co)Cartesian morphism is taken over a space that is larger than the fibre of either c or c'. This is why the "distance" f' is introduced as a kind of type converter. This means that the (inverse) translation is chosen not only over the objects that have the exact type but also those that "type-check".

We point this out is to recall that the categorical way of defining a concept, namely through a universal property, has to be morphism-oriented in order to yield good properties. The reader should experiment the weaker notion of (co)Cartesian morphism for which the universal property is required only over the fibres, and check which of the properties still hold. ♦

Split (co)fibrations allow us to wrap up the idea of translation between fibres with which we started in neat categorical clothing.

6.1.20 Proposition

Let φ:***D***→***C*** be a functor.

1. If φ is a split fibration, then it defines a functor $\boldsymbol{ind(\varphi)}:\boldsymbol{C}^{op}\to\boldsymbol{CAT}$ by mapping every ***C***-object c to its fibre $\boldsymbol{D}(c)$ and every morphism $f:c\to c'$ to the functor $f^{-1}:\boldsymbol{D}(c')\to\boldsymbol{D}(c)$ as above.

2. If φ is a split cofibration, then it defines a functor ***ind(φ)*:*C*→*CAT*** by mapping every ***C***-object c to its fibre ***D****(c)* and every morphism $f{:}c \to c'$ to the functor $f{:}\boldsymbol{D}(c) \to \boldsymbol{D}(c')$ as above.

Proof

Left as an exercise. ♦

Functors of the form $\boldsymbol{C}^{op} \to \boldsymbol{CAT}$ are called *indexed categories*, a structure widely applied in computing (see, for instance, [102]) precisely because it generalises what are usually called indexed sets, i.e. mechanisms for indexing given collections of objects with other kind of objects (indexes). We focus on indexed categories in Sect. 6.4.

One of the reasons to study fibrations is the close relationship that exists between their universal properties and those of the fibres.

6.1.21 Definition – Fibre Completeness

A cloven fibration $\varphi{:}\boldsymbol{D} \to \boldsymbol{C}$ is said to be *fibre-complete* if its fibres are complete categories and the inverse translation functors induced on the fibres preserve limits. ♦

6.1.22 Proposition

Let $\varphi{:}\boldsymbol{D} \to \boldsymbol{C}$ be a split fibration.

1. If φ is fibre-complete, then it lifts limits.
2. If in addition φ is amnestic, the lift is unique.

Proof

1. Consider a diagram $\delta{:}\boldsymbol{I} \to \boldsymbol{D}$ and a limit $\mu{:}c \to \delta;\varphi$ for the underlying ***C***-diagram. We are going to provide a step-by-step construction of a limit for δ that lifts μ.

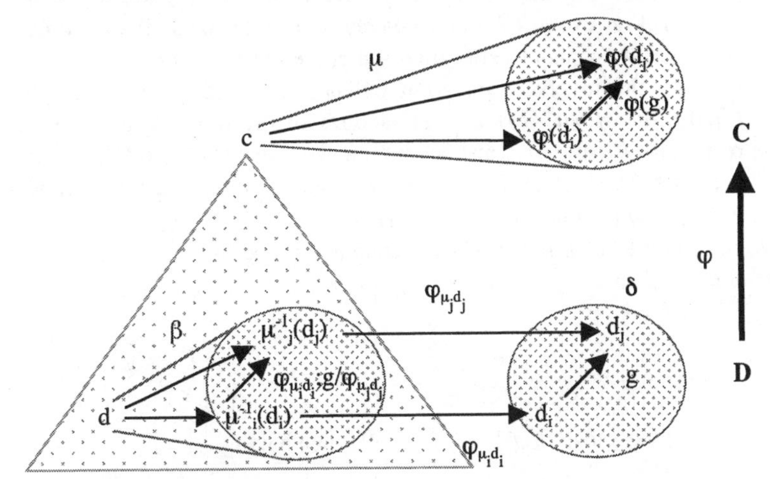

- First, we take the inverse image of δ induced by μ. We obtain a diagram within the fibre of c, related to the original one by the Cartesian morphisms $\varphi_{\mu_i d_i}$. Notice that every $g{:}d_i \rightarrow d_j$ is translated to $\varphi_{\mu_i d_i};g/\varphi_{\mu_j d_j}{:}\mu^{-1}(d_i) \rightarrow \mu^{-1}(d_j)$.
- Then, given that the fibre of c is complete, we compute the limit $\beta{:}d \rightarrow \mu^{-1}(\delta)$. We are going to prove that the cone obtained through the compositions $\beta_i;\varphi_{\mu_i d_i}$ is a limit for δ.
- The cone is commutative: consider $g{:}d_i \rightarrow d_j$ in δ; we have $\beta_i;\varphi_{\mu_i d_i};g= \beta_i;\varphi_{\mu_i d_i};g/\varphi_{\mu_j d_j};\varphi_{\mu_j d_j}= \beta_j;\varphi_{\mu_j d_j}$.
- The universal property of the cone is satisfied. Consider another commutative cone $\alpha{:}d' \rightarrow \delta$.

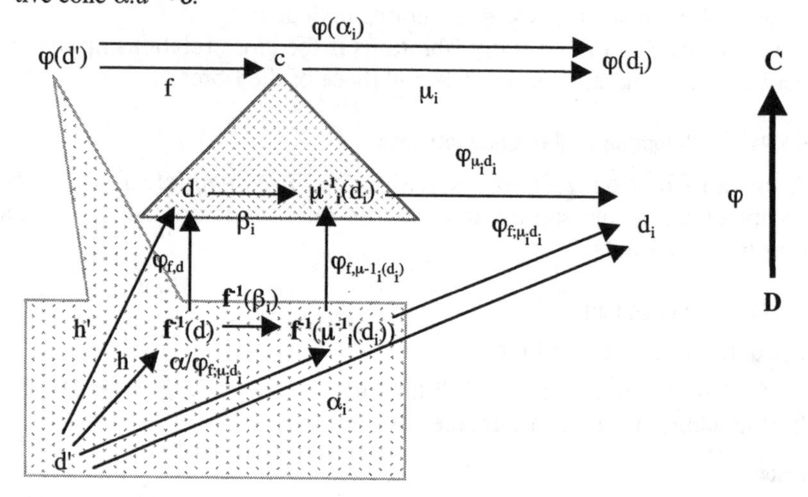

Because the image of α under φ is a commutative cone, the universal property of $\mu{:}c \rightarrow \delta;\varphi$ implies the existence of a unique $f{:}\varphi(d') \rightarrow c$ such that $f;\mu_i = \varphi(\alpha_i)$ for every $i \in I$; unfortunately, d' is not necessarily in the fibre of c, and hence we cannot use the universal properties of the limit.

- Because the inverse translation induced by f preserves limits, the cone $f^{-1}(\beta){:}f^{-1}(d) \rightarrow f^{-1}(\mu^{-1}(\delta))$ is itself a limit in the fibre of $\varphi(d')$; given that $\alpha_i/\varphi_{f;\mu,d}{:}d' \rightarrow (f;\mu)^{-1}(\delta)$ is also a commutative cone and $(f;\mu)^{-1}(\delta) = f^{-1}(\mu^{-1}(\delta))$ is a consequence of φ being split, we can infer the existence of a unique $h{:}d' \rightarrow f^{-1}(d)$ such that $h;f^{-1}(\beta_i) = \alpha_i/\varphi_{f;\mu_i d_i}$.

We now prove that $h;\varphi_{f,d}$ satisfies the required properties.

- $(h;\varphi_{f,d});(\beta_i;\varphi_{\mu_i d_i})$
  $= h;f^{-1}(\beta_i);\varphi_{f,\mu\text{-}1_i(d_i)};\varphi_{\mu_i d_i}$
  $= \alpha_i/\varphi_{f;\mu_i d_i};\varphi_{f,\mu\text{-}1_i(d_i)};\varphi_{\mu_i d_i}$
  $= \alpha_i/\varphi_{f;\mu_i d_i};\varphi_{f;\mu_i d}$
  $= \alpha_i$.
- Let $h'{:}d' \rightarrow d$ satisfy $h';(\beta_i;\varphi_{\mu_i d_i}) = \alpha_i$. We then have $\alpha_i = h'/\varphi_{f,d};\varphi_{f,d};\beta_i;\varphi_{\mu_i d_i} = h'/\varphi_{f,d};f^{-1}(\beta_i);\varphi_{f,\mu\text{-}1_i(d_i)};\varphi_{\mu_i d} = h'/\varphi_{f,d};f^{-1}(\beta_i);\varphi_{f;\mu_i d_i}$ which implies $h'/\varphi_{f,d};f^{-1}(\beta_i) = \alpha_i/\varphi_{f;\mu_i d_i}$, and hence $h'/\varphi_{f,d} = h$, and $h;\varphi_{f,d} = h'$.

2. Left as an exercise. ♦

6.1.23 Corollary

Let φ:***D***→***C*** be a split fibrationIf φ is fibre-complete and ***C*** is complete, then ***D*** is also complete. ♦

These results and their duals tell us how to compute universal constructions over (co)fibrations that are fibre-(co)complete: the diagram is projected to the underlying category and its (co)limit is computed. Then the original diagram is (co)translated to the fibre of the apex and its (co)limit is computed within the fibre.

6.1.24 Example – Colimits of Specification Diagrams

The fibres of $\boldsymbol{PRES_{LTL}}$, $\boldsymbol{SPRES_{LTL}}$ and $\boldsymbol{THEO_{LTL}}$ as concrete categories over ***SET*** are all complete and cocomplete as ordered sets. It is also easy to see that the (co)translations are (co)continuous. Taking into account the operations that define these universal constructions, we have the following procedure for calculating limits and colimits of a diagram δ with δ_i=<Σ_i,Φ_i> in these categories.

1. Calculate the limit σ:Σ→δ or colimit σ:δ→Σ of the underlying diagram of signatures.
2. Lift the result by computing the specification component according to the following rules:

	Limit	Colimit
$\boldsymbol{PRES_{LTL}}$	$\sigma:<\Sigma,\bigcap_{i\in I}\sigma_i^{-1}(c(\Phi_i))>\to\delta$	$\sigma:\delta\to<\Sigma,\bigcup_{i\in I}\sigma_i(\Phi_i)>$
$\boldsymbol{SPRES_{LTL}}$	$\sigma:<\Sigma,\bigcap_{i\in I}\sigma_i^{-1}(\Phi_i)>\to\delta$	$\sigma:\delta\to<\Sigma,\bigcup_{i\in I}\sigma_i(\Phi_i)>$
$\boldsymbol{THEO_{LTL}}$	$\sigma:<\Sigma,\bigcap_{i\in I}\sigma_i^{-1}(\Phi_i)>\to\delta$	$\sigma:\delta\to<\Sigma,c(\bigcup_{i\in I}\sigma_i(\Phi_i))>$

In order to illustrate these constructions, consider the specification of the vending machine given in Sect. 3.5. Therein, we see how the original specification (Par. 3.5.6) can be extended to include a mechanism for regulating the sale of cigars, which is captured by a morphism (Par. 3.5.11)

```
                     no cigars
vending machine  ──────────────▶  regulated vending machine
```

The need for this type of extensions arises whenever there is a change in the original requirements. Such changes may be very frequent in business domains that are very volatile, for instance, as a result of fierce competition, which forces companies to maintain a level of service that matches or beats the offer of their rivals. This prompts the need for mechanisms that allow systems to evolve in ways that localise the impact of changes. For instance, a better way of accounting for the need to regulate the sale of cigars is to interconnect the original vending machine with an external device (regulator). This would indicate that the required change does not need to be intrusive of the existing system in the sense that it is not necessary to change the way the existing system is implemented. Instead, we just need to implement the regulator and connect it to the system, even while it is running, i.e. without interruption of service.

The required regulator can be specified as follows:

```
specification   regulator is
signature       trigger, ted, tor
axioms          beg ⊃ (¬tor)
                trigger ⊃ (¬ted)Wtor
                tor ⊃ (¬ted)
```

The rationale is the following. When the trigger occurs, action *ted* (the one being regula*ted*) is blocked until the regula*tor* occurs.

As discussed in Chap. 4, we use diagrams to express systems as configurations of interconnected components, the (co)limits of which return the object that represents the system as a component, all the interconnections having been encapsulated. The regulated vending machine can be put together by synchronising action *trigger* of the regulator with action *coin* of the vending machine, and the regula*ted* action with *cigar*. The corresponding configuration diagram is:

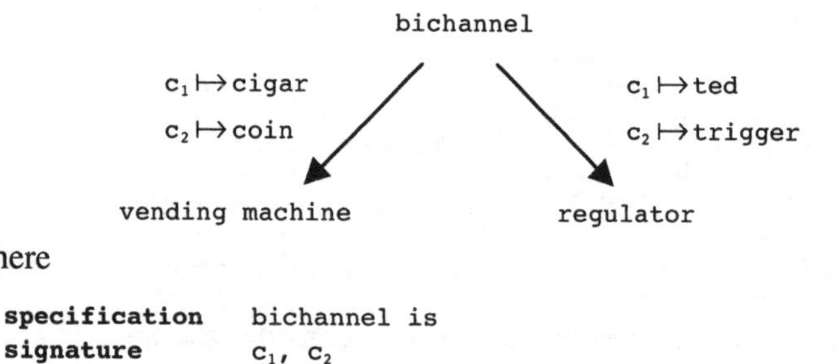

where

```
specification   bichannel is
signature       c1, c2
axioms
```

Notice that there is nothing in the regulator that names the vending machine: the regulator is intended to be a component that can be reused in different contexts. The same applies, a fortiori, to the vending machine because it was developed before the need to be regulated was determined. As a result, there is no implicit interaction between the two components. The interconnection through which the vending machine becomes regulated is completely externalised in the bichannel and the two morphisms. This is one basic difference between object and service-oriented development as already mentioned: whereas objects, through clientship, code up interactions in the way features are called, services cannot name other services because they are not specified with particular interactions in mind. Moreover, service integration is only performed when it is needed, at run-time, and for the services that will have been identified or selected *at that time* from the configuration.

According to the rules above, the pushout of this configuration diagram can be calculated in two steps. First, we compute the pushout of the underlying diagram of signatures. Because pushouts are determined only up to isomorphism, we choose the signature that matches that of the regulated vending machine.

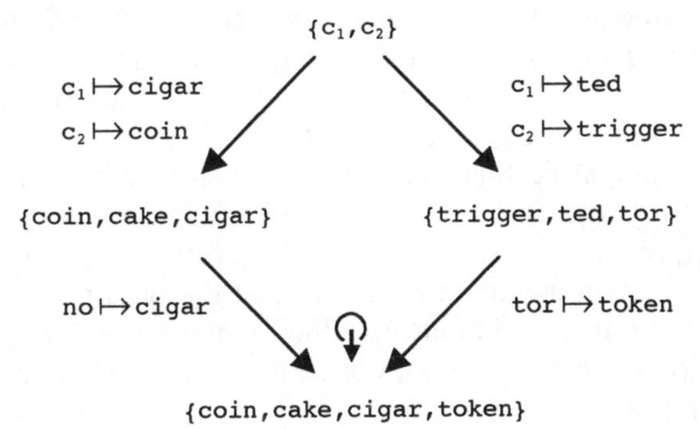

The second step consists in lifting the result back to the category of specifications. According to the rules given above, this can be achieved by choosing as axioms the translations of the axioms of the component specifications. From the vending machine we get:

```
beg ⊃ (¬cake∧¬cigar) ∧ (coin ∨ (¬cake∧¬cigar)Wcoin
coin ⊃ (¬coin)W(cake∨cigar)
(cake∨cigar) ⊃ (¬cake∧¬cigar)Wcoin
cake ⊃ (¬cigar)
```

And from the regulator we get:

```
beg ⊃ (¬token)
coin ⊃ (¬cigar)Wtoken
token ⊃ (¬cigar)
```

Notice that, because of the renaming imposed by the interconnection, the axioms are now in the shared language, i.e. they express properties of the same entities. For instance, the regulator now applies to coins and cigars. As a result, there is interference among the axioms from the different components, which makes new properties (the new requirements) emerge at the level of the resulting system. For instance, the vending machine now requires a token to be able to dispense cigars.

As mentioned in Chap. 1, this form of composition is the reason why the categorical approach brings software systems into the realm of general, complex systems, whose global behaviour is characterised in terms of properties or phenomena that emerge from components and interactions. This is a view that is now shared across different sciences, from biological to economics and sociological systems [71]. Furthermore, taking the union of the translations of the sets of axioms is logically equivalent to taking their conjunction. Hence, as claimed in the introductory remarks to Chap. 4 we are indeed complying with the "conjunction as composition" view used in [112]. Indeed, it is easy to see that this set of axioms is logically equivalent to the one we gave in Par. 3.5.11 for the regulated vending machine, which further shows that the lifting performed through the forgetful functor is not unique. Also notice that we recover the original extension (morphism), *no cigar*, as one of the cocone projections. Hence, basically, what we have done is factorise the extension by externalising the regulator that was coded over the original specification.

We should emphasise that, having performed the externalisation, the view of the system that interests us is the one given by the (configuration) diagram. The colimit construction is only useful as a semantics for the diagram, namely as a means of checking that the required properties will emerge from the interconnections. Further evolution of the system should be performed on the configuration, not on the global specification resulting from the colimit.

Other examples of the application of these techniques in computing can be found in the area of concurrency theory, namely in the works of G. Winskel [108], and F. Costa [21]. The idea is that operations on processes like synchronisation can be defined at the level of the actions that the processes can perform (their alphabet) through some algebraic operations and then can be lifted to the category of processes using a (co)fibration. An example in this area is given in Sect. 6.3. Finally, Part III is dedicated to a systematisation of this process of structuring complex systems through what are called software architectures.

6.2 Structured Objects and Morphisms

One of the best examples of the added expressive power that functors bring into the categorical discourse is the ability to work with more than one category. An exhaustive study of distinguished objects and arrows with respect to a functor can be found in [1]. In the following, we present some of the concepts that we have found to provide a good introduction, and thus motivate the reader to learn more about this topic.

6.2.1 Example – Realisations of a Specification

We argued in Par. 5.1.3 that the notion of satisfaction of specifications by programs can give rise to functors ***spec*:*PROG*→*SPEC*** that map programs to the strongest specification that they satisfy. The notion of satisfaction of a specification S by a program P can then be captured by the existence of a morphism σ:S→***spec***(P). Indeed, the morphism expresses the fact that the properties specified through S are entailed by the strongest specification of P. Such a "structured" morphism $\sigma:S\rightarrow \mathbf{spec}(P)$ accounts for representations of abstract concepts of the specification in terms of the syntax of the program. That is, the morphism reflects design decisions taken during the implementation of S in terms of P, say the choice of specific representations for the state of a system. Hence, there is an interest in manipulating morphisms of the form $\sigma:S\rightarrow \mathbf{spec}(P)$ as capturing the possible realisations of specifications in terms of programs. ♦

6.2.2 Definition – Structured Morphisms

Given a functor $\varphi:\mathbf{D}\rightarrow\mathbf{C}$, a *$\varphi$-structured morphism* is a $\mathbf{C}$-morphism of the form $f:c\rightarrow\varphi(d)$, where c is an object of $\mathbf{C}$ and d is an object of $\mathbf{D}$. ♦

Structured morphisms can be organised in categories that generalise the notion of comma category that is studied in Par. 3.3.2.

6.2.3 Definition – Comma Categories

1. Given a functor $\varphi:\mathbf{D}\rightarrow\mathbf{C}$ and an object $c:\mathbf{C}$, we define a category $c\downarrow\varphi$ that has for objects all pairs $<f:c\rightarrow\varphi(d),d>$, where f is a φ-structured morphism with domain c. The morphisms between $<f_1:c\rightarrow\varphi(d_1),d_1>$ and $<f_2:c\rightarrow\varphi(d_2),d_2>$ are all the $\mathbf{D}$-morphisms $h:d_1\rightarrow d_2$ that satisfy $f_1;\varphi(h)=f_2$. These are called under-cone categories in [22, p. 48].

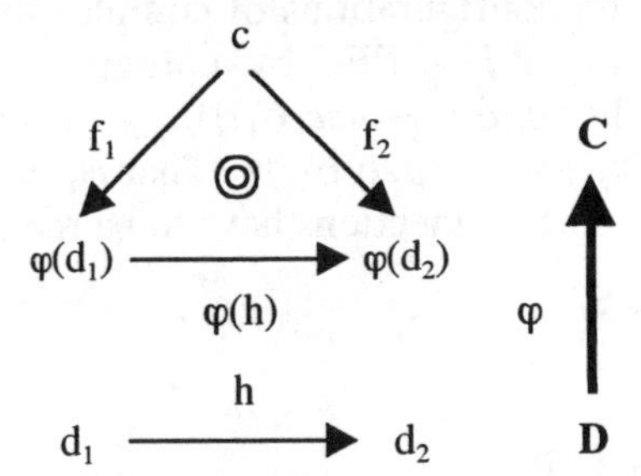

2. This construction can be further generalised to $\_\downarrow\varphi$ (or $\mathbf{C}\downarrow\varphi$) whose objects are φ-structured morphisms $<c,f:c\rightarrow\varphi(d),d>$. A morphism from $<c_1,f_1:c_1\rightarrow\varphi(d_1),d_1>$ to $<c_2,f_2:c_2\rightarrow\varphi(d_2),d_2>$ is a pair $<g,h>$ of a $\mathbf{C}$-morphism $g:c_1\rightarrow c_2$ and a $\mathbf{D}$-morphism $h:d_1\rightarrow d_2$ such that $f_1;\varphi(h)=g;f_2$.

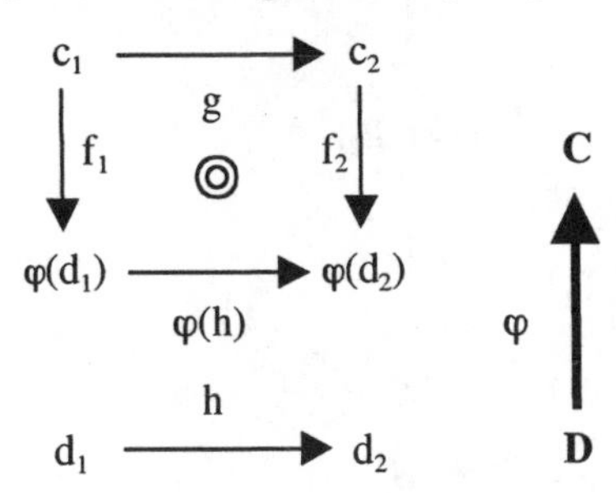

Proof

The proof that these constructions yield categories is left as an exercise. The reader is also invited to consult [80, p. 46–48] for further generalisations of the notion of comma category, including the one that justifies their name. ♦

6.2.4 Remark – Compositionality in Software Development

In Chap. 4, universal constructions are used to formalise notions of composition that occur in system development. These are applied to system specification in Par. 6.1.24, including an illustration of their role in supporting evolution through the interconnection, at runtime, of new components that can bring about new emergent properties.

However, ideally such constructions should apply to realisations and not to specifications alone in the sense that, when we compose specifications for which we have already correct implementations, we should be able to obtain a correct implementation for the resulting specification as a composition of the implementations of its components. This is usually called *compositionality* of system development. Following on [41], we are going to show how this form of compositionality can be formulated in this setting and derived from the existence of an abstraction functor.

First of all, we are now interested in extending the notion of realisation to diagrams as capturing configurations of complex systems: a realisation of a specification diagram $\delta:\boldsymbol{I}\rightarrow\boldsymbol{SPEC}$ by a program diagram $\delta':\boldsymbol{I}\rightarrow\boldsymbol{PROG}$ is an $|\boldsymbol{I}|$-indexed family $(\sigma_i:\delta(i)\rightarrow\boldsymbol{spec}(\delta'(i)))_{i\in|I|}$ of realisations such that, for every $f:i\rightarrow j$ in $\boldsymbol{I}$, $\delta(f);\sigma_j=\sigma_i;\boldsymbol{spec}(\delta'(f))$. That is, in addition to the individual components, the interconnections have to be realised as well.

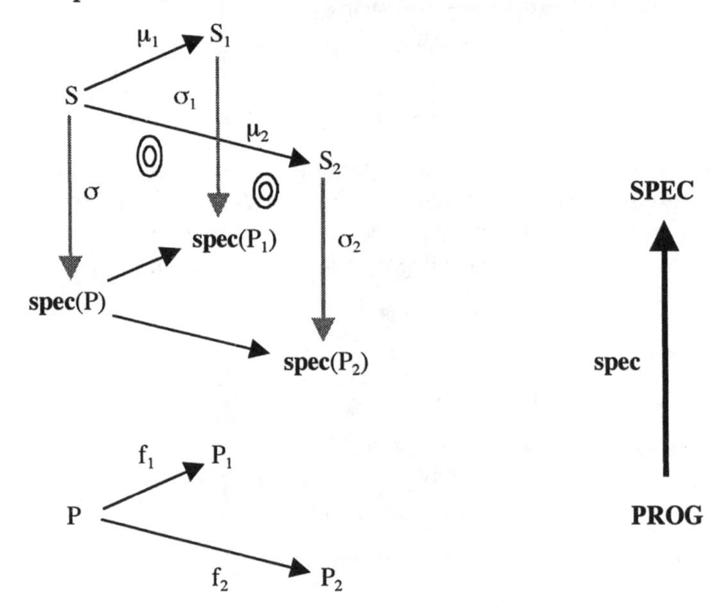

Compositionality of the relationship between programs and specifications can be expressed through the fact that the colimit of δ' – the diagram of programs – is a realisation of δ – the diagram of specifications – in an essentially unique way. We can actually prove that, as long as ***spec*** is a functor, this property holds regardless of the nature of the specifications and programs. Indeed, because functors preserve the commutativity of diagrams, the image of the program diagram commutes, i.e. in the case of the diagram in the figure,

$$\mathbf{spec}(f_1);\mathbf{spec}(f'_1)=\mathbf{spec}(f_2);\mathbf{spec}(f'_2).$$

This entails that

$$\mu_1;\sigma_1;\mathbf{spec}(f'_1)=\sigma;\mathbf{spec}(f_1);\mathbf{spec}(f'_1)=\sigma;\mathbf{spec}(f_2);\mathbf{spec}(f'_2)=\mu_2;\sigma_2;\mathbf{spec}(f'_2).$$

Because $\langle\mu'_1,\mu'_2,S'\rangle$ is a pushout of $\langle S,\mu_1,\mu_2\rangle$, we conclude that there exists a unique morphism $\sigma:S'\rightarrow\mathbf{spec}(P')$ such that $\sigma_1;\mathbf{spec}(f'_1)=\mu'_1;\sigma'$ and $\sigma_2;\mathbf{spec}(f'_2)=\mu'_2;\sigma'$. Notice that these two equations express the fact that the pairs $\langle\mu'_i;\mathbf{spec}(f'_i)\rangle$ are indeed morphisms of realisations. Also notice that ***spec*** is not required to preserve the pushout of programs!

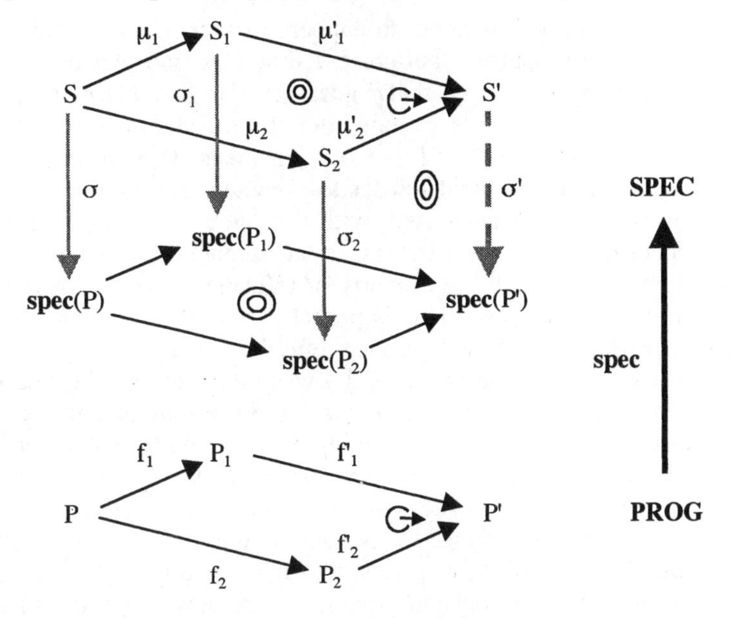

♦

Because compositionality is a property that is not exactly easy to achieve, as more than 20 years of research in the area has shown, this means that the existence of a functor relating two domains of system modelling expresses a very strong structural relationship between them. In [31], we gave examples of situations in which compositionality fails because some of the properties required of functors are not met by the way specifications relate to programs. Basically, at stake is the balance that must be struck between the nature of the properties that the abstraction map is able to derive, and the ability of program morphisms to preserve such properties. In [31] it was also shown that, in the absence of a functorial relationship, properties of systems may emerge that result from the need to regulate the interconnections between the components. This happens when the semantics of the programming language is not strong enough to ensure the preservation of the properties made observable through the specification language. This is similar to what happens in social systems in which regulators (e.g. the police) are necessary to enforce laws that are not "natural" but imposed by the society for coordinating the behaviour of individuals.

An example of such a situation can be given in terms of Eiffel class specifications and linear temporal logic. If we had chosen to restrict the behaviours of Eiffel classes to those that are normative in the sense that routines are only executed when their preconditions hold, the following sentence would be included in the strongest specification of every routine:

$$r \supset pre_r$$

That is, if the routine r is about to happen, its precondition holds. It is easy to see that this property is not preserved by class morphisms. Indeed, given a class morphism f, all that f guarantees is that $F(pre_r) \vdash pre_{F(r)}$. Hence, from $(F(r) \supset pre_{F(r)})$ we cannot infer the translation through F of $(r \supset pre_r)$, i.e. the property $(F(r) \supset F(pre_r))$. This means that an implementation of a class specification that refuses to execute every routine when the precondition fails cannot be reused, with the same semantics of refusal, when the specification is inherited into a larger one. In other words, inheritance does not preserve the property of refusing execution of routines for which preconditions fail. This is probably why the semantics of preconditions in Eiffel does not include such refusals.

Although the construction in Par. 6.2.4 was motivated by the extension of the notion of realisation to diagrams as compositions of complex systems, it admits a dual reading as extending composition to realisations, i.e. moving to the category $\_\downarrow \boldsymbol{spec}$ whose objects are precisely the realisations. Actually, this dual reading is supported by a dual visualisation of the figures above: whereas we analysed them from the point of view of "vertical" relationships (the realisations) between "horizontal structures" (the specification and the program configurations), we are now analysing them as horizontal relationships (a configuration) between vertical structures (the realisations of the component specifications).

Indeed, because, for every $f{:}i \rightarrow j$ in $\boldsymbol{I}$, $\delta(f);\sigma_j = \sigma_i;\boldsymbol{spec}(\delta'(f))$, δ and δ' define a diagram $<\delta,\delta'>{:}\boldsymbol{I} \rightarrow \_\downarrow \boldsymbol{spec}$ such that, for every i, $<\delta,\delta'>(i)$ is $<\delta(i),\sigma_i,\delta'(i)>$ and, for every $f{:}i \rightarrow j$, $<\delta,\delta'>(f)$ is $<\delta(f),\delta'(f)>$. Compositionality expresses the existence (and uniqueness) of a colimit for this diagram of realisations when δ and δ' admit colimits. Compositionality in the sense that we have just described is a property that can be stated and proved in the general setting of comma categories.

6.2.5 Definition/Proposition – Projection Creates Colimits

1. Given a functor $\varphi:\boldsymbol{D}\to\boldsymbol{C}$, we define two functors $\pi_C:\_\downarrow\varphi\to\boldsymbol{C}$ and $\pi_D:\_\downarrow\varphi\to\boldsymbol{D}$ by projecting objects and morphisms of the comma category $\_\downarrow\varphi$ to their $\boldsymbol{C}$ and $\boldsymbol{D}$ components, respectively.
2. The functor $<\pi_C,\pi_D>:\_\downarrow\varphi\to\boldsymbol{C}\times\boldsymbol{D}$ creates colimits.

Proof

1. Left as an exercise.
2. Consider a diagram $\delta:\boldsymbol{I}\to\_\downarrow\varphi$ and colimits $\mu:\delta;\pi_C\to c$ and $\beta:\delta;\pi_D\to d$ for its projections $\delta;\pi_C$ and $\delta;\pi_D$, respectively.

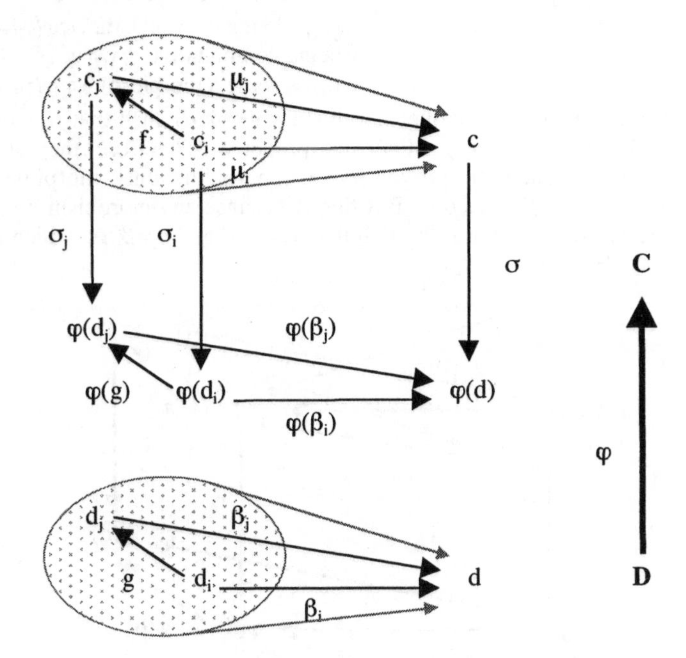

- Existence of a cocone in $\_\downarrow\varphi$ whose image is the pair of projections. Basically, we have to show that there is a morphism $\sigma:c\to\varphi(d)$ such that $\sigma_i;\varphi(\beta_i)=\mu_i;\sigma$.
  To prove this, consider the $\boldsymbol{C}$-cocone defined by $\{\sigma_i;\varphi(\beta_i)\}\to\varphi(d)$. This cocone is commutative because, given any morphism $f:c_i\to c_j$ in the base, if $g:d_i\to d_j$ is the morphism in $\boldsymbol{D}$ that, together with f, constitutes a morphism in $\_\downarrow\varphi$, we have that

$\beta_i=g;\beta_j$	the $\boldsymbol{D}$-cocone is commutative,
$\varphi(\beta_i)=\varphi(g);\varphi(\beta_j)$	functors preserve composition,
$f;\sigma_j=\sigma_i;\varphi(g)$	$<f,g>$ is a morphism in $\_\downarrow\varphi$,

$f;\sigma_j;\varphi(\beta_j)=\sigma_i;\varphi(g);\varphi(\beta_j)$ from the previous equality,
$f;\sigma_j;\varphi(\beta_j)=\sigma_i;\varphi(\beta_i)$ applying the second equality.

Hence, because $\mu:\delta;\pi_C\to c$ is a colimit, there is a (unique) morphism $\sigma:c\to\varphi(d)$ such that $\sigma_i;\varphi(\beta_i)=\mu_i;\sigma$.

- Uniqueness of the cocone. For the cocone to be in $\_\downarrow\varphi$, the equalities $\sigma_i;\varphi(\beta_i)=\mu_i;\sigma$ have to hold, and the fact that the cocone in **C** is a colimit implies that σ is the only morphism $c\to\varphi(d)$ that satisfies that property.
- The cocone is a colimit. Let $<\nu,\gamma>:\delta\to<c',\sigma',d'>$ be a commutative cocone. Its projections $\nu:\delta;\pi_C\to c'$ and $\gamma:\delta;\pi_d\to d'$ are also commutative. Therefore, because μ and β are colimits, there are unique $\tau:c\to c'$ and $\rho:d\to d'$ such that $\nu_i=\mu_i;\tau$ and $\gamma_i=\beta_i;\rho$. We want to prove, first of all, that $<\tau,\rho>$ is a morphism of $\_\downarrow\varphi$, i.e. $\tau;\sigma'=\sigma;\varphi(\rho)$. For that purpose, we are going to prove that the cocone defined by $(\sigma_i;\varphi(\gamma_i))$ is commutative: given an arbitrary $<f:c_i\to c_j,g:d_i\to d_j>$ in the base for a structured object $<c_j,\sigma_j,d_j>$,

$f;\sigma_j;\varphi(\gamma_j)$
$=f;\nu_i;\sigma'$ $<\nu_i,\gamma_i>$ being a morphism, $\sigma_j;\varphi(\gamma_j)=\nu_i;\sigma'$,
$=\nu_i;\sigma'$ ν being commutative, $f;\nu_i=\nu_i$,
$=\sigma_i;\varphi(\gamma_i)$ $<\nu_i,\gamma_i>$ being a morphism, $\sigma_i;\varphi(\gamma_i)=\nu_i;\sigma'$.

Hence, there is a unique morphism $\kappa:c\to\varphi(d)$ such that $\mu_i;\kappa=\sigma_i;\varphi(\gamma_i)$. But both $\tau;\sigma'$ and $\sigma;\varphi(\rho)$ satisfy that property. Therefore, they are equal. Finally, we have to prove that $<\tau,\rho>$ is the only morphism satisfying $<\nu_i,\gamma_i>=<\mu_i,\beta_i>;<\tau,\rho>$. But this is because any morphism $<\tau',\rho'>$ satisfying the same equation is such that $\nu_i=\mu_i;\tau'$ and $\gamma_i=\beta_i;\rho'$, which implies $\tau'=\tau$ and $\rho'=\rho$.

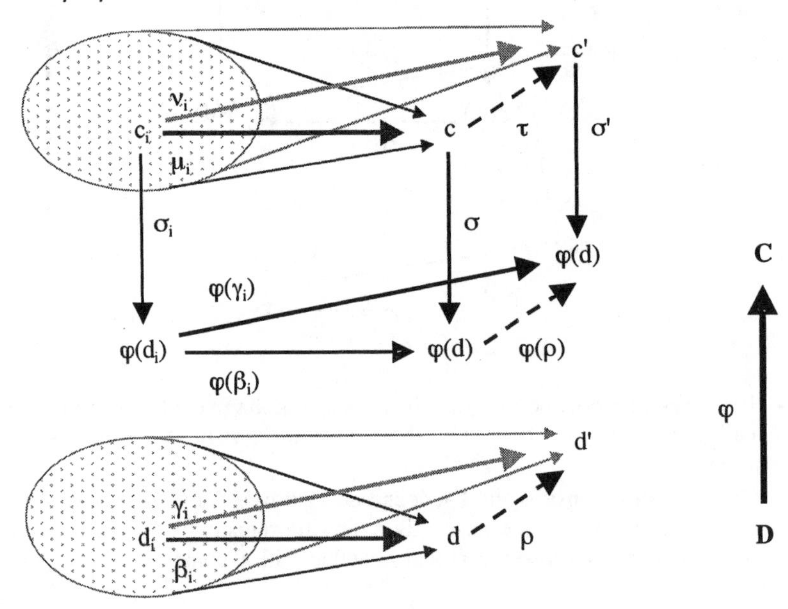

♦

6.3 Functor-Structured Categories

In this section, we present a construction that yields concrete categories of a very simple nature but with wide applicability.

6.3.1 Definition – Functor-Structured Categories

Let φ*:***C→SET** be a functor. We define the category ***spa(***φ***)*** whose objects are the pairs *<c,S>* where *c*:**C** and $S\subseteq\varphi(c)$ and whose morphisms *f:<c,S>→<d,T>* are the morphisms *f:c→d* of **C** such that $\varphi(f)(S)\subseteq T$.

Proof

As illustrated in Sect. 3.2, we have to prove that the composition law and the identity map inherited from **C** are applicable.

- Given morphisms *f:<c,S>→<d,T>* and *g:<d,T>→<e,R>*, we have
  1. $\varphi(f)(S)\subseteq T$ — because *f* is a morphism of ***spa(***φ***)***,
  2. $\varphi(g)(\varphi(f)(S))\subseteq\varphi(g)(T)$ — from 1,
  3. $\varphi(g)(T)\subseteq R$ — because *g* is a morphism of ***spa(***φ***)***,
  4. $\varphi(g)(\varphi(f)(S))\subseteq R$ — from 2, 3, and transitivity of inclusion,
  5. $(\varphi(f);\varphi(g))(S)\subseteq R$ — from 4 and function composition in ***SET***,
  6. $\varphi(f;g)(S)\subseteq R$ — from 5 and the properties of functors.
- The case of the identity map is trivial: given a pair *<c,S>*, the identity on *c* is such that $\varphi(id_c)(S)=id_{\varphi(c)}(S)=S$. ♦

Functor-structured categories, sometimes also called *spa-categories*, are studied in detail in [1], where some additional terminology is introduced. The objects of such categories are usually called φ-spaces and their morphisms φ-maps.

6.3.2 Example – Processes

In Par. 3.2.1 and the discussion that follows it, we regard a pointed set as a signature or alphabet of a process: the proper elements of the set denote actions or events in which the process can get involved, and the designated element denotes an action of the environment, i.e. an action in which the process is not involved. We can associate with every pointed set $A_\perp$, the set of possible trajectories over $A_\perp$: ***tra***$(A_\perp)=\{\lambda:\omega\to A\}$. That is, a trajectory for an alphabet is an infinite sequence of actions; it represents one possible behaviour in a given environment. For instance, in the case of the alphabet $<\{produce,store,\perp_P\},\perp_P>$ of a *producer*, the set of all behaviours in which *produce* and *store* succeed each other is given by:

```
(⊥P*produce⊥P*store)∞
```

Notice that we make explicit the occurrence of the designated event because we are capturing behaviours that take place in given environments. Hence, finite behaviours can be represented by infinite sequences that, after a certain point, consist of the designated element, i.e. consist only of environment steps:

$(\bot_P{}^*\text{produce}\bot_P{}^*\text{store})^*\bot_P{}^\infty$

This association defines a functor $\boldsymbol{tra}$:$\boldsymbol{SET}_\bot \rightarrow \boldsymbol{SET}$ if we extend it to morphisms as follows: $\boldsymbol{tra}(f{:}A_\bot \rightarrow B_\bot)(\lambda)=\lambda;f$, i.e. $\boldsymbol{tra}(f)(\lambda)(i)=f(\lambda(i))$. That is, every morphism of pointed sets induces a translation between the corresponding trajectories by pointwise application of the function between the base sets. We usually denote $\boldsymbol{tra}(f)$ by f^ω.

We can now define the category ***PROC*** of processes as ***spa(tra)***. A process consists of a pair $<A_\bot,\Lambda>$ where $\Lambda \subseteq \boldsymbol{tra}(A_\bot)$, i.e. a process consists of an alphabet and a set of trajectories that capture its behaviour. A process morphism $f{:}<A_{\bot_1},\Lambda_1> \rightarrow <A_{\bot_2},\Lambda_2>$ is a morphism $f{:}A_{\bot_1} \rightarrow A_{\bot_2}$ between the underlying pointed sets such that $f^\omega(\Lambda_1) \subseteq \Lambda_2$, i.e., such that every trajectory of the source is translated to a trajectory of the target.

As mentioned in Par. 3.2.1, a morphism $f{:}P_1 \rightarrow P_2$ identifies process P_2 as a component of process P_1: it identifies, for every action of P_1, the participation of P_2 in that action; if the event of P_1 is mapped to the designated event of P_2, this means that P_2 does not participate in it, making it an environment step for P_2. Notice that the preservation of the designated element, as enforced through the morphism, is consistent with the view of P_2 as a component of P_1: an environment step for P_1 must necessarily be an environment step for P_2. Hence, P_1 identifies part of the environment of P_2. The condition on the trajectories requires that every possible behaviour of P_1 (the system) be mapped to one of the allowed behaviours of P_2 (the component). That is, the system cannot exhibit behaviours that are not allowed by the component. However, it is possible that behaviours of the component do not show up in the system, e.g. because interactions with other components within the system prevent them from occurring. ◆

Functor-structured categories provide examples of some of the kinds of categories discussed in Sect. 6.1.

6.3.3 Proposition

Let $\varphi{:}\boldsymbol{C} \rightarrow \boldsymbol{SET}$ be a functor.

1. The functor $v{:}\,\boldsymbol{spa}(\varphi) \rightarrow \boldsymbol{C}$ that forgets the ***SET***-component of each object defines $\boldsymbol{spa}(\varphi)$ as a concrete category over ***C***.
2. The forgetful functor $v{:}\boldsymbol{spa}(\varphi) \rightarrow \boldsymbol{C}$ is both a split fibre-complete fibration and a split fibre-cocomplete cofibration. Given $f{:}c \rightarrow c'$ in ***C***, $f{:}<c,\varphi(f)^{-1}(S')> \rightarrow <c',S'>$ is the Cartesian morphism for $<c',S'>$ and $f{:}<c,S> \rightarrow <c',\varphi(f)(S)>$ is the co-Cartesian morphism for $<c,S>$.

Proof

Left as an exercise. ◆

6.3.4 Example – Processes

We call ***alph*** the forgetful functor defined by Proposition 6.3.3(1) on ***PROC***: it projects processes and their morphisms to the underlying alphabets (pointed sets). Notice how the (co)Cartesian morphisms have an intuitive interpretation in ***PROC***. The Cartesian morphism returns, for a given process and alphabet morphism having the process as target, the least-deterministic system over that alphabet in which the process can fit, through the morphism, as a component: any additional behaviours of the system would violate the allowed behaviours of the given process.

For instance, consider the alphabet of the *producer/consumer* system defined in Par. 4.3.8, whose proper events are: *store|retrieve, produce|consume, produce, consume*.

The Cartesian morphism relative to the following behaviour of the *producer* component

$$(\bot_P\text{*produce}\bot_P\text{*store})^\infty$$

defines the following behaviour for the system

$$(\{\bot_s,\text{consume}\}\text{*}\{\text{produce|consume},\text{produce}\}\{\bot_s,\text{consume}\}\text{*store|retrieve})^\infty$$

That is, we obtain the system trajectories in which the *producer* component alternates between executing *produce* and *store*. Notice how the environment of *producer* is captured by one of $\bot_s$ or *consume*. In other words, an environment step for the component is either performed by the environment of the whole system $\bot_s$ or by the *consumer* component without interacting with *producer*.

On the other hand, the co-Cartesian morphism returns, for a given process and alphabet morphism having the process as source, the most deterministic process over the target alphabet that it admits, through the morphism, as a component. Any additional behaviours that the component may have will not be able to be observed in the given system.

For instance, consider a system that consists of all trajectories in which only one exchange is performed between *consumer* and *producer*:

$$\bot_s\text{*produce}\bot_s\text{*store|retrieve}\bot_s\text{*consume}\bot_s{}^\infty$$

The Cartesian morphism relative to the *consumer* component returns:

$$\bot_c\text{*retrieve}\bot_c\text{*consume}\bot_c{}^\infty$$

Any other trajectory in the language of *consumer* will not show up during the execution of the given system. Notice that the Cartesian morphism relative to this behaviour of the *consumer* component defines the following behaviour for the system

$\{\perp_s, \mathtt{produce}\}^*$
$\quad \mathtt{store|retrieve}$
$\quad\quad \{\perp_s, \mathtt{produce}\}^*$
$\quad\quad\quad \{\mathtt{produce|consume}, \mathtt{consume}\}$
$\quad\quad\quad\quad \{\perp_s, \mathtt{produce}\}^\infty$

This process is less deterministic than the one we started from because it captures only restrictions imposed by the *consumer*. This means that the original system behaviour can only emerge from the interactions of the *consumer* with other components of the system. Indeed, the behaviours

$$\perp_s{}^*\mathtt{produce}\perp_s{}^*\mathtt{store|retrieve}\perp_s{}^*\mathtt{consume}\perp_s{}^\infty$$

belong to the intersection defined by the Cartesian morphisms relative to both components:

$$\perp_c\mathtt{retrieve}\perp_c{}^*\mathtt{consume}\perp_c{}^\infty$$

for the *consumer* and, for the *producer*

$$(\perp_P{}^*\mathtt{produce}\perp_P{}^*\mathtt{store})^\infty$$

We show in Par. 6.3.5 that this intersection, which captures parallel composition, is computed by a limit (pullback). This is yet another instance of the "conjunction as composition" idea [112] and Goguen's dogma [57].

In summary, the Cartesian and co-Cartesian morphism allow us to gauge the roles that a given process can play through a morphism as a system or as a component. ♦

The characterisation provided in Par. 6.3.3 for (co)Cartesian morphisms allows us to derive some useful properties.

6.3.5 Corollary

Let $\varphi{:}\boldsymbol{C}\to\boldsymbol{SET}$ be a functor, and $\upsilon{:}\boldsymbol{spa}(\varphi)\to\boldsymbol{C}$ the forgetful functor induced by the ***spa***-construction.[1]

1. υ lifts limits uniquely. Any limit $\mu{:}c\to\delta;\upsilon$ for $\delta{:}\boldsymbol{I}\to\boldsymbol{spa}(\varphi)$ with $\delta_i=<c_i,S_i>$ is lifted uniquely to $\mu{:}<c,S>\to\delta$ where $S=\bigcap_{i\in I}\varphi(\mu_i)^{-1}(S_i)$.
2. υ lifts colimits uniquely. Any colimit $\mu{:}\delta;\upsilon\to c$ for $\delta{:}\boldsymbol{I}\to\boldsymbol{spa}(\varphi)$ with $\delta_i=<c_i,S_i>$ is lifted uniquely to $\mu{:}\delta\to<c,S>$ where $S=\bigcup_{i\in I}\varphi(\mu_i)(S_i)$. ♦

6.3.6 Corollary

Let $\varphi{:}\boldsymbol{C}\to\boldsymbol{SET}$ be a functor.

1. If $\boldsymbol{C}$ is complete so is $\boldsymbol{spa}(\varphi)$, all limits being concrete (Par. 6.1.9).
2. If $\boldsymbol{C}$ is cocomplete so is $\boldsymbol{spa}(\varphi)$, all colimits being concrete. ♦

---

[1] For readability, we shall omit the forgetful functor when referring to objects and morphisms of $\boldsymbol{C}$ that are projected from $\boldsymbol{spa}(\varphi)$.

Therefore, we can apply the recipe for calculating limits and colimits in ***spa(φ)*** discussed in Sect. 6.1:

1. Project the diagram to the base category and calculate its (co)limit.
2. Lift the result by computing the intersection of the inverse images of the set components (for a limit) or the union of the direct images of the set components (for the colimit).

It is useful to see how this construction can be instantiated to particular universal constructions:

- Lifting of the initial and terminal objects:
  If 0_C is an initial object of $\boldsymbol{C}$, then $<0_C,\emptyset>$ is initial for ***spa(φ)***..
  If 1_C is a terminal object of $\boldsymbol{C}$, $<1_C,\varphi(1_C)>$ is terminal for ***spa(φ)***.
- Universal constructions in three steps: projection of the ***spa(φ)***-diagram to $\boldsymbol{C}$, universal construction performed in $\boldsymbol{C}$, lifting the result back to ***spa(φ)***.

Product:

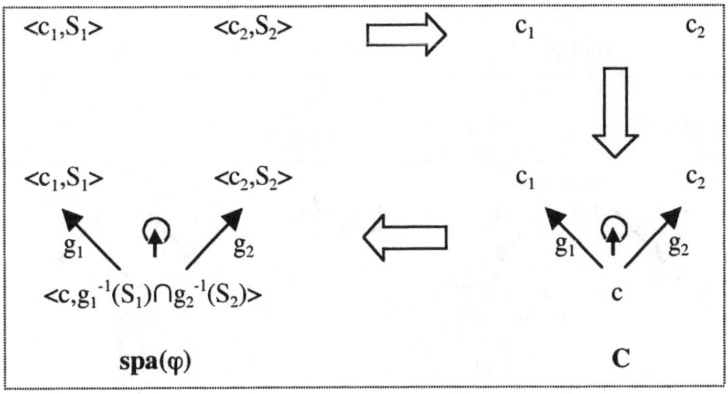

$<c_1,S_1>$ $<c_2,S_2>$ ⇒ c_1 c_2

$<c_1,S_1>$ $<c_2,S_2>$ ⇐ c_1 c_2

g_1 g_2 g_1 g_2

$<c,g_1^{-1}(S_1)\cap g_2^{-1}(S_2)>$ c

spa(φ) **C**

Sum:

$<c_1,S_1>$ $<c_2,S_2>$ ⇒ c_1 c_2

$<c_1,S_1>$ $<c_2,S_2>$ ⇐ c_1 c_2

g_1 g_2 g_1 g_2

$<c,g_1(S_1)\cup g_2(S_2)>$ c

spa(φ) **C**

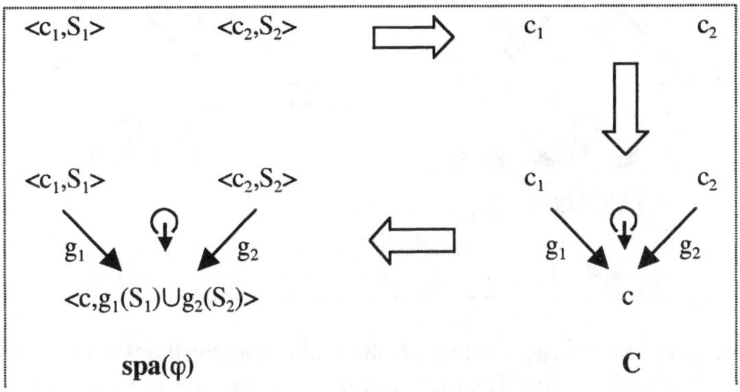

Pullback:

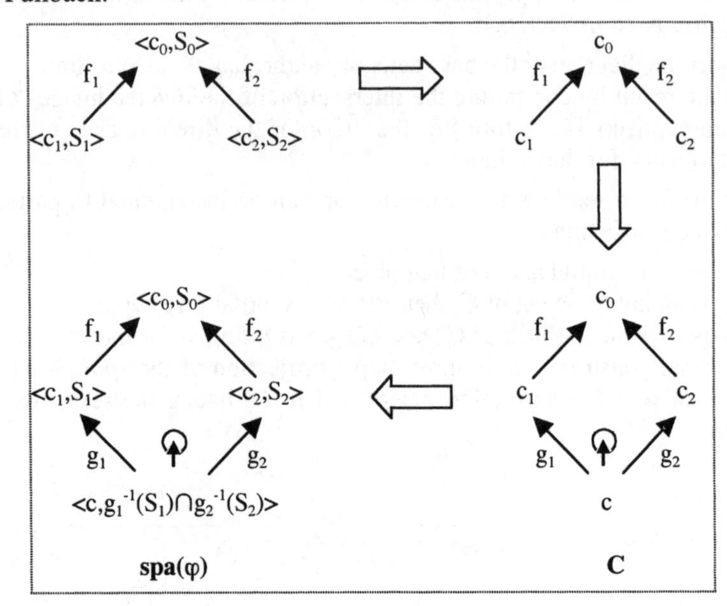

Pushout:

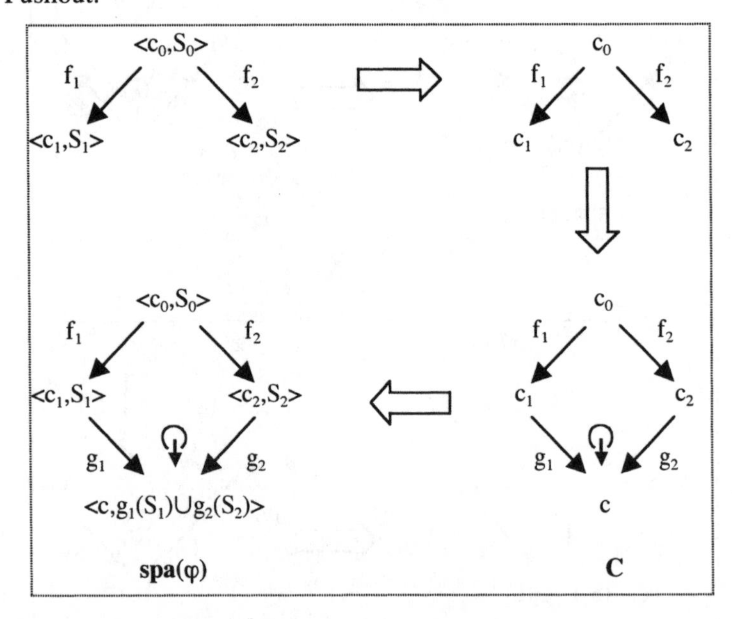

Notice that the lifting of the product and the pullback are exactly the same: the "shared" object is only used for determining the ***C***-component of the apex of the pullback and does not interfere with the calculation of the

SET component. This is another example of the kind of separation that can be achieved between "coordination" and "computation" as discussed in Sect. 5.2. We return to these aspects in Sect. 7.5, but this is a simple illustration of the principle.

6.3.7 Example – Parallel Composition of Processes

When applied to processes, these constructions provide us with a mathematical semantics for typical operations of process calculi. Recalling the universal constructions studied in Chap. 4 for pointed sets, we have:

1. The terminal process is $<\{\perp_\emptyset\},\{\perp_\emptyset{}^\omega\}>$. Its alphabet contains only the witness for actions of the environment, and its behaviour reflects exactly that – it witnesses life go by. That is, we have an idle process.
2. The initial process is $<\{\perp_\emptyset\},\emptyset>$. Its alphabet is exactly the same as for the terminal process, but its behaviour is completely different – it does nothing, not even witnessing life go by! That is, we have a blocking process that deadlocks any system to which it is interconnected.
3. As seen in Par. 4.2.7, products of pointed sets model the interleaving of alphabets in the sense that they compute the set of all the synchronisation pairs of actions between the components plus the individual actions themselves.

   When we take into account the behaviour of processes, products return the infinite sequences of such parallel actions that, once projected into the components, result in behaviours of the components. That is, the product of processes $<A_1,\Lambda_1>$ and $<A_2,\Lambda_2>$ is obtained by computing the product $<A,g_1,g_2>$ of the alphabets and by taking as set of behaviours the intersection of the inverse images of the sets of behaviours of the components. This intersection consists of the sequences $\lambda:\omega\rightarrow A$ such that $g_1{}^\omega(\lambda)\in\Lambda_1$ and $g_2{}^\omega(\lambda)\in\Lambda_2$. Hence, what we obtain is the traditional trace-based semantics of parallel composition.

   Notice the difference between the idle and the blocking process. When put in parallel with another process, the idle process is "absorbed", i.e. the result of the parallel composition is the other process. This is because its behaviour is already present in any other process (except the blocking one). Indeed, the product of a terminal object with any other object is, up to isomorphism, that object. Hence, the idle process does nothing and lets the others do as they please.

   On the contrary, the blocking process "absorbs" any other process with which it is put in parallel: it does nothing and does not let the others do anything. This is because the product of the initial object with any other process *P* returns a process with the alphabet of *P* but with an empty behaviour. As claimed at beginning of the book, category theory is all about the social behaviour of objects...

4. Pullbacks allow us to select only the behaviours that satisfy certain synchronisation requirements. We can see in Par. 4.3.8 that such requirements are expressed through the morphisms that connect the components to the "channel" that interconnects them. Each action of the channel acts as a point for rendez vous synchronisation in the sense that the actions that participate in a rendez vous are not allowed to occur in isolation, i.e. are not part of the alphabet of the resulting process. Concerning the resulting behaviour, its computation is exactly as for products: it consists of the sequences of actions determined by the pullback of alphabets that are projected into behaviours of the components. Hence, the synchronisation is achieved at the level of the alphabets. In summary, pullbacks model parallel composition with synchronisation.
5. What we have just observed about pullbacks can be generalised to limits in general. The limit of the diagram of alphabets internalises all the interconnections established via the morphisms as synchronisation sets: each component may be involved with at most one action in a synchronisation set. That is, internal synchronisations cannot be established and a given component may not participate in a given rendez vous. When we take into account the behaviour of the processes involved, limits return the infinite sequences of such synchronisation sets that are projected into allowed behaviours of the components. ♦

6.4 The Grothendieck Construction

In Sect. 6.1 every split fibration $\varphi{:}\boldsymbol{D}\rightarrow\boldsymbol{C}$ is shown to define a functor $\boldsymbol{ind(\varphi)}{:}\boldsymbol{C}^{op}\rightarrow\boldsymbol{CAT}$ that maps every object of $\boldsymbol{C}$ to its fibre. In the case of a split cofibration, we obtain $\boldsymbol{ind(\varphi)}{:}\boldsymbol{C}\rightarrow\boldsymbol{CAT}$. The objects of $\boldsymbol{C}$ can be seen as indexes or types that are used for classifying the objects of $\boldsymbol{D}$. The functor φ makes the type assignments and the functor $\boldsymbol{ind(\varphi)}$ groups the objects according to their types.

On the morphism side, the functor $\boldsymbol{ind(\varphi)}$ tells us how to translate between fibres when moving from one type to another through a type morphism. For fibrations, this translation is contravariant (like when taking inverse images), whereas for cofibrations the translation is covariant (like when taking direct images).

For instance, if $\boldsymbol{C}$ is the inheritance hierarchy of an object-oriented system, and $\boldsymbol{D}$ models the population of objects, we can think of a cofibration that assigns to every object the type of which it is a direct instance. For every inheritance morphism, the co-Cartesian morphisms define how each object of the child class is also an instance of the parent type (which feature renaming applies, etc). The $\boldsymbol{CAT}$-based functor returns, for each type, the population of direct instances associated with the type and, for each inheritance morphism, the functor that maps instances of the child to instances of the parent.

The fact that we work with a fibration or a cofibration, and that the associated *CAT*-based functor is contravariant or covariant, is not very important. After all, it is all a matter of arrow direction. However, as pointed out in other parts of this book, it is important that the direction of the arrows is chosen so as to be "natural" in the context in which they are going to be used. For instance, in the case of inheritance, hierarchies are traditionally depicted as graphs with the arrows pointing to the parents, and it would be counterintuitive to formalise them with morphisms going in the opposite direction. This is why, in this section, one direction is preferred for definitions and results, but examples are given in their "natural" direction. The reader should detail the dual constructions as an exercise.

We have also mentioned that indexed categories are very much part of the categorical folklore of computing [e.g., 12, 22, 102]. We follow [102] more closely because it shares with us the same kind of "engineering" inspiration that we set ourselves to promote. The reader can also find therein many more examples and a much more extended coverage of this topic.

6.4.1 Definition – Indexed Categories

An *indexed category* ι over a category $\boldsymbol{I}$ (of indexes) is a functor of the form $\iota:\boldsymbol{I}^{op}\rightarrow\boldsymbol{CAT}$. ♦

6.4.2 Example – Comma categories

In Par. 3.2.2 we show how, given a category $\boldsymbol{C}$ and an object $a{:}\boldsymbol{C}$, we can define the category of objects under a, that is $a{\downarrow}\boldsymbol{C}$. Given a morphism $r{:}a'\rightarrow a$, there is a natural way of translating objects under a to objects under a': given $f{:}a\rightarrow x$, we map it to $(r;f){:}a'\rightarrow x$.

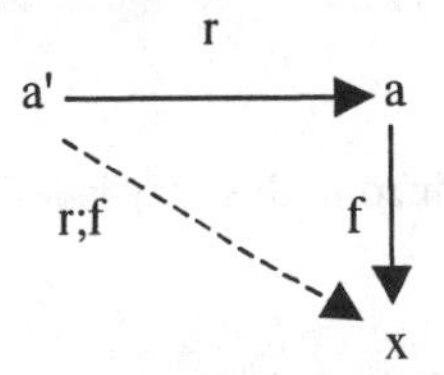

It is easy to prove that this translation is indeed a functor $a{\downarrow}\boldsymbol{C}\rightarrow a'{\downarrow}\boldsymbol{C}$ so that we obtain an indexed category $\_{\downarrow}\boldsymbol{C}{:}\boldsymbol{C}^{op}\rightarrow\boldsymbol{CAT}$. ♦

6.4.3 Exercise

Complete the previous example by doing the proofs and generalising it to the comma categories of the form $a{\downarrow}\varphi$ as defined in Par. 6.2.3. ♦

Once we look at functors $\iota:\boldsymbol{I}^{op}\rightarrow\boldsymbol{CAT}$ as performing an indexing of certain classes of objects, it seems natural to think about the amalgamation of all these classes into a single population, using the indexes to distinguish between the objects according to their provenance. This operation is traditionally called the "Grothendieck construction" [12, 22] in honour of A. Grothendieck, or "flattening" [102].

6.4.4 Definition – Flattening an Indexed Category

Any indexed category $\iota{:}\boldsymbol{I}^{op}\rightarrow\boldsymbol{CAT}$ defines a category $\boldsymbol{FLAT(\iota)}$ as follows:

1. The objects of $\boldsymbol{FLAT(\iota)}$ are all the pairs $<i,a>$ where $i{:}\boldsymbol{I}$ is an index and $a{:}\iota(i)$ is an object of type i.
2. The morphisms $<i,a>\rightarrow<j,b>$ are all the pairs $<\sigma,f>$, where $\sigma{:}i\rightarrow j$ is a morphism of indexes and $f{:}a\rightarrow\iota(\sigma)(b)$ is a morphism in $\iota(i)$.
3. Composition is defined componentwise: given $<\sigma,f>{:}<i,a>\rightarrow<j,b>$ and $<\mu,g>{:}<j,b>\rightarrow<k,c>$, we define $<\sigma,f>;<\mu,g>=<\sigma;\mu,f;\iota(\sigma)(g)>$.
4. The identity for an object $<i,a>$ is $<id_i;id_a>$.

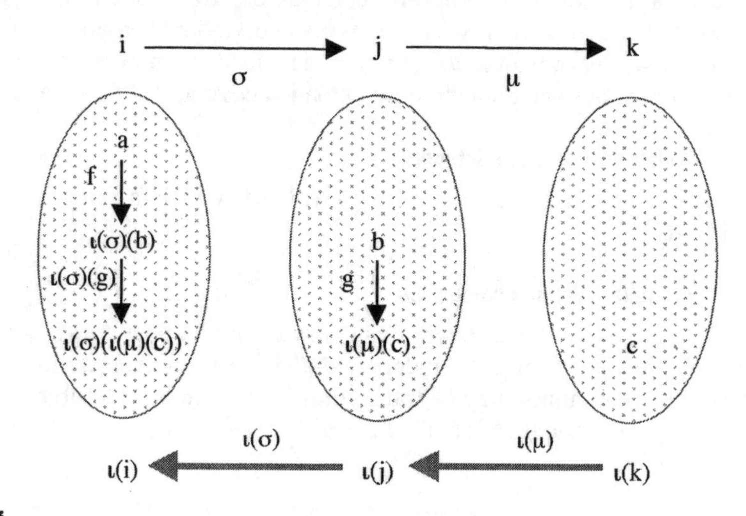

Proof

The proof that a category is defined in this way is left as an exercise. ♦

6.4.5 Example

The flattening of $\_\downarrow\boldsymbol{C}{:}\boldsymbol{C}^{op}\rightarrow\boldsymbol{CAT}$ is the arrow category of $\boldsymbol{C}$. Its objects are all triples $<x,f{:}x\rightarrow y,y>$. The morphisms $<x,f{:}x\rightarrow y,y>\rightarrow<z,g{:}z\rightarrow w,w>$ are the pairs $<h,k>$ such that $h{:}x\rightarrow z$, $k{:}y\rightarrow w$ and $h;g=f;k$. The equation results from the requirement that k be a morphism between $f{:}x\rightarrow y$ and $h;g{:}x\rightarrow w$ in $x\downarrow\boldsymbol{C}$, which is the translation of $g{:}z\rightarrow w$ defined by h.

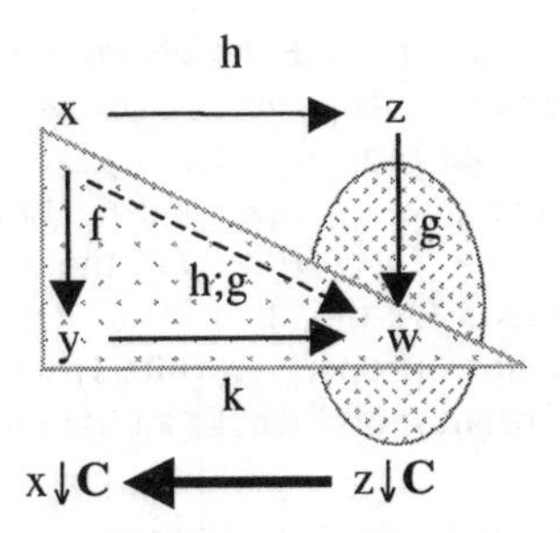

◆

6.4.6 Exercise

Prove that the category $\_\downarrow\varphi$ defined in Par 6.2.3 "is" the flattening of the indexed category of Exercise 6.4.3. ◆

The flattened indexed category comes equipped with some structure.

6.4.7 Proposition

Consider an indexed category $\iota{:}I^{op}\rightarrow CAT$.

1. We define a functor $fib(\iota){:}FLAT(\iota)\rightarrow I$ by projecting objects and morphisms to their I-components.
2. This functor is faithful, defining $FLAT(\iota)$ as being concrete over I.
3. This functor is a split fibration: the Cartesian morphism associated with $\sigma{:}i\rightarrow j$ and $<j,b>$ is $<\sigma,id_{\iota(\sigma)(b)}>:<i,\iota(\sigma)(b)>\rightarrow<j,b>$. That is, we simply use the translation mechanism of the indexed category to perform the required lift.

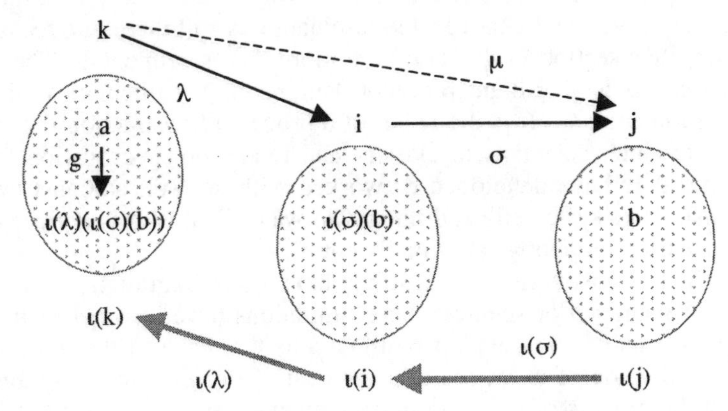

Proof

We leave the proof of the first two properties to the reader.

3. We have to prove that the proposed Cartesian morphisms satisfy the required properties.

- First of all, it is easy to see that the morphism is well defined: $\iota(\sigma)(b)$ is indeed indexed by i. Moreover, its projection over $\boldsymbol{I}$ is σ as required.
- Consider now the universal property. Let $<\mu,g>:<k,a>\rightarrow<j,b>$ and $\lambda:k\rightarrow i$ satisfy $\mu=\lambda;\sigma$. Because $g:a\rightarrow\iota(\mu)(b)$ and $\iota(\mu)(b)=\iota(\lambda;\sigma)(b)=\iota(\lambda)(\iota(\sigma)(b))$, we obtain $<\lambda,g>:<k,a>\rightarrow<i,\iota(\sigma)(b)>$. On the one hand, this morphism is projected to λ as required. On the other hand, $<\mu,g>=<\lambda,g>;<\sigma,id_{\iota(\sigma)(b)}>$ as required. The uniqueness property results from the fact that the second component of the Cartesian morphism is an identity and the projection over $\boldsymbol{I}$ is faithful.

The fact that the fibration is split is trivially checked. ◆

This result allows us to derive properties of flattened indexed categories as corollaries of general results about fibrations.

6.4.8 Proposition

Consider an indexed category $\iota:\boldsymbol{I}^{op}\rightarrow\boldsymbol{CAT}$. If $\boldsymbol{I}$ is complete, $\iota(i)$ is complete for every $i{:}\boldsymbol{I}$, and every translation functor $\iota(\sigma)$ preserves limits, then $\boldsymbol{FLAT}(\iota)$ is complete. ◆

Examples of indexed categories, their properties, and some of their uses in computing are given in the next section.

6.5 Institutions

I am proud to say that I grew up into category theory when trying to learn the concept of institution put forward by Goguen and Burstall in the early 1980s! However, and although the fascination is still there, the reasons for including this section in this book are more than sentimental. The theory of institutions is a brilliant piece of “engineering mathematics”, both in form and in content. It is the result of a process of abstraction that has allowed computer scientists to classify and relate the specification formalisms that they have developed or worked with, as well as construct new formalisms in a systematic and integrated way. That is, it is a piece of science, one that computing can proudly boast.

Like every abstraction, institutions have their limitations. But, as in every other branch of science, these limitations have inspired many other computer scientists to develop ramifications that have further enlightened the path of those of us who are in the business of systematising and supporting the process of software development. Hence, our own account of the basics of institutions is left here both to exemplify some of the constructions that we developed so far in the book, and as a source of inspiration for the reader.

Some readers and, it is to be hoped, users of this book will complain that this topic comes too late. For instance, it does not require the material that we introduced so far in this section. While this is clearly true, the reason is, as for all the other examples, that we deliberately placed institutions where we think they "belong". The point, once again, is that the purpose of this section is not to promote institutions, but rather illustrate the constructions that we have been defining with the best examples we know. The community has been longing for a book on the theory of institutions, but this section is no substitute for it.

We start by defining not the original notion of institution as it appears in [62], but a variant that we developed in [45] called π-institution. This is not for self-promotion, but because it will allow us to capitalise on material introduced in previous sections. We defined in Sect. 3.6 the notion of closure system as consisting of a pair $<L,c>$, where L is a set and $c:L \rightarrow L$ is a total function satisfying the properties of a closure operator: reflexivity, idempotence and monotonicity. The idea is that the elements of L are the well-formed formulae that result from a specific choice of vocabulary symbols and a given grammar. The closure operator captures the notion of consequence of the logic at stake, instantiated to that particular language.

The idea of (π-)institutions is to add to this abstract view of logic the idea that the notion of consequence is independent of the specific choice of vocabulary symbols that determine the language L. More precisely, it makes explicit the distinction between logical and nonlogical symbols in a language: whereas the former (usually called the *connectives*of the logic) are captured by the grammar, the latter are given by (typed) sets. The notion of consequence is then constrained to be invariant under changes of nonlogical symbols in order to make it depend only on the connectives. That is, we group together several of these closure systems that we regard as being different instantiations of the same logic through different choices of nonlogical symbols, and take that set as defining that logic.

Basically, in Sect. 3.5, when defining linear temporal logic, we introduced the notion of signature as a means of modelling the instantiations of the general structures of temporal logic, and made explicit the grammar that defines the connectives. For every specific signature, we obtain a closure system whose language is determined by the application of the grammar of temporal logic to the signature, and whose closure operator is obtained through a notion of consequence derived from the traditional Kripke semantics of temporal logic. The idea that all these closure systems are related in a way that makes them instantiations of the same logic can be captured by the following property.

6.5.1 Proposition

The mapping ***prop*** defined in Par. 3.5.2 and the translations induced by signature morphisms as defined in Par. 3.5.7 extend to a functor *ltl:SET→CLOS*.

Proof

The crux of the proof is in showing that signature morphisms map to morphisms of closure systems. But this is a direct consequence of the presentation lemma (Par. 3.5.10). The reader is invited to fill in the details. ♦

Through the functor that maps closure systems to their underlying languages, ***ltl*** maps each signature to the set of well-formed formulae over that signature. The fact that we have a functor means that changes of vocabulary induce corresponding translations at the level of the languages, thus capturing the idea of "uniformity" that we tend to associate with a grammar. The same applies to the closure operator. The fact that changes of vocabulary induce morphisms between the corresponding closure systems captures the flavour of uniformity and continuity that we associate with a logic. Naturally, we can make these "flavours" concrete by making the grammar explicit rather than implicit. It all depends on the level of abstraction at which one wants to work. The reader interested in these aspects should consult [63].

6.5.2 Definition – Π-Institutions

A *π-institution* consists of a pair <***SIGN***,***clos***>, where ***SIGN*** is a category (of signatures), and ***clos***:***SIGN***→***CLOS*** is a functor. ♦

6.5.3 Remark

An equivalent definition of a π-institution, given in [45], consists of:

- A category ***SIGN***.
- A functor ***gram***:***SIGN***→***SET***.
- For every Σ:***SIGN***, a relation $\vdash_\Sigma : 2^{\mathbf{gram}(\Sigma)} \times \mathbf{gram}(\Sigma)$ satisfying the following properties:
  - For every $p \in \mathbf{gram}(\Sigma)$, $p \vdash_\Sigma p$.
  - For every $p \in \mathbf{gram}(\Sigma)$ and $\Phi_1, \Phi_2 \subseteq \mathbf{gram}(\Sigma)$, if $\Phi_1 \subseteq \Phi_2$ and $\Phi_1 \vdash_\Sigma p$ then $\Phi_2 \vdash_\Sigma p$.
  - For every $p \in \mathbf{gram}(\Sigma)$ and $\Phi_1, \Phi_2 \subseteq \mathbf{gram}(\Sigma)$, if $\Phi_1 \vdash_\Sigma p$ and $\Phi_2 \vdash_\Sigma p'$ for every $p' \in \Phi_1$, then $\Phi_2 \vdash_\Sigma p$.
  - For every $\sigma:\Sigma \to \Sigma'$, $p \in \mathbf{gram}(\Sigma)$ and $\Phi \subseteq \mathbf{gram}(\Sigma)$, $\Phi \vdash_\Sigma p$ implies $\mathbf{gram}(\sigma)(\Phi) \vdash_{\Sigma'} \mathbf{gram}(\sigma)(p)$.

Notice that the functor ***gram*** is the composition of ***clos*** with the forgetful functor that maps closure systems to the underlying languages. The closure operator itself is derived from the consequence relation: for every $\Phi \subseteq \mathbf{gram}(\Sigma)$, $c_\Sigma(\Phi) = \{p \in \mathbf{gram}(\Sigma) : \Phi \vdash_\Sigma p\}$. On the other hand, every closure operator defines a consequence relation: $\Phi \vdash_\Sigma p$ iff $p \in c_\Sigma(\Phi)$.

The definition given in [45] further requires the consequence relation to be compact, a property that we do not need for the constructions that we are present in this book. ♦

6.5.4 Remarks

1. Given the equivalence between the two definitions, we use the consequence relation and the grammar function without notifying the reader.
2. Because the notation can become quite cumbersome, we often omit the reference to the grammar functor when applied to morphisms and write $\sigma(p)$ instead of $\boldsymbol{gram}(\sigma)(p)$, like we did for temporal logic in Sect. 3.5. ♦

When defining a π-institution, the consequence relation or the closure operator can be presented in many different ways. One of the possible ways is the one we adopted in the definition of linear temporal logic: by providing a notion of model and a satisfaction relation. This is what institutions, as defined by Goguen and Burstall [62], consist of.

6.5.5 Definition – Institutions

An *institution* is a quadruple $\langle \boldsymbol{SIGN}, \boldsymbol{gram}, \boldsymbol{mod}, \models \rangle$ where:

- $\boldsymbol{SIGN}$ is a category.
- $\boldsymbol{gram}: \boldsymbol{SIGN} \rightarrow \boldsymbol{SET}$ is a functor.
- $\boldsymbol{mod}: \boldsymbol{SIGN}^{op} \rightarrow \boldsymbol{CAT}$ is a functor.
- For every $\Sigma: \boldsymbol{SIGN}$, $\models_\Sigma : \boldsymbol{mod}(\Sigma) \times \boldsymbol{gram}(\Sigma)$ satisfies for every $\sigma: \Sigma \rightarrow \Sigma'$, $p \in \boldsymbol{gram}(\Sigma)$ and $M' \in \boldsymbol{mod}(\Sigma')$, $\boldsymbol{mod}(\sigma)(M') \models_\Sigma p$ iff $M' \models_{\Sigma'} \boldsymbol{gram}(\sigma)(p)$.

The functor $\boldsymbol{mod}$ provides, for every signature, the category of models that can be used to interpret the language defined by $\boldsymbol{gram}$ over that signature. Signature morphisms induce translations between these classes of models in the opposite direction. The idea is that, as seen in Par. 3.5.9 for linear temporal logic, sentences in the language of the source signature can be interpreted in a model for the target signature through the interpretation of their translations. The condition that is required on the satisfaction relation, which is usually called the *satisfaction condition*, states precisely this property – that satisfaction is invariant under change of notation.

When reasoning about institutions, we normally use the simplified notation for the $\boldsymbol{gram}$ functor that we mentioned for π-institution. We also adopt a similar simplification for the $\boldsymbol{mod}$ functor: for every morphism $\sigma: \Sigma \rightarrow \Sigma'$ and $M' \subseteq \boldsymbol{mod}(\Sigma')$, we normally write $M'|_\sigma$ instead of $\boldsymbol{mod}(\sigma)(M')$. For every signature Σ, $\Phi \subseteq \boldsymbol{gram}(\Sigma)$ and $M \in \boldsymbol{mod}(\Sigma)$, we usually write $M \models_\Sigma \Phi$, meaning that $M \models_\Sigma p$ for every $p \in \Phi$. ♦

6.5.6 Proposition

Every institution $\langle \boldsymbol{SIGN}, \boldsymbol{gram}, \boldsymbol{mod}, \models \rangle$ presents the π-institution $\langle \boldsymbol{SIGN}, \boldsymbol{gram}, \vdash \rangle$, where, for every signature Σ, $p \in \boldsymbol{gram}(\Sigma)$ and $\Phi \subseteq \boldsymbol{gram}(\Sigma)$, $\Phi \vdash_\Sigma p$ iff for every $M \in \boldsymbol{mod}(\Sigma)$, $M \models_\Sigma \Phi$ implies $M \models_\Sigma p$.

Proof

The proof that a π-institution is obtained in this way offers no difficulties and is left as an exercise. ♦

Another way of presenting the closure operator or consequence relation of a π-institution is through the notion of proof, choosing an inference system for the logic. See [85] for such proof-theoretic presentations and their integration in a more comprehensive categorical account of logical systems that the author calls "General Logics".

6.5.7 Remark – Models Defined Via a Split (Co)Fibration

The model functor of an institution is defined in Par. 6.5.5 directly as an indexed category, signatures providing the indexes. As seen in Par. 6.1.20, indexed categories can be presented as split (co)fibrations, which is how, in many cases, the model-theory of some institutions is more naturally defined. That is, it is often more intuitive, not to say "practical", to provide directly either a split fibration ***sign**: **MODL→SIGN*** or a split cofibration ***sign**: **MODL→SIGN**^{op}* for some category ***MODL*** of models, and take ***ind(sign)*** as the model functor of the required institution. This happens when we already have a notion of model that corresponds to the domain of interpretation intended for a given specification formalism, and we want to use it as the "semantics functor" of the institution. An example can be given through linear temporal logic.

6.5.8 Example – Linear Temporal Logic Over *PROC*

Temporal signatures were interpreted in Par. 3.5.3 over infinite sequences of sets of atomic propositions corresponding to synchronisation sets of actions. Yet, from the point of view of modelling concurrent systems, such infinite sequences do not represent abstractions of full behaviour; they usually represent one single behaviour among many that may also be observed on a given system. Hence, from the point of view of providing a domain of interpretation for a formalism supporting concurrent system specification, sets of such trajectories are more meaningful in the sense that they can be taken to represent *full* behaviours. Put another way, if we take single sequences as models of full behaviour, we are restricting the expressive power of the formalism to a much smaller class of systems: those that are deterministic and run in a deterministic environment.

We can see in Sect. 6.3 how such a trace-based model of process behaviour can be organised in the functor-structured category ***spa(tra)*** that we called ***PROC***, where ***tra**:**SET**$_\perp$→**SET*** is the functor that generates and translates between sets of traces. Hence, it seems intuitive that we investigate the use of this category for providing the models of an institution of linear temporal logic. However, the split cofibration defined by ***spa(tra)*** is over ***SET**$_\perp$*, not over ***SET**^{op}* as required for the temporal signatures. Indeed, processes were defined over arbitrary alphabets of actions, whereas in the case of temporal logic, propositions are interpreted over traces of synchronisation sets. Hence, some adaptation is required. There are two equivalent ways in which this adaptation can be performed.

The first approach is to work with the subcategory of ***spa(tra)*** that involves only alphabets of synchronisation sets. Powersets can be regarded as pointed sets, the empty set providing the distinguished element. More precisely, we show in 3.3.2 that, by choosing the morphisms $2^B \to 2^A$ to be the inverses of the functions $A \to B$, we define a subcategory of $\boldsymbol{SET_\perp}$ that we called ***POWER***. Therefore, we can particularise the ***spa*** construction that yields processes to the functor ***ptra*: *POWER*→*SET*** that yields the set of infinite traces taken over sets of actions. Finally, there is a straightforward (and powerful, as seen in Chap. 7) way of relating $\boldsymbol{SET^{op}}$ to ***POWER***: through the contravariant functor $\boldsymbol{2^-}$ that maps every set A to its powerset 2^A and every function $A \to B$ to the map $2^B \to 2^A$ that computes inverse images according to that function. Hence, we can particularise ***tra*** to the functor $\boldsymbol{ftra{:}SET^{op} \to SET = 2^-;tra}$ and use the split cofibration defined by ***spa(ftra)*** over $\boldsymbol{SET^{op}}$ as the model functor $\boldsymbol{ind(spa(ftra)){:}SET^{op} \to CAT}$ of an institution of linear temporal logic. See Par. 6.1.20(2) for details pn the construction. This is the institution that associates every temporal signature (set of actions) with the category of processes whose alphabet consists of the subsets of the signature (synchronisation sets of actions).

An alternative approach is to restrict not the (co)fibration that generates the indexed category but the indexed category generated by the original (co)fibration, i.e. to work with $\boldsymbol{2^-;ind(spa(tra))}$. This is simpler because it does not require the definition of a new (co)fibration. For the same reason, this approach is less satisfying because it is less structural in the sense that it does not characterise explicitly the class of models that are chosen. In the case of linear temporal logic, both approaches lead to the same indexed category. Indeed, taking the indexed category $\boldsymbol{SET_\perp \to CAT}$ generated by ***spa(tra)***, we can compose it with the contravariant powerset functor $\boldsymbol{SET^{op} \to SET_\perp}$ to define the required model functor $\boldsymbol{SET^{op} \to CAT}$.

It remains to define the satisfaction relation of this institution. Given that models over a signature Σ consist of sets of traces $\Lambda \subseteq (2^\Sigma)^\omega$, satisfaction of a temporal proposition by a process is defined as satisfaction by all the traces of the process as defined in Par. 3.5.3. The satisfaction condition results from the properties proved in Par. 3.5.9. Indeed, the reader should check that the notion of reduct defined therein coincides with the functor between fibres defined by the cofibration as in Par. 6.1.16. ◆

It is easy to see that the institution defined over this indexed category yields the same π-institution as the one that can be defined directly by applying Par. 6.5.2 to the functor defined in Par. 6.5.1 (which is based on a semantic interpretation over single sequences as given in Par. 3.5.3). This is because the ***PROC***-based institution and the one that uses single trajectories as in Par. 3.5.3 give rise to the same consequence relation. However, we shall see that they satisfy different structural properties, which shows that, by forgetting the model-theory to retain just the consequence relations, π-institutions are indeed more abstract than institutions.

6.5.9 Remark – When a Split (Co)Fibration Is Not Available

The availability of a split cofibration for the definition of the model functor is very useful because it allows us to work with reduct functors that operate on the chosen semantic models, and therefore compare the structures that models and theories provide for specification. However, it may happen that we have a notion of model in mind that can still be captured by a functor ***sign**: **MODL**→**SIGN**^{op}*, but one that is not a cofibration. The problem here is the definition of a suitable translation between models.

This problem can be solved by not working directly over the fibres defined by the functor, but with what we could call "generalised" models: those that are provided by structured morphisms as defined in Par. 6.2.2. That is, the idea is, for every signature Σ, to work with models of the form $<\sigma:\Sigma\to \mathbf{sign}(M),M>$ so that ***mod**(Σ)* is $\Sigma\!\downarrow\!\mathbf{sign}$, making ***mod*** the functor $\_\!\downarrow\!\mathbf{sign}:\mathbf{SIGN}^{op}\to\mathbf{CAT}$ mapping morphisms $\mu:\Sigma\to\Sigma'$ to the functors $\Sigma'\!\downarrow\!\mathbf{sign}\to\Sigma\!\downarrow\!\mathbf{sign}$ that map generalised models $<\sigma:\Sigma'\to\mathbf{sign}(M'),M'>$ to $<\mu;\sigma:\Sigma\to\mathbf{sign}(M'),M'>$. (The reader is encouraged to fill in the way the functor works on morphisms between models.) In a way, what this construction does is compensate for the lack of (co)Cartesian morphisms by providing an explicit "adaptor" that maps the signature to the language of the model. As a result, instead of translating directly between models, reducts operate "syntactically" at the level of these adaptors.

The reader may be wondering whether this is not just a contrived way of bringing structured morphisms into the discussion and a good example of "abstract nonsense": an accusation often made to categorical techniques when developed without a clear application in mind, in a self-justified process of "structure for structure"… Certainly this is not true in this case (nor in any other case in this book); these generalised models are quite common in modal logic (e.g. [67])! Indeed, models for a modal logic, normally called frames or *Kripke structures*, can be regarded as concrete categories over a base category of possible worlds. The adaptors that we defined correspond to interpretation functions that, for each atomic proposition, return the set of possible worlds in which the proposition is true. A generalised model, consisting of a Kripke frame and an interpretation function, is usually called a model in modal logic.

In the case of linear temporal logic, this means that interpretation structures are no longer traces $\lambda\in(2^{\Sigma})^{\omega}$ as defined in Par. 3.5.3 – where we mentioned that they are *canonical* Kripke structures for linear, discrete, propositional logic [110] – but triples $<W,\lambda\in W^{\omega},v:\Sigma\to 2^{W}>$ that correspond to the more traditional notion of models. In this case W is a set of possible worlds (events), the pair $<W,\lambda\in W^{\omega}>$ defines a linear frame

(Kripke structure), and the function v returns the set of worlds (events) in which each atomic proposition (action) holds (occurs).

In the case that interests us, ***MODL*** is the original ***spa(tra)***, but ***sign*** is not the original cofibration but its composition with the powerset functor $\boldsymbol{SET}_\perp \rightarrow \boldsymbol{SET}^{op}$ (in fact, as shown in Par. 7.3.6, the left adjoint of the powerset functor used in Par. 6.5.8). This means that reducts are not operated by the cofibration, and hence we lose some of the structural properties of processes. More details on these constructions can be found in [32, 33]. ♦

In logic, models are considered to be the "duals" of theories. In institutions, this duality can be explored and made explicit in several ways. We can start by realising that the theories of the underlying closure systems can be organised in a category that provides itself a split (co)fibration over the category of signatures. Indeed, we proved in Sect. 6.1 that the category $\boldsymbol{THEO}_{LTL}$ of linear temporal theories, as defined in Par. 3.5.4, is both a split fibration and a split cofibration over ***SET***. The extension to any π-institution is trivial.

6.5.10 Proposition – The (Co)Indexed Category of Theories

1. Every π-institution ***<SIGN,clos>*** defines a functor $\boldsymbol{theo}:\boldsymbol{SIGN} \rightarrow \boldsymbol{CAT}$ as an extension of ***clos*** by mapping every signature Σ to the category $\boldsymbol{THEO}_{clos(\Sigma)}$ of the theories over $\boldsymbol{clos}(\Sigma)$, and every signature morphism $\sigma:\Sigma \rightarrow \Sigma'$ to the functor $\boldsymbol{theo}(\sigma):\boldsymbol{THEO}_{clos(\Sigma)} \rightarrow \boldsymbol{THEO}_{clos(\Sigma')}$ that is defined by $\boldsymbol{theo}(\sigma)(\Phi)=c'(\boldsymbol{clos}(\sigma)(\Phi))$.
2. Every π-institution ***<SIGN,clos>*** defines a contravariant functor $\boldsymbol{theo}^{-1}:\boldsymbol{SIGN}^{op} \rightarrow \boldsymbol{CAT}$ by mapping every signature Σ to the category $\boldsymbol{THEO}_{clos(\Sigma)}$ of the theories over $\boldsymbol{clos}(\Sigma)$, and every signature morphism $\sigma:\Sigma' \rightarrow \Sigma$ to the functor $\boldsymbol{theo}^{-1}(\sigma):\boldsymbol{THEO}_{clos(\Sigma)} \rightarrow \boldsymbol{THEO}_{clos(\Sigma')}$ defined by $\boldsymbol{theo}^{-1}(\sigma)(\Phi)=\boldsymbol{clos}(\sigma)^{-1}(\Phi)$.

Proof

Notice that, because the category of theories over a closure system is, in fact, a preorder, the functors $\boldsymbol{theo}(\sigma)$ and $\boldsymbol{theo}^{-1}(\sigma)$ do not have to be defined on morphisms. However, we are required to prove:

1. Given theories $\Phi \leq \Gamma$ in $\boldsymbol{THEO}_{clos(\Sigma)}$, $\boldsymbol{theo}(\sigma)(\Phi) \leq \boldsymbol{theo}(\sigma)(\Gamma)$. This holds because ***clos*** is a functor. More precisely, given $\Phi \subseteq \Gamma$ in $\boldsymbol{THEO}_{clos(\Sigma)}$, we have $\boldsymbol{clos}(\sigma)(\Phi) \subseteq \boldsymbol{clos}(\sigma)(\Gamma)$ because $\boldsymbol{clos}(\sigma)$ is a function between the languages of the two closure systems, which implies $c'(\boldsymbol{clos}(\sigma)(\Phi)) \subseteq c'(\boldsymbol{clos}(\sigma)(\Gamma))$ because closure operators are monotonic with respect to set inclusion.
2. Given theories $\Phi \leq \Gamma$ in $\boldsymbol{THEO}_{clos(\Sigma)}$, $\boldsymbol{theo}^{-1}(\sigma)(\Phi) \leq \boldsymbol{theo}^{-1}(\sigma)(\Gamma)$. Given $\Phi \subseteq \Gamma$, we have $\boldsymbol{clos}(\sigma)^{-1}(\Phi) \subseteq \boldsymbol{clos}(\sigma)^{-1}(\Gamma)$ because $\boldsymbol{clos}(\sigma)$ is a normal function between the languages of the two closure systems, and the inverse image of a closed set is itself closed.

The reader is invited to complete the proof as an exercise. ♦

The (co)flattening of ***theo*** as a (co)indexed category provides us with a split (co)fibration $\boldsymbol{THEO}_{<SIGN,clos>} \rightarrow \boldsymbol{SIGN}$, where $\boldsymbol{THEO}_{<SIGN,clos>}$, the flattened category, consists of pairs $<\Sigma,\Phi>$, where Σ is a signature and Φ is a closed set of sentences of ***gram***(Σ), i.e. Φ is a theory of ***clos***(Σ). A morphism $\sigma:<\Sigma,\Phi>\to<\Sigma',\Phi'>$ is a signature morphism $\sigma:\Sigma\to\Sigma'$ such that $c_{\Sigma'}(\sigma(\Phi))\subseteq\Phi'$. Theories were originally defined in (π-)institutions in this way. Notice that it does not matter whether one flattens the indexed or the coindexed category: the categories obtained are dual. This is because $c_{\Sigma'}(\sigma(\Phi))\subseteq\Phi'$ holds iff $\Phi\subseteq\sigma^{-1}(\Phi')$.

This construction can be generalised to (strict) presentations.

6.5.11 Definition – Theories/Presentations in a ∏-Institution

Given a π-institution $<$***SIGN,clos***$>$, we define the following categories:

1. $\boldsymbol{THEO}_{<SIGN,clos>}$, the category of theories, has for objects the pairs $<\Sigma,\Phi>$, where Σ is a signature and Φ is a closed set of sentences of ***gram***(Σ). A morphism $\sigma:<\Sigma,\Phi>\to<\Sigma',\Phi'>$ is a signature morphism $\sigma:\Sigma\to\Sigma'$ such that $\sigma(\Phi)\subseteq\Phi'$.
2. $\boldsymbol{PRES}_{<SIGN,clos>}$, the category of theory presentations, has for objects the pairs $<\Sigma,\Phi>$, where Σ is a signature and $\Phi\subseteq$***gram***(Σ). A morphism $\sigma:<\Sigma,\Phi>\to<\Sigma',\Phi'>$ is a signature morphism $\sigma:\Sigma\to\Sigma'$ such that $\sigma(c_\Sigma(\Phi))\subseteq c_{\Sigma'}(\Phi')$.
3. $\boldsymbol{SPRES}_{<SIGN,clos>}$, the category of strict theory presentations, has for objects the pairs $<\Sigma,\Phi>$, where Σ is a signature and $\Phi\subseteq$***gram***(Σ). A morphism $\sigma:<\Sigma,\Phi>\to<\Sigma',\Phi'>$ is a signature morphism $\sigma:\Sigma\to\Sigma'$ such that $\sigma(\Phi)\subseteq\Phi'$. ♦

These categories satisfy the properties that we have already proved for their linear temporal logic instantiations.

6.5.12 Proposition

Consider a π-institution $<$***SIGN,clos***$>$..

1. The categories $\boldsymbol{PRES}_{<SIGN,clos>}$, $\boldsymbol{SPRES}_{<SIGN,clos>}$ and $\boldsymbol{THEO}_{<SIGN,clos>}$ define split fibrations and split cofibrations through the functor ***sign*** that projects their objects and morphisms to the corresponding signature components. Given a signature morphism $\sigma:\Sigma\to\Sigma'$,

	Cartesian morphism for $<\Sigma',\Phi'>$	Cocartesian morphism for $<\Sigma,\Phi>$
PRES	$\sigma:<\Sigma,\sigma^{-1}(c(\Phi))>\to<\Sigma',\Phi'>$	$\sigma:<\Sigma,\Phi>\to<\Sigma',\sigma(\Phi)>$
SPRES	$\sigma:<\Sigma,\sigma^{-1}(\Phi)>\to<\Sigma',\Phi'>$	$\sigma:<\Sigma,\Phi>\to<\Sigma',\sigma(\Phi)>$
THEO	$\sigma:<\Sigma,\sigma^{-1}(\Phi)>\to<\Sigma',\Phi'>$	$\sigma:<\Sigma,\Phi>\to<\Sigma',c(\sigma(\Phi))>$

2. All these (co)fibrations are fibre-(co)complete. The following procedure calculates limits and colimits of a diagram δ with $\delta_i=<\Sigma_i,\Phi_i>$ in these categories:

- Calculate the limit $\sigma{:}\Sigma \rightarrow \delta$ or colimit $\sigma{:}\delta \rightarrow \Sigma$ of the underlying diagram of signatures.
- Lift the result by computing the ***SIGN***-component according to the following rules:

	Limit	**Colimit**
PRES	$\sigma{:}<\Sigma, \bigcap_{i\in I} \sigma_i^{-1}(c(\Phi_i))> \rightarrow \delta$	$\sigma{:}\delta \rightarrow <\Sigma, \bigcup_{i\in I} \sigma_i(\Phi_i)>$
SPRES	$\sigma{:}<\Sigma, \bigcap_{i\in I} \sigma_i^{-1}(\Phi_i)> \rightarrow \delta$	$\sigma{:}\delta \rightarrow <\Sigma, \bigcup_{i\in I} \sigma_i(\Phi_i)>$
THEO	$\sigma{:}<\Sigma, \bigcap_{i\in I} \sigma_i^{-1}(\Phi_i)> \rightarrow \delta$	$\sigma{:}\delta \rightarrow <\Sigma, c_\Sigma(\bigcup_{i\in I} \sigma_i(\Phi_i))>$

3. If ***SIGN*** is (co)complete, so are $\boldsymbol{PRES}_{<SIGN,clos>}$, $\boldsymbol{SPRES}_{<SIGN,clos>}$ and $\boldsymbol{THEO}_{<SIGN,clos>}$. ♦

This symmetry between theories and models, as captured through indexed categories, can be used to provide every π-institution with a canonical notion of model that makes it an institution.

6.5.13 Proposition – Institution Presented by a ∏-Institution

Every π-institution $<$***SIGN*,*gram***, $\vdash>$ defines the institution $<$***SIGN*,*gram*,*theo***$^{-1}$, $\models>$, where for every signature Σ, $p \in \boldsymbol{gram}(\Sigma)$ and $\Phi \in \boldsymbol{theo}^{-1}(\Sigma)$, $\Phi \models_\Sigma p$ iff $p \in \Phi$.

Proof

We just have to prove that the satisfaction condition holds. Let $\Sigma{:}\boldsymbol{SIGN}$, $\models_\Sigma{:}\boldsymbol{mod}(\Sigma) \times \boldsymbol{gram}(\Sigma)$ satisfy for every morphism $\sigma{:}\Sigma \rightarrow \Sigma'$, $p \in \boldsymbol{gram}(\Sigma)$ and $\Phi' \in \boldsymbol{theo}^{-1}(\Sigma')$, $\boldsymbol{theo}^{-1}(\sigma)(\Phi') \models_\Sigma p$ iff $\sigma^{-1}(\Phi') \models_\Sigma p$ iff $p \in \sigma^{-1}(\Phi')$ iff $\sigma(p) \in \Phi'$ iff $\Phi \models_\Sigma \sigma(p)$. ♦

What this results says is that, whereas having a model-theory that corresponds to an intended domain of interpretation for the language of a logic may be a convenient way of defining a π-institution, once we are given a π-institution there is not much point in searching for a model-theory for it. Its theories are up to the job at no conceptual expense.

Notice that the consequence relation induced by this notion of satisfaction is $\Phi \vdash_\Sigma p$ iff, for every $\Gamma \in \boldsymbol{theo}^{-1}(\Sigma)$, $\Phi \subseteq \Gamma$ implies $p \in \Gamma$, which is equivalent to $p \in c(\Phi)$. That is, we recover the π-institution from which we started. Hence, institutions do give us a more concrete level of abstraction at which properties of specifications can be discussed.

Duality between models and theories can also be explored by lifting some of the signature-based constructions to theory-based ones.

6.5.14 Proposition

Let $<$***SIGN*,*gram*,*mod***, $\models>$ be an institution.

1. The functor ***mod:SIGN***^{op}***→CAT*** extends to ***tmod:THEO***^{op}***→CAT*** by assigning to every theory $<\Sigma,\Phi>$ the full subcategory of ***mod***(Σ) that consists of the models that satisfy Φ.
2. When ***mod*** is generated by a cofibration ***sign:MODL→SIGN***^{op}, we can lift it to ***theo:MODL→THEO***^{op} by associating with every model the theory that consists of all the sentences that are true for that model.

Proof

We have to prove that a functor is, indeed, defined in point 1. This is because, given a theory morphism $\sigma:<\Sigma,\Phi>\rightarrow<\Sigma',\Phi'>$ and a model M' of $<\Sigma',\Phi'>$, $M'|_{\sigma}$ satisfies Φ. The reader is invited to work out the proof of the second property. ♦

6.5.15 Remark – Theories Over Generalised Models

We saw that functors ***sign: MODL→SIGN***^{op} give rise to model functors of the form ***_↓sign:SIGN***^{op}***→CAT*** that are typical of Modal Logic. In an institution defined over such a model functor, the notions of theory and the constructions that we have presented around them for arbitrary (π-)institutions still apply. For instance, ***_↓sign:SIGN***^{op}***→CAT*** extends to ***tmod:THEO***^{op}***→CAT*** as defined in Par. 6.5.14. However, in this particular case, it is interesting to check if, or when, ***sign:MODL→SIGN***^{op} can lift to ***theo:MODL→THEO***^{op} so that ***tmod*** is, in fact, ***_↓theo***.

The intuition for this interest is that we should be able to associate with every model M a canonical specification ***theo***(M) so that every morphism T→***theo***(M) identifies T as a specification of M and, dually, M as a realisation of T. In this case, the models of T correspond to all possible refinements of T into "programs", something that we already discussed in Par. 6.2.1. Identifying ***MODL*** with the behaviours of programs (or processes), this corresponds to the idea that programs can themselves be regarded as specifications, and that the models of a specification can be identified with the programs that, in some sense, complete the specification.

Although it is always possible to define a mapping that associates a theory ***theo***(M) with every model M – the obvious candidate assigns to ***theo***(M) the signature ***sign***(M) and all the sentences that are true in it $\{p\in$***gram***$($***sign***$(M)): <id,M>\models p\}$ – it is not always possible to extend it to a functor that lifts ***sign***, i.e. such that ***theo***$(h:M\rightarrow M')=$***sign***(h). In order to motivate why this is so, assume that we do have ***theo:MODL→THEO***^{op} as a functor and consider a morphism $h:M'\rightarrow M$. Because ***theo*** is a functor, we have ***theo***(h):***theo***$(M)\rightarrow$***theo***(M'). Hence, if M is a model of a theory T, i.e. if we have a morphism $\sigma:T\rightarrow$***theo***(M), then M' is also a model of T through the morphism σ;***theo***(h). In particular this implies that, for every signature Σ and $p\in$***gram***(Σ), if $<\sigma,M>\models_{\Sigma}p$ then $<\sigma;$***sign***$(h),M'>\models_{\Sigma}p$. This property, however, is not universal: only some institutions (logics) satisfy it.

There are two important aspects about this property. The first is that, as proved in [33], it guarantees that ***sign***: ***MODL***$\rightarrow$***SIGN***^{op} lifts to ***theo***: ***MODL***$\rightarrow$***THEO***^{op} as defined, and that ***tmod*** is, in fact, $\_\downarrow$***theo***. The second is that, once again, it is not an artificial "fabrication": it captures what, in modal logic, is known as "the p-property" [67]. That is, there are well-known principles of modal logic that can be captured in the categorical framework that we have defined, meaning that we are, indeed, addressing structural properties of logic as we have known them. Once again, a more detailed discussion of the p-property and its impact in specification can be found in [32, 33]. ♦

6.5.16 Exercise

Check that the institution of linear temporal logic as defined in Par. 6.5.9 over general Kripke structures satisfies the p-property. ♦

The reader will have noticed that, in an institution, the model functor is defined over ***CAT*** instead of ***SET***. This is because, some times, models come equipped with a non-trivial notion of morphism and, as argued in Par. 6.5.9, it is not always possible to reflect in the logic the structure that they induce on models.

However, so far, the constructions that we discussed do not depend on the notion of morphism between models, and hence do not reflect the structure of the interpretation domain that are chosen for the specification formalisms. We are now going to illustrate a situation in which such structures are of interest.

6.5.17 Definition – Initial/Terminal Semantics

We say that an institution $<\mathit{SIGN},\mathit{gram},\mathit{mod},\models>$ has *initial* (resp. *terminal*) *semantics* provided that, for every theory $<\Sigma,\Phi>$, the category ***tmod***$(<\Sigma,\Phi>)$ has an initial (resp. terminal) object. ♦

A non-trivial (but simple) example of an institution with terminal semantics is linear temporal logic.

6.5.18 Example – Terminal Semantics of Linear Temporal Logic

The institution of linear temporal logic as defined in Par. 6.5.8 over the cofibration ***spa(ftra)*** has terminal semantics. This is because, given any theory $<\Sigma,\Phi>$, the category ***tmod***$(<\Sigma,\Phi>)$ consists of sets of traces ordered by inclusion and is closed under intersection. ♦

Notice that linear temporal logic interpreted over single sequences does not have terminal semantics: the (discrete) category that represents the set of all the sequences that satisfy a given theory does not have a sequence that represents the whole set unless the set is singular.

6.5.19 Exercise

Check that the institution of linear temporal logic as defined in Par. 6.5.9 over general Kripke structures has terminal semantics. Does it coincide with the one obtained in Par. 6.5.18? ♦

Properties such as having an initial or terminal semantics can be used for differentiating between different institutions that present the same π-institution. It is interesting to note that what we have called the "canonical" institution for a given π-institution, the one that uses theories as models, has both initial and terminal semantics. The initial model of every theory is the theory itself and the terminal model is the inconsistent theory, i.e. the full language. That is, we do not extract much information from the structure of models, which was only to be expected because we do not extract any information from the model-theory itself.

This example can also be used to transmit a certain "moral" message, namely that the pursuit for an institution with "good" properties like the ones above is, most of the times, a trivial one in the sense that it is usually possible to engineer a model-theory that satisfies one's requirements. As already said, the real purpose of these properties is to measure the relationship that the institution provides between specifications and the domain as abstracted through the models. Hence, one needs to determine the nature of the models relative to the domain of interpretation and not to the properties, otherwise what we get is "abstract nonsense".

7 Adjunctions

7.1 The Social Life of Functors

Yes, even functors are entitled to have their own social lives. And they can be quite rich, too. In this book, we will remain at the level of what is basic and indispensable for covering this last topic of our introduction to category theory: adjunctions. For that purpose, we will provide a short introduction to what are normally accepted to be "the" morphisms between functors: natural transformations. The best way of understanding what natural transformations consist of, and can be used for, is to look at functors as views that one has from one category into another and to formulate the properties that characterise the "preservation" of such views.

7.1.1 Example – Two Views of Eiffel Class Specifications

We have already seen how one can look into Eiffel classes from the point of view of temporal logic specification: this is the view that is provided through the functor ***spec*** that we defined in Par. 5.1.3. This functor accounts for both the pre/postconditions of methods and the class invariants. However, one is often interested in a higher-level view of the behaviour of object classes that is concerned with the global properties that can be observed from their interfaces, typically their functions and routines. For this particular view, the actual specification of the functionality of the methods is of little relevance; they account for "how" these global properties are achieved rather than just "what" they are. Therefore, it makes sense that we define another functor $\boldsymbol{obsv}$: $\boldsymbol{CLASS\_SPEC \rightarrow PRES_{FOLTL}}$ that offers the more abstract point of view. Formally, these observable properties can be defined as sentences that involve only routines and functions. Hence, given an Eiffel class specification $e=<\Sigma,P,I>$ we define

$$\boldsymbol{obsv}(e)=<\Sigma,\{\phi \in LTL(fun(\Sigma) \cup rou(\Sigma)) \mid \Phi \vdash_{\Sigma} \phi\}>,$$

where Φ is the set of axioms of $\boldsymbol{spec}(e)$. The reader is invited to check that this mapping defines a functor, i.e. that class inheritance induces an interpretation between the corresponding global properties.

Given two such views of class specifications, how can we relate them? Clearly, such a relationship has to be established on the basis of morphisms that, for each class specification e, relate $\boldsymbol{obsv}(e)$ and $\boldsymbol{spec}(e)$. By definition, observable properties are derivable from the full class properties, so there is a morphism of presentations between each ***obsv(e)*** and ***spec(e)***. However, in order to respect the structure that morphisms (inheritance between class specification) induce on both views, the way observable properties relate through inheritance must be "the same" as the full specifications relate between them. ◆

7.1.2 Definition – Natural Transformations

Given two functors $\psi:\boldsymbol{D}\to\boldsymbol{C}$ and $\varphi:\boldsymbol{D}\to\boldsymbol{C}$, a natural transformation τ from ψ to φ, denoted by $\psi\xrightarrow{\tau}\varphi$ or $\tau:\psi\xrightarrow{\bullet}\varphi$, is a function that assigns to each object d of $\boldsymbol{D}$ a morphism $\tau_d:\psi(d)\to\varphi(d)$ of $\boldsymbol{C}$ such that, for every morphism $f:d\to d'$ of $\boldsymbol{D}$, the following square commutes. In this case we say that τ_d is *natural in d* or that the *naturality condition* holds:

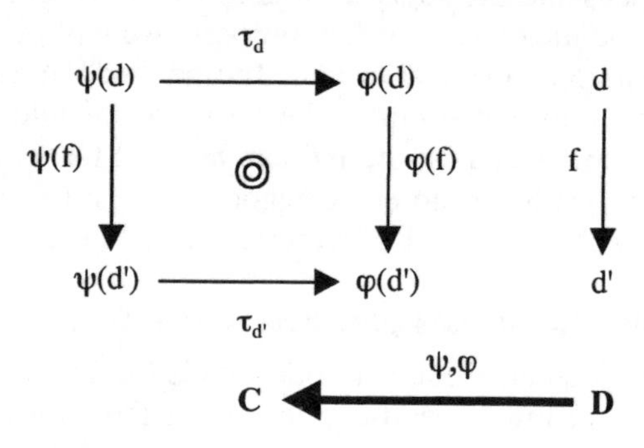

◆

7.1.3 Exercise

Work out the example in full by defining and proving all properties. ◆

The most obvious example of a natural transformation is the identity.

7.1.4 Definition – Identities

Given a functor $\psi:\boldsymbol{D}\to\boldsymbol{D}$, the identity natural transformation id_ψ assigns to each object d of $\boldsymbol{D}$ the identity morphism $id_{\psi(d)}$. ◆

In the rest of this section, we analyse some of the mechanisms and properties that are available for using natural transformations when reasoning about how given functors relate to each other. The main results that we need concern the way natural transformations compose and the mechanisms we have to act on them.

7.1.5 Definition – Dual of a Natural Transformation

Consider two functors $\psi:\boldsymbol{D}\to\boldsymbol{C}$ and $\varphi:\boldsymbol{D}\to\boldsymbol{C}$, and a natural transformation $\psi\xrightarrow{\tau}\varphi$. We define $\varphi^{op}\xrightarrow{\tau^{op}}\psi^{op}$ by $\tau^{op}{}_d=\tau_d$. ◆

Notice that a morphism $\varphi^{op}(d)\to\psi^{op}(d)$ in $\boldsymbol{C}^{op}$ corresponds exactly to a morphism $\psi(d)\to\varphi(d)$ in $\boldsymbol{C}$.

A useful class of operations on natural transformations is induced by functors into the sources, or from the targets, of the categories involved. For instance, supposing that we have a way (functor) to relate the domain of a viewpoint to another one (say between ***PRES***$_{FOLTL}$ and ***PROC*** as in Par. 7.3.14), it makes sense to compose it with a natural transformation to provide a new one that extends the former to the second domain, e.g. to provide a mapping between the processes that capture the observable and the full view of object behaviour.

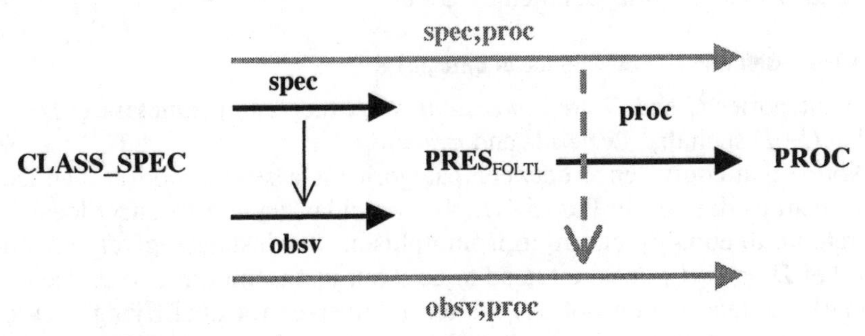

7.1.6 Definition – External Composition

Consider a natural transformation $\psi \xrightarrow{\tau} \varphi$ between functors $\psi{:}D{\to}C$ and $\varphi{:}D{\to}C$.

1. Given $\rho{:}E{\to}D$ we define $\rho;\psi \xrightarrow{\rho;\tau} \rho;\varphi$ by $(\rho;\tau)_e=\tau_{\rho(e)}$.
2. Given $\rho{:}C{\to}B$ we define $\psi;\rho \xrightarrow{\tau;\rho} \varphi;\rho$ by $(\tau;\rho)_d=\rho(\tau_d)$. ♦

These are external operations on a given natural transformation. An internal law can also be defined that allow us to compose views into more complex ones in the sense that they bridge over sequences of viewpoints.

7.1.7 Definition – Internal Composition

Consider functors $\psi,\varphi,\kappa{:}D{\to}C$ and natural transformations $\psi \xrightarrow{\tau} \varphi$ and $\varphi \xrightarrow{\nu} \kappa$. The composition $\psi \xrightarrow{\tau;\nu} \kappa$ is defined by $(\tau;\nu)_d=\tau_d;\nu_d$.

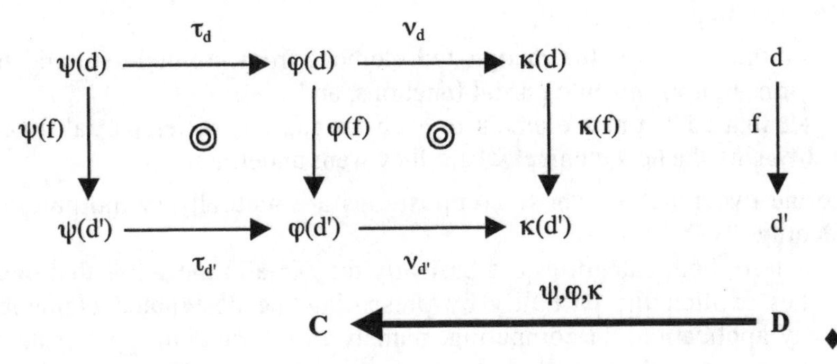

♦

7.1.8 Exercise

Prove that the composition of natural transformations is well defined, is associative and that the identities are units for the composition law. ♦

7.1.9 Definition – Natural Isomorphism

A natural transformation $\psi \xrightarrow{\tau} \varphi$ is said to be a *natural isomorphism* provided that each τ_d is an isomorphism, in which case ψ and φ are said to be *naturally isomorphic*, denoted by $\psi \cong \varphi$. ♦

7.1.10 Definition – Equivalence of Categories

Two categories $\boldsymbol{C}$ and $\boldsymbol{D}$ are *equivalent* when they admit functors $\psi{:}\boldsymbol{D}\rightarrow\boldsymbol{C}$ and $\varphi{:}\boldsymbol{C}\rightarrow\boldsymbol{D}$ such that $\psi;\varphi \cong \boldsymbol{id}_D$ and $\varphi;\psi \cong \boldsymbol{id}_C$. ♦

Notice that equivalence between categories is a weaker notion than isomorphism as defined in Par. 5.1.7. In particular, an equivalence does not operate up to equality but up to isomorphism. For instance, given any object d of $\boldsymbol{D}$, $\varphi(\psi(d))$ does not need to be d but just isomorphic to d. Hence, to mark the fact, we do not use the term "inverse" for qualifiying each of these functors with respect to the other but "pseudo-inverse". This is the terminology used, for instance, in [12].

For instance, for any institution, ***PRES*** and ***THEO*** are usually not isomorphic (the same theory may admit many presentations), but they are equivalent (all the presentations that define the same theory are isomorphic). This example shows that a category may be equivalent to one of its strict subcategories.

7.1.11 Exercise

What about ***SPRES***? How does it relate to ***PRES*** and ***THEO***? ♦

Another example of an equivalence concerns two categories for which we have already highlighted many relationships.

7.1.12 Example – Equivalence Between *PAR* and $SET_\perp$

The mappings

- $-\perp$ that removes the designated element from pointed sets and transforms morphisms into partial functions, and
- $+\perp$ that adds a new element to each set and completes partial functions by using the new element where they were undefined

define two functors whose compositions are naturally isomorphic to the identity. ♦

Indeed, both categories are basically the same in the sense that one just makes explicit the partiality by presenting the designated element. In many applications in computing, namely in the modelling of system behaviour as we illustrated with processes and specifications (theories), one tends to switch between one category and the other depending on whether we wish to attribute a meaning to the designated element, like for processes where it models steps performed by the environment, or be less bureaucratic (more pragmatic) and keep it just implicit. In fact, this is what we did in the graphical representations of the examples of universal constructions on processes; to simplify the notation, we omitted the designated element from the alphabets and represented the morphisms as partial functions. The old duality between syntax and semantics also tends to play a role: for semantic domains, like processes, it is useful to handle the designated element; but for design languages, like CommUnity (Chap. 8), it is often more "practical" to work instead with partial functions. The equivalence tells us how we can switch between the two views.

Note that the two categories are not isomorphic: the new element that is used to complete a set is not necessarily the one that was forgotten when that set is obtained from a pointed one. In fact, it is interesting for the reader to come up with a "real" definition of the functor that adds new elements to sets to make pointed sets: Which elements does it add? Is there a canonical way of performing this completion?

The fact that we do not have an isomorphism is more significant for the "social life" of the categories themselves than for the structures that they endow internally on their objects (which is basically the same). We give an example of what we mean by this in Par. 7.3.13.

7.2 Reflective Functors

Keeping the promise of writing for the community of software scientists and practitioners, who are not necessarily as mathematically oriented as most of the other books on category theory assume, we abstain from the traditional introduction to adjunctions, such as the construction of free monoids and other mathematical structures, or Galois connections. Instead, and besides giving examples closer to system development, we follow the method adopted in previous chapters. That is, we give and use concepts introduced about subcategories to motivate the definition of "similar" properties of functors, based on the fact that the inclusion of a subcategory in another one defines a functor.

The generalisation from subcategories into functors that interests us for adjunctions concerns reflections and coreflections as introduced in Pars. 3.3.8–3.3.12. Therein, we discussed (co)reflections as a specialised class of "secretaries" through which all the interactions can be factorised. The generalisation consists, once again, in replacing the inclusion by a functor; secretaries become "interpreters", i.e. some kind of "adjuncts" through which all the communication with the other side of the functor is handled.

7.2.1 Definition – Reflections

Let φ:***D***→***C*** be a functor.

1. Let c be a ***C***-object. A φ-reflection for c is a ***C***-morphism o:c→$\varphi(d)$ for some ***D***-object d such that, for any ***C***-morphism f:c→$\varphi(d')$ where d' is a ***D***-object, there is a unique ***D***-morphism f':d→d' such that $f=o;\varphi(f')$, i.e. the ***C***-diagram commutes:

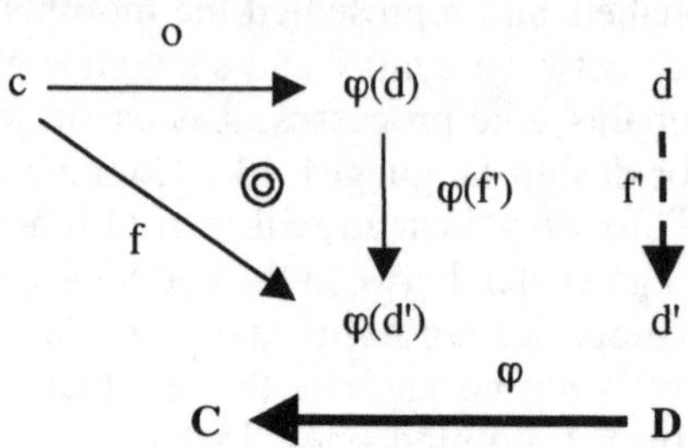

2. The functor φ is said to be *reflective* iff every ***C***-object admits a φ-reflection. We tend to denote functors that are reflective with the special arrow •⟶. ◆

That is, given an object c of ***C***, we are looking for the "best" object of ***D*** that can handle its relationships "across the border", i.e. with other objects of ***D*** through the functor φ. The morphism o can be seen as the protocol that c needs to have with its "interpreter", or the "distance" that remains to be bridged in ***D***.

Notice that φ-reflections are φ-structured morphisms in the sense of Par. 6.2.2. The following proposition provides a useful characterisation of reflections.

7.2.2 Proposition

1. Given a functor φ:***D***→***C*** and a ***C***-object c, the φ-reflections for c are the initial objects of the category $c\downarrow\varphi$.
2. A functor φ:***D***→***C*** is reflective iff, for every ***C***-object c, the category $c\downarrow\varphi$ has initial objects. ◆

7.2.3 Exercise

Prove 7.2.2 and conclude that φ-reflections for an object c are essentially unique, i.e. two φ-reflections for c are isomorphic and, if f:c→$\varphi(d)$ is a φ-reflection for c and h:d→d' is an isomorphism, then $f;\varphi(h)$ is also a φ-reflection for c. ◆

The notion of reflector for the inclusion functor defined by a reflective subcategory put forward in Par. 5.1.13 can also be generalised to an arbitrary reflective functor.

7.2.4 Definition/Proposition – Reflectors

Let $\varphi:\boldsymbol{D}\rightarrow\boldsymbol{C}$ be a reflective functor. We define a functor $\rho:\boldsymbol{C}\rightarrow\boldsymbol{D}$ as follows:

- Every $\boldsymbol{C}$-object c has a φ-reflection arrow $\eta_c:c\rightarrow\varphi(d)$. We define $\rho(c)=d$.
- Consider a morphism $h:c\rightarrow c'$. The composition $h;\eta_{c'}$ is such that the definition of φ-reflection arrow for c guarantees the existence and uniqueness of a morphism $h':\rho(c)\rightarrow\rho(c')$ such that $h;\eta_{c'}=\eta_c;\varphi(h')$. We define $\rho(h)=h'$.

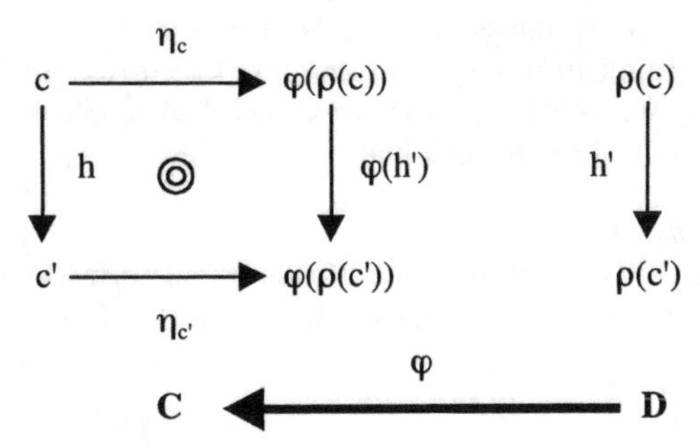

This functor is called a *reflector for* φ.

Proof

The proof is trivially generalised from the one given in Par. 5.1.13 and is left as an exercise. ♦

7.2.5 Definition/Proposition – Reflection Unit

Let $\varphi:\boldsymbol{D}\rightarrow\boldsymbol{C}$ be a reflective functor and $\rho:\boldsymbol{C}\rightarrow\boldsymbol{D}$ a reflector. The definition of ρ directly provides a natural transformation $\boldsymbol{id_C}\xrightarrow{\eta}\rho;\varphi$. We call it the *unit* of the reflection. ♦

By duality, we obtain the notion of *coreflective* functor and coreflector for a coreflective functor, generalising what was defined for coreflective subcategories. We tend to denote functors that are coreflective with the special arrow ◆⟶.

7.2.6 Proposition

Let $\varphi:\boldsymbol{D}\rightarrow\boldsymbol{C}$ be a reflective functor. Every reflector $\rho:\boldsymbol{C}\rightarrow\boldsymbol{D}$ for φ is coreflective and admits φ as a coreflector. Moreover, given any $\boldsymbol{D}$-object d, its ρ-coreflection $\varepsilon_d:\rho(\varphi(d))\rightarrow d$ satisfies $\eta_{\varphi(d)};\varphi(\varepsilon_d)=id_{\varphi(d)}$.

Proof

Let d be a $\boldsymbol{D}$-object. The universal properties of $\eta_{\varphi(d)}$ ensure the existence and uniqueness of a morphism $\varepsilon_d:\rho(\varphi(d))\rightarrow d$ such that $\eta_{\varphi(d)};\varphi(\varepsilon_d)=id_{\varphi(d)}$.

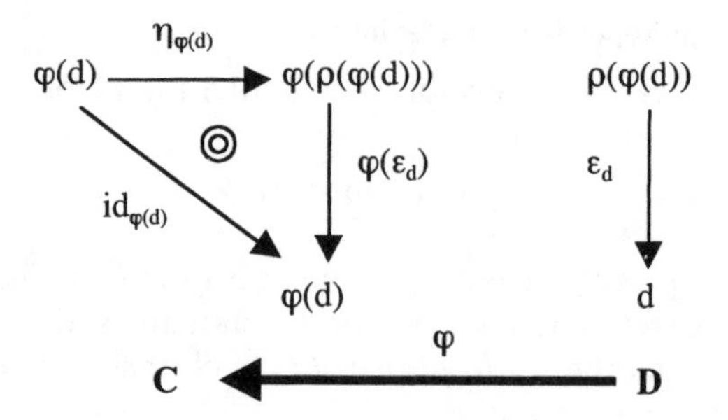

◆

It is easy to see that ε_d is, indeed, a ρ-coreflection for d. Let $g{:}\rho(c)\to d$ be a $\boldsymbol{D}$-morphism. We are going to prove that $g\#=\eta_c;\varphi(g){:}c\to\varphi(d)$ satisfies $\rho(g\#);\varepsilon_d=g$. Because there is only one morphism $h{:}\rho(c)\to d$ such that $\eta_c;\varphi(h)=g\#$, we are going to prove that $\rho(g\#);\varepsilon_d$ satisfies the equation:

$\eta_c;\varphi(\rho(g\#);\varepsilon_d)$

$= \eta_c;\varphi(\rho(g\#));\varphi(\varepsilon_d)$

$=g\#;\eta_{\varphi(d)};\varphi(\varepsilon_d)$ properties of natural transformations,

$=g\#$ properties of ε_d.

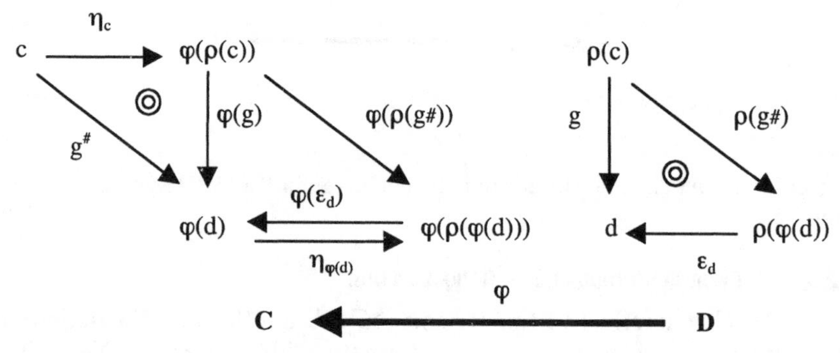

Moreover, $g\#$ is the only morphism $g'{:}\ c\to\varphi(d)$ that satisfies $g=\rho(g');\varepsilon_d$, as the equality implies $\eta_c;\varphi(g)=\eta_c;\varphi(\rho(g'));\varphi(\varepsilon_d)=g';\eta_{\varphi(d)};\varphi(\varepsilon_d)=g'$. ◆

7.2.7 Corollary

Consider a reflective functor $\varphi{:}\boldsymbol{D}\to\boldsymbol{C}$ and its reflector $\rho{:}\boldsymbol{C}\to\boldsymbol{D}$.

1. From Proposition 7.2.6 we derive a natural transformation $\varphi;\rho\xrightarrow{\varepsilon}\boldsymbol{id_D}$ that we call the *counit* of the reflection. The two natural transformations (unit and counit) satisfy

$$\varphi \xrightarrow{\varphi;\eta} \varphi;\rho;\varphi \xrightarrow{\varepsilon;\varphi} \varphi = \varphi \xrightarrow{id_\varphi} \varphi$$

$$\rho \xrightarrow{\eta;\rho} \rho;\varphi;\rho \xrightarrow{\rho;\varepsilon} \rho = \rho \xrightarrow{id_\rho} \rho$$

2. Every morphism $f{:}c\to\varphi(d)$ can be mapped to $f\#=\rho(f);\varepsilon_d{:}\rho(c)\to d$, which is the unique morphism $\rho(c)\to d$ that makes the $\boldsymbol{C}$-triangle commute.

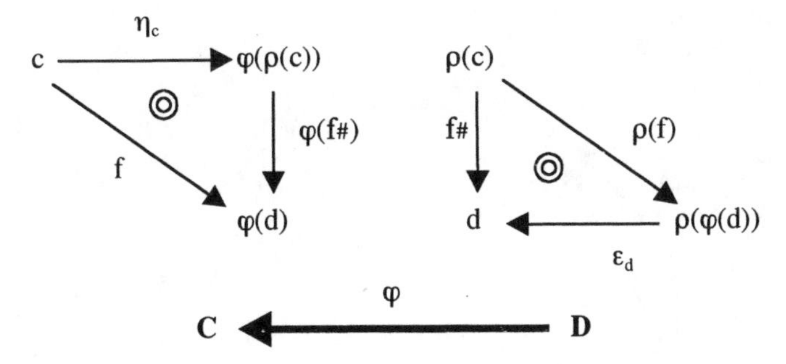

Every morphism $g:\rho(c)\rightarrow d$ can be mapped to $g\#=\eta_c;\varphi(g):c\rightarrow\varphi(d)$, which is the unique morphism $c\rightarrow\varphi(d)$ that makes the $\boldsymbol{D}$-triangle commute.

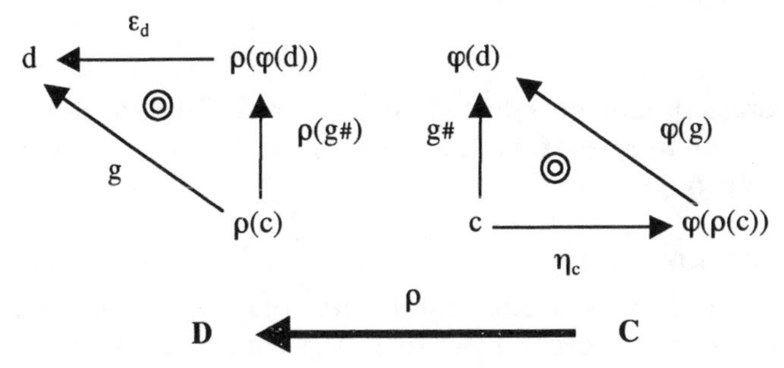

These mappings define a bijection that is "natural" in $\boldsymbol{C}$ and $\boldsymbol{D}$ in the sense that it satisfies, for every $h:c'\rightarrow c$ and $k:d\rightarrow d'$,

$$\rho(h);f\#;k=(h;f;\varphi(k))\# \text{ and } h;g\#;\varphi(k)=(\rho(h);g;k)\#$$

Proof

We leave the proof as an exercise. Notice that, for instance,

$$f\#=\eta_c;\varphi(f\#)=\eta_c;\varphi(\rho(f);\varepsilon_d)=\eta_c;\varphi(\rho(f));\varphi(\varepsilon_d)=f;\eta_{\varphi(d)};\varphi(\varepsilon_d)=f$$ ♦

Some useful properties of reflective functors are as follows.

7.2.8 Proposition

1. Reflective functors compose, i.e. if $\psi:\boldsymbol{E}\rightarrow\boldsymbol{D}$ and $\varphi:\boldsymbol{D}\rightarrow\boldsymbol{C}$ are reflective then so is $\psi;\varphi:\boldsymbol{E}\rightarrow\boldsymbol{C}$.
2. Reflective functors preserve limits.

Proof

Left as an exercise. We give a brief illustration of point 2 for pullbacks. We start with a pullback diagram in $\boldsymbol{D}$ that we translate to $\boldsymbol{C}$. If we now consider a commutative cone $<c\rightarrow\varphi(d_i)>$, we can lift it back to $\boldsymbol{D}$ through the reflection as a commutative cone $<\rho(c)\rightarrow d_i>$. From the properties of the pullback, we are given a morphism $\rho(c)\rightarrow d$ of commutative cones that translates, once again, to $\boldsymbol{C}$ as a morphism of commutative cones. Uniqueness can be easily checked.

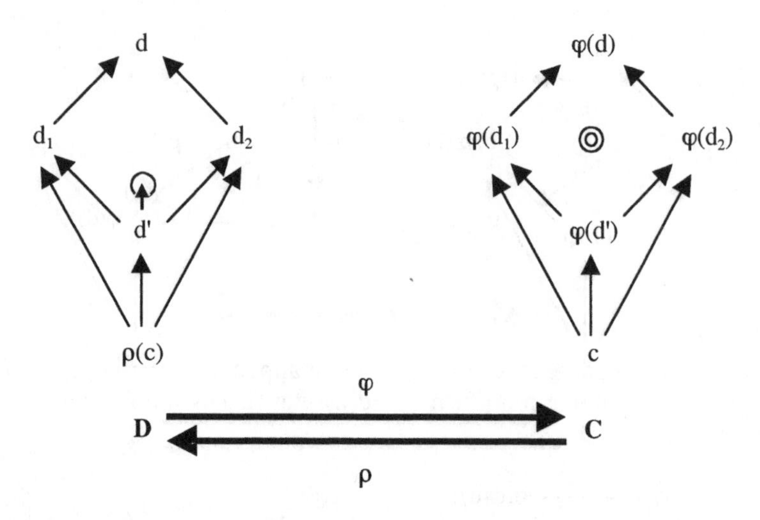

♦

The characterisation of reflective functors provided through Proposition 7.2.2 allows us to give examples among some of the constructions analysed in Chap. 6.

7.2.9 Corollary

1. Let φ:***C***→***SET*** be a functor. The functor υ:***spa***(φ)→***C*** that forgets the ***SET***-component of each object (6.3.3) is both reflective and coreflective. The υ-reflection of any ***C***-object *c* is id_c:*c*→υ(<*c*,∅>), and its coreflection is id_c: υ(<*c*,φ(*c*)>)→*c*.
2. Consider an indexed category ι:$\boldsymbol{I}^{op}$→***CAT***. As defined in Par. 6.4.7, the functor ***fib***(ι):***FLAT***(ι)→***I*** that projects objects and morphisms to their ***I***-components is reflective iff, for every index *i*, $\iota(i)$ has an initial object $0_{\iota(i)}$, and is coreflective iff, for every index *i*, $\iota(i)$ has a terminal object $1_{\iota(i)}$. The ***fib***(ι)-reflection of any index *i*, when it exists, is given by id_i:*i*→***fib***(ι)(<*i*,$0_{\iota(i)}$>), and its ***fib***(ι)-coreflection, when it exists, is given by id_i:***fib***(ι)(<*i*,$1_{\iota(i)}$>)→*i*. ♦

Another obvious example of (co)reflective functors concerns, of course, (co)reflective subcategories,

7.2.10 Corollary

1. If ***D*** is a reflective subcategory of a category ***C***, then the inclusion functor is reflective.
2. If ***D*** is a coreflective subcategory of a category ***C***, then the inclusion functor is coreflective. ♦

We can also generalise the results given in Par. 3.3.10 that relate full (co)reflective subcategories with properties of the (co)unit.

7.2.11 Proposition

Consider a reflective functor $\varphi: \boldsymbol{D} \rightarrow \boldsymbol{C}$, and let ε be its counit.

1. φ is faithful iff, for every $\boldsymbol{D}$-object d, ε_d is epi.
2. φ is full and faithful iff, for every $\boldsymbol{D}$-object d, ε_d is an isomorphism. ♦

7.2.12 Exercise

Complete the constructions and proofs of Pars. 7.2.6 and 7.2.7 to get acquainted with these newly acquired tools. ♦

7.3 Adjunctions

Readers who are acquainted with category theory will have noticed that we are not only following a different path to the topic, even if it turns out not to be that different from [1], but also departing from the standard terminology (if one really exists) for adjunctions. The reason is that the terminology that we introduce is a natural continuation of the one we used for subcategories (which is standard, or at least complies with [80], which comes more or less to the same point). What we have called a *φ-reflection for d* is called in [1] a *φ-universal arrow for d* (or with domain d), and a reflective functor is called therein an *adjoint*. The prefix *co* is used in [1] exactly in the same way so that a φ-coreflection is a φ-couniversal arrow, and a coreflective functor is coadjoint.

Although we prefer the terminolgy that we introduced in the previous sections, there are also good reasons for using the terminology introduced in [1]: adjoints and coadjoints arise in *adjunctions*. In this section, we are going to introduce the standard terminology on adjunctions (i.e. [80]) because it *really* is standard. Adjunctions are an intrinsic part of the vocabulary of category theory. What is hardly standard is the way to approach and define the notion of adjunction. This is where, as authors, we can allow ourselves a little illusion of originality.

7.3.1 Definition – Adjunctions

An *adjunction* from a category $\boldsymbol{C}$ to another category $\boldsymbol{D}$ consists of

- Two functors $\varphi: \boldsymbol{D} \rightarrow \boldsymbol{C}$ and $\rho: \boldsymbol{C} \rightarrow \boldsymbol{D}$.
- Two natural transformations $\boldsymbol{id}_{\boldsymbol{C}} \xrightarrow{\eta} \rho;\varphi$, $\varphi;\rho \xrightarrow{\varepsilon} \boldsymbol{id}_{\boldsymbol{D}}$ satisfying

$$\varphi \xrightarrow{\varphi;\eta} \varphi;\rho;\varphi \xrightarrow{\varepsilon;\varphi} \varphi = \varphi \xrightarrow{\mathrm{id}\varphi} \varphi$$

$$\rho \xrightarrow{\eta;\rho} \rho;\varphi;\rho \xrightarrow{\rho;\varepsilon} \rho = \rho \xrightarrow{\mathrm{id}\rho} \rho$$

We use the notation $\rho \underset{\eta}{\overset{\varepsilon}{-}} \varphi$ for such an adjunction, in which case:

- φ can be called: the *right adjoint*, the *adjoint*, the *forgetful functor*.

- ρ can be called: the *left adjoint*, the *coadjoint*, the *free functor*.
- η is called the *unit*.
- ε is called the *counit*. ♦

Concerning the terminology, the left/right classification is quite widespread; (co)adjoints are used in [1] as already mentioned. Classifying the functors as forgetful/free can be very helpful when the roles that they play are obvious. This is precisely the case of the adjunctions that result from reflective subcategories, functor-structured categories and indexed categories as illustrated below: the (right) adjoint usually "forgets" part of the structure of objects that the left/coadjoint is able to freely generate the additional structure.

7.3.2 Proposition

Every equivalence defines two adjunctions. ♦

7.3.3 Proposition

For every adjunction $\rho \overset{\varepsilon}{\underset{\eta}{\dashv}} \varphi$, $\varphi^{op} \overset{\eta^{op}}{\underset{\varepsilon^{op}}{\dashv}} \rho^{op}$ is also an adjunction. ♦

7.3.4 Proposition

Given functors $\varphi{:}\boldsymbol{D}\rightarrow\boldsymbol{C}$, $\rho{:}\boldsymbol{C}\rightarrow\boldsymbol{D}$, $\psi{:}\boldsymbol{E}\rightarrow\boldsymbol{D}$, $\gamma{:}\boldsymbol{D}\rightarrow\boldsymbol{E}$, if ρ is a left adjoint of φ and γ is a left adjoint of ψ, then $\rho;\gamma$ is a left adjoint of $\psi;\varphi$. ♦

There are many alternative ways of characterising adjunctions. It can even be hard to find two books that adopt the same characterisation as the defining one. However, they all involve, in some way or the other, but not arbitrarily, the properties analysed in Par. 7.2.7.

7.3.5 Proposition

An adjunction from a category $\boldsymbol{C}$ to another category $\boldsymbol{D}$ can be obtained from:

- Two functors $\varphi{:}\boldsymbol{D}\rightarrow\boldsymbol{C}$ and $\rho{:}\boldsymbol{C}\rightarrow\boldsymbol{D}$.
- A bijection between morphisms $c\rightarrow\varphi(d)$ and $\rho(c)\rightarrow d$ that is natural in $\boldsymbol{C}$ and $\boldsymbol{D}$.

Proof

Much of the proof is sketched in Par. 7.2.7. We leave it as an exercise. ♦

This characterisation is useful for the following example in particular.

7.3.6 Proposition

The powerset functor $\boldsymbol{2}^{-}{:}\boldsymbol{SET}^{op}\rightarrow\boldsymbol{SET}_{\perp}$ that maps every set to its powerset as a pointed set, the empty set being the designated element, and every function to its inverse image, defines an adjunction from $\boldsymbol{SET}_{\perp}$ to $\boldsymbol{SET}^{op}$.

Its left adjoint computes powersets of proper elements (i.e. excluding the designated element) and inverse images.

Proof

The reader is invited to carry out the proof as it is very instructive, if not challenging, due to the fact that it operates on a contravariant functor! The natural bijection is defined by associating functions of the form $f:A \rightarrow 2^B$ with $g:B \rightarrow 2^A$ by $a \in g(b)$ iff $b \in f(a)$. ♦

An adjunction is a very strong relationship between two categories: it allows us to give canonical approximations of objects in one domain with respect to the structural properties that are captured in the other domain. Examples of the use of adjunctions abound, even in the particular aspects of computing science that interest us in this book. One that we have worked out and presented in [36] concerns the synthesis of programs from specifications.

7.3.7 Example – Synthesis of Programs from Specifications

Par. 5.1.3 shows how a satisfaction relationship between programs and specifications can be defined from a functor ***spec*:*PROG*→*SPEC*** that maps every program to the maximal set of properties that it satisfies. In Par. 6.2.1, every morphism *S*→***spec***(*P*) is called a possible realisation of the specification *S* by the program *P*. Par. 6.1.24 illustrates how systems can be evolved by interconnecting components that make new required properties to emerge. Par. 6.2.4 shows how the process of assembling a system from smaller components, including the interconnection of new components, can be addressed in a compositional way by addressing realisations and not just individual specifications or programs.

We are now interested in incorporating into the picture the ability to synthesise programs from specifications. This is intended as a means of supporting the process of compositional evolution that we have been addressing. According to this principle, the addition of a component to a system should not require the recalculation (or the resynthesis) of the whole system, but only of the new component and its interconnections to the previous system. To illustrate our purpose, consider once again the vending machine as defined in Par. 3.5.6. In Par. 6.1.24, we develop the specification of a regulator and show that, once interconnected with the vending machine, the new system does not allow arbitrary sales of cigars. but requires the insertion of a special token before a cigar can be selected. Assuming that the original vending machine is implemented and running, and that we are in possession of a realisation of the regulator, possibly by using the synthesis method of [83], we want to be able to synthesise the interconnections between the two programs (the running vending machine and the realisation of the regulator) in order to obtain a realisation of the specification diagram.

In summary, given realisations of component specifications (either obtained through traditional transformational methods, or synthesised directly from the specifications, or reused from previous developments), we would like to be able to synthesise the interconnections between the programs in such a way that the program diagram realises the specification diagram. That is, given specifications S_1 and S_2 (newly) interconnected via morphisms $\varphi_1:S\to S_1$ and $\varphi_2:S\to S_2$, and realisations $<\sigma_1,P_1>$, $<\sigma_2,P_2>$ of S_1 and S_2, respectively, one would like to be able to synthesise a realisation $<\sigma,P>$ of S and interconnecting morphisms $\mu_1:P\to P_1$ and $\mu_2:P\to P_2$ such that $\sigma;\boldsymbol{spec}(\mu_i)=\varphi_i;\sigma_i$ (i=1,2).

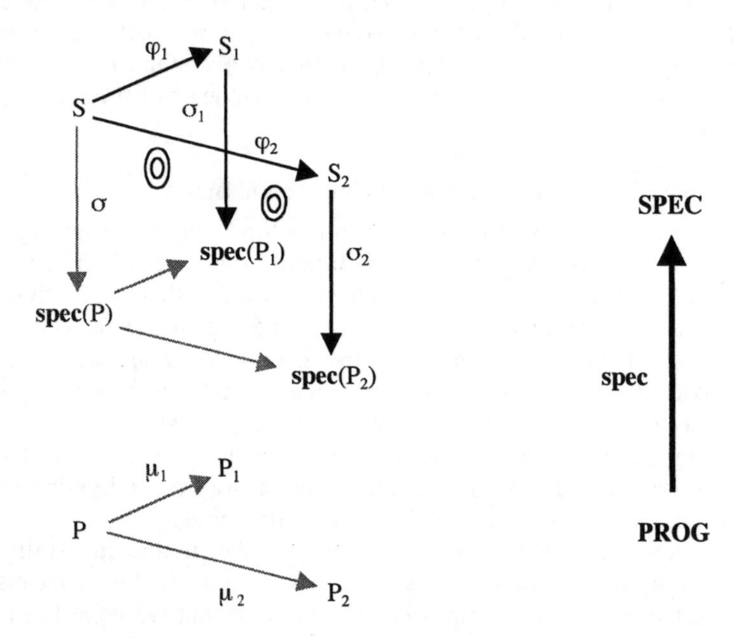

This general statement of what it means to synthesise interconnections makes it clear that it is necessary to synthesise both the middle program P and the morphisms μ_i that are required to interconnect the given programs. Because, in the general case, any object can be used in an interconnection, this suggests, rather obviously, that a functor ***synt***: ***SPEC→PROG*** is required that is somehow related to ***spec***. One possible such relationship is for ***spec*** to be the inverse (in the sense of Par. 5.1.7) of ***synt***, but this is a rather strong property because it would require the two categories of programs and specifications to be isomorphic. Clearly, if this were to be the case, we could hardly claim that we were dealing with two different levels of abstraction. Hence, it makes sense to look for weaker properties of the relationship between the two functors.

It seems clear that, more than programs, synthesis must return realisations of the given specifications. That is, for every specification S, ***synt(S)*** must be provided together with a morphism η_S*:S→***spec(synt(S))** that establishes ***synt(S)*** as one of the possible realisations of *S*. Hence, η_S expresses a correctness criterion for ***synt***. Moreover, ***synt*** must respect interconnections in the following sense: given a specification diagram ***δ**:* ***I→SPEC***, it is necessary that the program diagram ***δ;synt*** be a realisation of ***δ*** through $(\sigma_i{:}\delta(i)\rightarrow spec(\delta'(i)))_{i\in|I|}$ as defined in Par. 6.2.4. That is, we must have, for every *f:i→j* in ***I***, $\delta(f);\sigma_j=\sigma_i;spec(\delta'(f))$. But these are the ingredients that define a natural transformation. Hence, ***synt*** must be provided together with a natural transformation $\eta{:}1_{SPEC}\rightarrow synt;spec$.

Consider now the synthesis of interconnections themselves. Given an interconnection of specifications σ*:S→***spec(P)**, we should be able to synthesise σ'*:***synt(S)***→P* in such a way that the interconnection is respected, i.e. $\sigma=\eta_S;spec(\sigma')$. This is equivalent to defining a (natural) bijection between the morphisms *S→***spec(P)** and the morphisms ***synt(S)→P***. But this is precisely the property that characterises the existence of an adjunction between ***SPEC*** and ***PROG***. Hence, synthesis of interconnections can be characterised by the existence of a reflector (left adjoint) ***synt*** for ***spec***. Notice that Proposition 7.2.2 characterises the synthesis functor precisely in terms of the existence, for every specification, of a "minimal" realisation in the sense that all other programs that implement the specification simulate it.

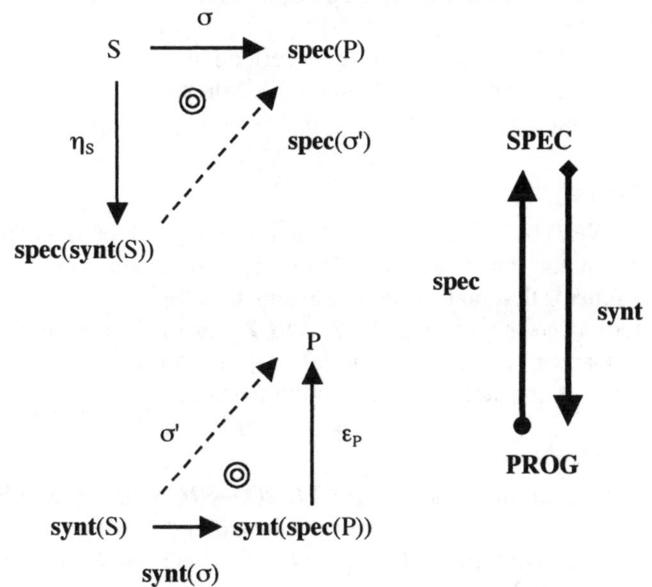

Notice that the counit of the adjunction ε_P*:***synt(spec(P))→P** is not necessarily an isomorphism because ***spec(P)*** may not be powerful enough to fully characterise *P* (we cannot guarantee that the specification domain is expressive enough to capture the semantics of P in full). Hence, we are not even in the presence of an equivalence.

The direction of the counit reflects the fact that if we synthesise from the strongest specification of a program P, we obtain a program that cannot be stronger than P. Hence, the morphism ε_P provides a sort of "universal adaptor" between the program synthesised from ***spec(P)*** and P itself.

Although weaker than the existence of an inverse or a pseudo-inverse, the existence of a left adjoint to the functor ***spec***: ***PROG→SPEC*** is quite a strong property. This is not surprising because the ability to synthesise any specification is itself, in intuitive terms, a very strong property. In the literature, examples of synthesis of finite state automata from temporal logic specifications can be found, both from propositional linear temporal logic as above [83] and from branching time logic [30]. However, their generalisation to a full systems view is difficult. We shall see in Sect. 7.5 that we can go a longer way in the context of formalisms that separate "coordination" from "computation".

Another important property that results from the properties of reflective functors is as follows.

7.3.8 Proposition – Adjunctions and Reflections

1. Every reflective functor defines an adjunction in which it plays the role of right adjoint.
2. In every adjunction, the adjoint φ is reflective with the co-adjoint ρ as reflector, and ρ is coreflective with coreflector φ. ♦

This result allows us to derive from Par. 7.2.9 two useful adjunctions.

7.3.9 Corollary

1. Let φ:***C→SET*** be a functor. The functor υ: ***spa(***φ***)→C*** has for left adjoint the functor that maps each ***C***-object c to $<c,\emptyset>$, and for right adjoint the functor that maps each ***C***-object c to $<c,\varphi(c)>$.
2. The functor ***alph*** that maps ***PROC*** to $\boldsymbol{SET_\perp}$ by forgetting behaviours has both a left and right adjoint. The left adjoint maps each alphabet $A_\perp$ to the process $<A_\perp,\emptyset>$ and the right adjoint to $<A_\perp,\boldsymbol{tra}(A_\perp)>$. ♦

7.3.10 Corollary

In any π-institution, the functor ***sign:THEO→SIGN*** that maps theories to their underlying signatures has both a left and right adjoint. The left adjoint maps each signature to the theory $<\Sigma,c_\Sigma(\emptyset)>$, and the right adjoint maps it to $<\Sigma,\boldsymbol{gram}(\Sigma)>$. ♦

Basically, both results tell us how to map back and forth between processes and alphabets, and between theories and signatures.

7.3.11 Exercise

Work out direct proofs for Pars. 7.3.9 and 7.3.10, and interpret the meaning of the natural transformations. Check how far Par. 7.3.10 extends to presentations and strict presentations. ♦

From Pars. 7.2.10 and 7.3.8 we get another class of adjunctions.

7.3.12 Corollary

1. Every reflective subcategory defines an adjunction in which the inclusion functor is the right adjoint.
2. Every coreflective subcategory defines an adjunction in which the inclusion is the left adjoint. ♦

7.3.13 Example – Adjunctions Between *SET*, *PAR* and $SET_\perp$

The fact that, as seen in Par. 3.3.11, ***SET*** is a coreflective subcategory of ***PAR*** tells us that the inclusion has a right-adjoint. As also seen in Par. 3.3.11, this right adjoint (called $+\perp$ in Par. 7.1.12) is the one that performs the traditional "elevation" of partial into total functions by extending sets with an "undefined" element, or "bottom", that serves as image for the elements in which the partial functions are undefined. This construction may well remind the reader of one of the functors over which the equivalence between ***PAR*** and $\boldsymbol{SET_\perp}$ was built in Par. 7.1.12: the one that bears the same name. That is, we have the same kind of construction – the "elevation" – being performed over two different categories. Are they related?

There is a "natural" way in which every pointed set can be viewed as a (normal) set: just forget the "added structure", i.e. the fact that it has a designated element. Note that this does not mean "through away the designated element", which is what the functor $-\perp$ (the pseudo-inverse of $+\perp$ in the equivalence) does. Going back to Par. 3.2.1, we are mapping pointed sets$<A,\perp_A>$ to the underlying set A and morphisms between two pointed sets $f{:}<A,\perp_A>\rightarrow<B,\perp_B>$ to the corresponding total function $f{:}A\rightarrow B$. This mapping defines $\boldsymbol{SET_\perp}$ as a concrete category over ***SET***. The two elevations of ***PAR*** are related by this forgetful functor. The elevation to ***SET*** is simply the result of forgetting that there is a designated element in the elevation to $\boldsymbol{SET_\perp}$, which is captured by the following commutative diagram:

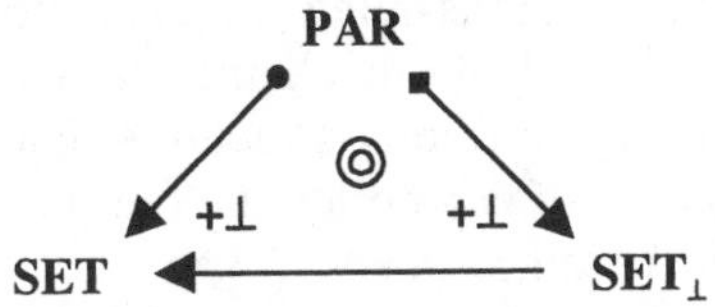

Notice that the elevation to $\boldsymbol{SET_\perp}$ is explicitly recorded into a structure that is added to sets, whereas the elevation to ***SET*** is merely a representation or encoding. This difference is well captured in the fact that it gives rise to an equivalence in the first case, but "just" a reflective functor in the second. The elevation from $\boldsymbol{PAR}$ to $\boldsymbol{SET}_\perp$ is also reflective, but the fact that it is a coreflection is more interesting. If we complete the diagram with the adjunctions that we have already built

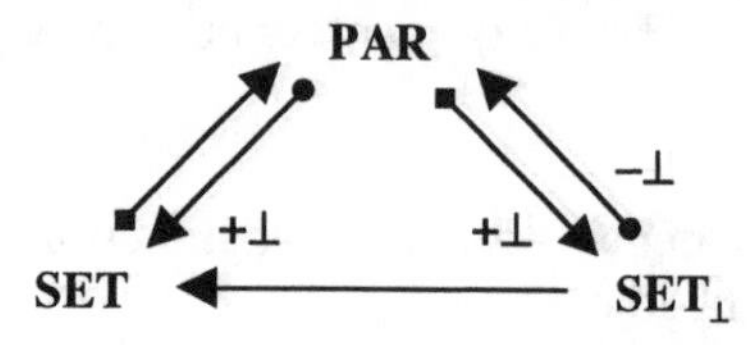

we can see that the functor that forgets the designated elements admits a right adjoint that again performs another kind of elevation, this time superposing a designated element to every set. This is just the elevation of partial functions being performed on total ones as a particular case.

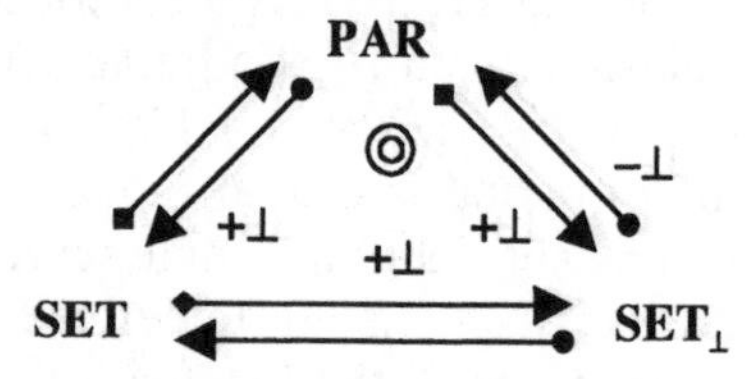

Notice that we obtain a commutative diagram of adjunctions (i.e. of reflective and coreflective functors), but not of the functors in general. For instance, the diagram of $+\perp$ is clearly not a commutative one!

It is also important to point out, in what follows the remarks made in Par. 7.1.12, that because $\boldsymbol{PAR}$ and $\boldsymbol{SET}_\perp$ are equivalent, we tend to look at them as being "the same", but they may bear quite different relationships to other categories like, in this case, $\boldsymbol{SET}$. For instance, the "elevations" go in opposite directions, one from $\boldsymbol{PAR}$, the other into $\boldsymbol{SET}_\perp$; one is reflective and the other coreflective. What is more interesting is that these are "technical" differences. Conceptually, both $\boldsymbol{PAR}$ and $\boldsymbol{SET}_\perp$ provide a coreflective representation for "normal" sets; the representations are different because, in spite of being equivalent, the two categories offer different structures and, hence, require different encodings of what is, essentially, the same kind of relationship.

This example also shows how diagrams of adjunctions can be useful to understand how different domains relate to each other. They provide a kind of "roadmap" or "classification scheme" that is essential for being able to "navigate" among the different structures that one tends to find in the literature. For instance, we can enrich the previous diagram of adjunctions with the one that we obtained in Par. 7.3.6:

$$\mathbf{SET} \underset{+\perp}{\rightleftarrows} \mathbf{PAR} \underset{-\perp}{\overset{+\perp}{\approx}} \mathbf{SET}_\perp \underset{2^-}{\overset{2^-}{\rightleftarrows}} \mathbf{SET}^{op}$$

The composition of these adjunctions gives us the well-known adjunction between ***SET*** and SET^{op} performed by the powerset functor.

This kind of roadmap was used in [97, 109] for formalising relationships between models of concurrency like transition systems, synchronisation trees, event structures, etc., in what constitutes one of the most striking examples of the expressive power of adjunctions. Each such model is endowed with a notion of morphism that captures a form of simulation as a behaviour-preserving mapping. Typical operations of process calculi are captured as universal constructions as exemplified in Par. 4.3.8 for ***PROC***. Reflections and coreflections[1] are used for expressing the way one model is embedded in another: one of the functors in the adjunction embeds the more abstract model in the other, while the other functor abstracts away from some aspect of the representation.

Instead of reproducing an example of such uses of adjunctions, which on its own would hardly capture the richness of the classification that is developed for different kinds of concurrency models in [97, 109], we present a related kind of application: a duality between process models and specifications. More precisely, a duality between ***PROC*** and linear temporal logic specifications given by $THEO_{LTL}$ as presented in [32] to show how both semantics domains – theories and models – can be made part of the same roadmap.

7.3.14 Example – Processes versus Specifications

We start by recalling that ***PROC*** is concrete over $SET_{\perp}$ through ***alph*** and that ***THEO*** is concrete over ***SET*** through ***sign***. In fact, we proved in Pars. 7.3.9 and 7.3.10 that these functors are both reflective and coreflective. Moreover, we showed in Par. 7.3.6 that the contravariant powerset functor $2^{-}:SET^{op}\rightarrow SET_{\perp}$ is reflective. We now show that 2^{-} can be lifted to $proc:THEO^{op}\rightarrow PROC$ as a reflective functor that makes the diagram (of reflective functors) commute.

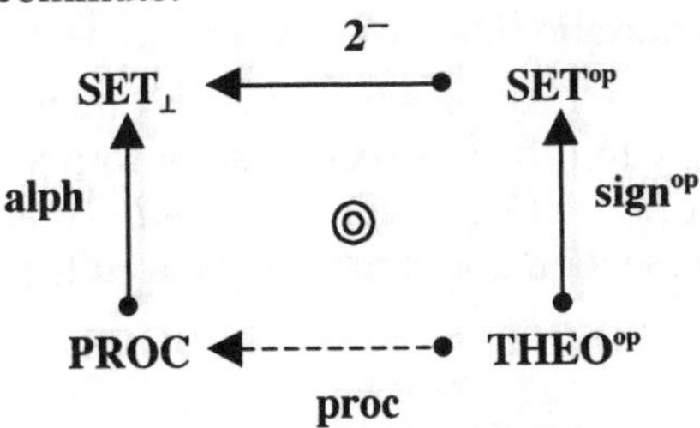

Let $<\Sigma,\Phi>$ be a theory. Requiring the diagram to commute fixes the choice of the alphabet for $proc(<\Sigma,\Phi>)$: the powerset 2^{Σ} considered as a

[1] Further terminological confusion arises with respect to [109], where a coreflection is an adjunction in which the reflective functor (the adjoint) is a full embedding, i.e. the straight generalisation of a full reflective subcategory.

pointed set. We are going to choose for its behaviour the set $\{\lambda \in (2^\Sigma)^\omega \mid \lambda \models \Phi\}$, i.e. the least deterministic process that satisfies the properties given by the theory. It is not difficult to prove that we do obtain a functor.

If we consider now a process $<A_\perp,\Lambda>$, the category $<A_\perp,\Lambda>\downarrow$***proc*** consists of the specifications that are validated by the behaviours in Λ after a suitable translation (which we called generalised models in Par. 6.5.9). This category has an initial object: the set of all sentences in the language of 2^A that are validated by the behaviours of Λ translated by the unit of the powerset adjunction. It is the largest, not the smallest, because we are working with a contravariant functor. Hence, ***proc*** is indeed reflective, with its reflector assigning to $<A_\perp,\Lambda>$ the theory $<2^A,\{p \mid \Lambda;\eta_A \models p\}>$.

Note that because, as shown in Par. 7.2.8, reflective functors preserve limits, colimits of specifications are mapped by ***proc*** to limits of the corresponding processes – again a form of *compositionality*. This says that composition of specifications as given by colimits of configuration diagrams captures parallel composition of the corresponding processes taking into account their interactions. In other words, compositionality expresses precisely the view that "conjunction as composition" [112] applies to development steps, not just specifications or designs. ♦

The reader is encouraged to consult [32, 33] for a wider discussion of the relationships between these two categories, namely in the context of what are called categorical institutions in [85].

7.3.15 Exercises

1. Work out the full proof of Example 7.3.14.
2. Relate this result to Par. 6.5.18.
3. Since, in the diagram in Par. 7.3.14, ***sign***^{op} and $\mathbf{2}^-$ are reflective and ***alph*** is coreflective, why didn't we take the composition of ***sign***^{op} with $\mathbf{2}^-$ and the coreflector of ***alph*** to obtain an adjunction from ***PROC*** to ***THEO***^{op}?
4. What kind of generalisation into institutions can we hope for? ♦

Can we extend this ability of adjunctions to relate different semantic models of concurrency to different specification formalisms as captured by the categories of theories defined by institutions? To answer this question, we present in the next section a summary of the results published in [5].

7.4 Adjunctions in Institutions

We start with an example in order to motivate the structures that are involved in mapping between specification formalisms as captured by institutions. The typical temporal logics that have been used for the specification of reactive systems are based on linear time, of which the one we have been working with is an example. However, sometimes, branching time is more justified. For instance, it is well known that verification techniques over branching time logic can be more effective. The expressive power of branching time logic can also be useful, especially in relation to progress properties related to required non-determinism, like responsiveness. So, we would like to have ways of mapping between specifications in these different logics that support translations back and forth between them, through which one can take advantage of the best features of each.

As an example, consider the specification of a "user-friendly" vending machine that, once it accepts a coin, will make a cake and a cigar available. Notice that this is not a property of the vending machine as specified in Par. 3.5.6; the specification given therein allows behaviours in which, for instance, after a coin is accepted, no cakes and no cigars are delivered! One could think that adding a property like $(coin \supset F(cake \vee cigar))$ would solve the problem, but it is easy to see that this requirement does not capture the availability of a cake or a cigar for the customer to choose. It requires that, in every behaviour, the acceptance of a coin is followed by the delivery of a cigar or a cake. Hence, it admits as an implementation a machine that only delivers cigars! Moreover, it forces the customer to take either the cake or the cigar among the options that are given, which does not make sense when that activity is not initiated by the machine. All this is because the trace-based semantics is not expressive enough to model choice; for that purpose, branching structures are required.

A logic in which such properties can be easily expressed is the branching time logic *CTL**. This logic is said to be branching because operators are provided that quantify over the possible future behaviours from the current state.

7.4.1 Definition – CTL* as an Institution

The branching temporal logic institution *CTL** is defined as follows:

- Its category of signatures is ***SET***.
- We define two classes of propositions (the state propositions ϕ_s and the path propositions ϕ_p) for a temporal signature Σ

$$\phi_s ::= a \mid (\neg\phi_s) \mid (\phi_s \supset \psi_s) \mid (A\phi_p)$$
$$\phi_p ::= \phi_s \mid (\neg\phi_p) \mid (\phi_p \supset \psi_p) \mid (\phi_p U \psi_p)$$

- The set of branching temporal propositions $\mathbf{gram}_{CTL*}(\Sigma)$ is the set of state propositions.
- Every signature morphism $f{:}\Sigma \to \Sigma'$ induces the translation function $\mathbf{gram}(f){:}\mathbf{gram}_{CTL*}(\Sigma) \to \mathbf{gram}_{CTL*}(\Sigma')$ defined as follows:

$$\mathbf{gram}(f)(\phi_s) ::= f(a) \mid \neg\mathbf{gram}(f)(\phi_s) \mid \mathbf{gram}(f)(\phi_s) \supset \mathbf{gram}(f)(\psi_s) \mid A\mathbf{gram}(f)(\phi_p)$$
$$\mathbf{gram}(f)(\phi_p) ::= f(\phi_s) \mid \neg\mathbf{gram}(f)(\phi_p) \mid \mathbf{gram}(f)(\phi_p) \supset \mathbf{gram}(f)(\psi_p) \mid \mathbf{gram}(f)(\phi_p) U \mathbf{gram}(f)(\psi_p)$$

- The model functor $\boldsymbol{mod}_{CTL*}$ is defined as follows:
  - For every signature Σ, a branching Σ-model is a triple $<W,R,V>$ with R a total relation on W and $V{:}\Sigma\to 2^W$.
  - Given a branching model $M=<W,R,V>$ we denote by $L(M)$ the set of all infinite sequences $\omega\to 2^\Sigma$ of the form$\lambda;V^{-1}$ where $\lambda{:}\omega\to W$ is such that $\lambda(i)R\lambda(i+1)$ for every $i\in\omega$. We denote by $L(M,s)$ the subset of $L(M)$ that is generated by the sequences $\lambda_s{:}\omega\to W$ such that $\lambda_s(0)=s$, i.e. $L(M,s)$ contains the subset of paths that begin at state s.
  - Let $M_1=<W_1,R_1,V_1>$ and $M_2=<W_2,R_2,V_2>$ be Σ-models. A morphism from M_1 to M_2 is a map $f{:}W_1\to W_2$ such that:
    1. sR_1t implies $f(s)R_2f(t)$.
    2. $f(s)R_2u$ implies the existence of $t\in W_1$ such that sR_1t and $f(t)=u$.
    3. $s\in V_1(a)$ iff $f(s)\in V_2(a)$.
  - Let $f{:}\Sigma_1\to\Sigma_2$ be a morphism. If $< W_2,R_2,V_2>$ is a Σ_2-model, then $<W_2,R_2,f;V_2>$ is a Σ_1-model called the *f-reduct* of $<W_2,R_2,V_2>$.
- The satisfaction relation is as follows: the truth of a Σ-proposition ϕ in $M=<W,R,V>$ at state $s\in W$ (which we write $(M,s)\models_\Sigma\phi)$ is inductively defined as for LTL except for the operator $\boldsymbol{A}$ for which

  $$(M,s)\models_\Sigma \boldsymbol{A}\phi \text{ iff, for every } \lambda_s\in L(M,s),\ \lambda_s\models^o{}_\Sigma\phi.$$
- The branching temporal proposition ϕ is said to be true in M, which we denote by $M\models_\Sigma\phi$, iff $(M,s)\models_\Sigma\phi$ at every state s of W. ♦

Notice how the new operator $\boldsymbol{A}$ quantifies over all possible paths that start from the current state.

The reader is invited to check that the satisfaction condition holds.

7.4.2 Proposition – Satisfaction Condition

Let $f{:}\Sigma_1\to\Sigma_2$ be a signature morphism. For every $M\in|\boldsymbol{mod}_{CTL*}(\Sigma_2)|$ and $\phi\in\boldsymbol{gram}_{CTL*}(\Sigma_1)$, $M\models_{\Sigma_2}\boldsymbol{gram}(f)(\phi)$ iff $\boldsymbol{mod}_{CTL*}(f)(m)\models_\Sigma\phi$. ♦

As done in previous chapters, we often use f instead of $\boldsymbol{gram}(f)$.

7.4.3 Corollary

*CTL** as defined in Par. 7.4.1 is an institution. ♦

As an example of a specification in CTL*, consider the user-friendly vending machine.

```
specification   user-friendly vending machine is
signature       coin, cake, cigar
axioms          beg ⊃ A(¬cake∧¬cigar)
                     ∧ A(coin ∨ (¬cake∧¬cigar)Wcoin)
                coin ⊃ A((¬coin)W(cake∨cigar))
                coin ⊃ (EXcake ∧ EXcigar)
                (cake∨cigar) ⊃ A((¬cake∧¬cigar)Wcoin)
                cake ⊃ (¬cigar)
```

The operator E is the dual of A: it expresses the existence of a path from the current state in which the given property holds. Notice the use of the conjunction in $(coin \supset (EXcake \wedge EXcigar))$; it requires the machine to give the customer the choice; hence, for instance, if the machine runs out of cakes, it may not accept coins even if cigars are still available.

It is clear from the definition of $CTL*$ that this logic "incorporates" LTL in the sense that it can express at least as much as LTL. A theory in LTL expresses properties about all possible behaviours of a system taken as infinite sequences of states. In $CTL*$ this quantification can be made explicit through the operator A. Hence, it should be straightforward to map a theory of LTL to a theory of $CTL*$ by qualifying every proposition with A.

This syntactic transformation between the two languages respects the translations defined by signature morphisms, i.e. is "natural" on signatures. Indeed, it is captured by a natural transformation.

7.4.4 Definition/Proposition

The family of functions $\alpha_\Sigma{:}\mathbf{gram}_{LTL}(\Sigma)\to\mathbf{gram}_{CTL*}(\Sigma)$ defined by $\alpha_\Sigma(\phi)=A\phi$ is a natural transformation from $\mathbf{gram}_{LTL}$ to $\mathbf{gram}_{CTL*}$. ♦

There is also a way in which this translation can be claimed to be "correct". Every branching structure gives rise to a linear one in a natural way.

7.4.5 Definition/Proposition

Let β_Σ map every branching model $M=\langle W,R,V\rangle$ to the linear model $L(M)$ defined in Par. 7.4.1. Given branching Σ-models $M=\langle W,R,V\rangle$ and $M'=\langle W',R',V'\rangle$, and a p-morphism $f{:}M\to M'$, let $\beta_\Sigma(f)$ be the inclusion $L(M)\rightarrowtail L(M')$ that is induced by the properties of the morphism. The mapping β_Σ is a functor $\mathbf{mod}_{CTL*}(\Sigma)\to\mathbf{mod}_{LTL}(\Sigma)$ and the family $\langle\beta_\Sigma\rangle_{\Sigma\in|SIGN|}$ defines a natural transformation $\mathbf{mod}_{CTL*}\to\mathbf{mod}_{LTL}$. ♦

That is, we generate from every branching structure M the linear structure $L(M)$ that consists of all possible paths through M. The syntactic and semantic transformations agree in the sense of the proposition below.

7.4.6 Definition/Proposition

If $M=\langle W,R,V\rangle$ is a branching Σ-model and $\phi\in\mathbf{gram}_{LTL}(\Sigma)$, then $M\models_\Sigma A\phi$ iff $\beta_\Sigma(M)\models_\Sigma\phi$. ♦

This relationship between the syntax and semantics of the given institutions allows us to define the intended functor between the corresponding categories of theories.

7.4.7 Definition/Proposition

The mapping $\mathcal{T}{:}\mathbf{THEO}_{LTL}\to\mathbf{THEO}_{CTL*}$ defined by $\mathcal{T}(\langle\Sigma,\Gamma\rangle)=\langle\Sigma,c(A\Gamma)\rangle$ is a functor. ♦

By $A\Gamma$ we are denoting the set $\{A\phi \mid \phi \in \Gamma\}$ and by c the closure operator of $CTL*$. This functor allows us to translate any specification (theory) in LTL to a specification in $CTL*$. This translation is "canonical" in the sense of the proposition below.

7.4.8 Proposition

The functor $\mathcal{T}: \mathbf{THEO}_{LTL} \rightarrow \mathbf{THEO}_{CTL*}$ is coreflective, that is, admits a right adjoint. ◆

It is not difficult to guess the nature of the right adjoint. Because $\mathcal{T}$ computes direct images through α, its right adjoint computes inverse images, i.e. the adjunction is given by a generalisation to closure operators of the well-known Galois connection between direct and inverse images of sets. The unit of the adjunction is given by the inclusion $\Gamma \subseteq \alpha_\Sigma^{-1}(c(\alpha_\Sigma(\Gamma)))$.

7.4.9 Proposition

The mapping $\mathcal{U}: \mathbf{THEO}_{CTL*} \rightarrow \mathbf{THEO}_{LTL}$ given by $\mathcal{U}(<\Sigma,\Gamma>) = <\Sigma, \alpha_\Sigma^{-1}(\Gamma)>$ is a coreflector (right adjoint) of $\mathcal{T}$.

Proof (of Propositions 7.4.7 and 7.4.8)

The functor $\mathcal{U}$ is a coreflector of $\mathcal{T}$ iff for every theory $<\Sigma,\Gamma>$ of LTL, the pair $(<\Sigma, \alpha_\Sigma^{-1}(c(\alpha_\Sigma(\Gamma)))>, id_\Sigma)$ is a reflection. Consider $f: <\Sigma,\Gamma> \rightarrow <\Sigma', \alpha_{\Sigma'}^{-1}(\Gamma')>$. We have to prove that there is a unique $CTL*$ morphism $f': <\Sigma, c(\alpha_\Sigma(\Gamma))> \rightarrow <\Sigma',\Gamma'>$ such that $id_\Sigma; f' = f$. Unicity is automatically guaranteed by this equation. All that remains is the proof that f is a theory morphism $<\Sigma, c(\alpha_\Sigma(\Gamma))> \rightarrow <\Sigma',\Gamma'>$ in $CTL*$. Let $\alpha_\Sigma(\Gamma) \vdash_\Sigma \phi$. We have to prove that $f(\phi) \in \Gamma'$. Because $\alpha_\Sigma(\Gamma) \vdash_\Sigma \phi$ we have $f(\alpha_\Sigma(\Gamma)) \vdash_{\Sigma'} f(\phi)$ (a consequence of the satisfaction condition of $CTL*$). But $f(\alpha_\Sigma(\Gamma)) = \alpha_{\Sigma'}(f(\Gamma))$ because α is a natural transformation. Hence, $f(\phi) \in \alpha_{\Sigma'} c(f(\Gamma))$. On the other hand, $f(\Gamma) \subseteq \alpha_{\Sigma'}^{-1}(\Gamma')$ because f was taken as a theory morphism $<\Sigma,\Gamma> \rightarrow <\Sigma', \alpha_{\Sigma'}^{-1}(\Gamma')>$. Hence, $f(\phi) \in \Gamma'$. ◆

The coreflector "forgets" the branching nature of time in $CTL*$ by retaining only those propositions ϕ for which $A\phi$ is a theorem in $CTL*$, i.e. it retains those truths that hold for every possible path.

The existence of the adjunction means that, in order to prove that a $CTL*$-theory BT provides an interpretation (refinement) of an LTL-theory LT, it is equivalent to prove $LT \subseteq \mathcal{U}(BT)$ or $\mathcal{T}(LT) \subseteq BT$. For practical purposes, the inclusion $\mathcal{T}(LT) \subseteq BT$ is easier to prove because it can be lifted to presentations. Indeed, if we take the category $\mathbf{PRES}_{LTL}$ of the theory presentations of LTL, the adjunction between presentations and theories as given in Par. 3.6.4 allows us to extend the adjunction between $\mathbf{THEO}_{CTL*}$ and $\mathbf{THEO}_{LTL}$ to one between $\mathbf{THEO}_{CTL*}$ and $\mathbf{PRES}_{LTL}$. Hence, the inclusion $\mathcal{T}(LT) \subseteq BT$ can be proved at the level of a presentation of LT. The converse, however, does not hold because although there is also an adjunction between $\mathbf{THEO}_{CTL*}$ and $\mathbf{PRES}_{CTL*}$, the right adjoint does not go in the same direction as $\mathcal{U}$.

The actual relationship between $\boldsymbol{THEO}_{CTL*}$ and $\boldsymbol{THEO}_{LTL}$ is stronger than what we proved. The proof above showed us that every *LTL*-theory $<\Sigma,\Gamma>$ is included in $<\Sigma,\alpha_\Sigma^{-1}(c(\alpha_\Sigma(\Gamma)))>$ but, in fact, they are equal. That is, when translated back from its image in *CTL**, an *LTL*-theory does not gain any theorems. This result can be proved by noticing that every linear structure can be generated by a branching one, i.e. the natural transformation β consists of surjective mappings, which gives us the faithfulness of the right-adjoint as in Par. 7.2.11. We will generalise this result below, but it is important to realise that this means that the translation from *LTL* to *CTL** is "conservative", i.e. the representation of *LTL* in *CTL** is faithful.

Notice that, in the proof above, no use was made of the syntactic transformation itself. Only the fact that α is a natural transformation was used, which indicates that the relationship between *LTL* and *CTL** can be generalised to other institutions.

In order to perform the generalisation, let us first analyse what in the example above can be cast directly in categorical terms. The basic ingredients in our example were:

- A natural transformation $\alpha{:}\boldsymbol{gram}_{LTL}\rightarrow\boldsymbol{gram}_{CTL*}$.
- A natural transformation $\beta{:}\boldsymbol{mod}_{CTL*}\rightarrow\boldsymbol{mod}_{LTL}$.
- The invariance condition $M\models_\Sigma a_\Sigma(\phi)$ iff $b_\Sigma(M)\models_\Sigma\phi$.

These are exactly the ingredients found in institution morphisms [62] and institution maps [85].

7.4.10 Definition – Institution Morphisms

Let $\iota=<\boldsymbol{SIGN},\boldsymbol{gram},\boldsymbol{mod},\models>$ and $\iota'=<\boldsymbol{SIGN}',\boldsymbol{gram}',\boldsymbol{mod}',\models'>$ be institutions. An institution morphism $\rho{:}\iota\rightarrow\iota'$ is a triple $<\Phi,\alpha,\beta>$ where:

- $\Phi{:}\boldsymbol{SIGN}\rightarrow\boldsymbol{SIGN}'$ is a functor.
- $\alpha{:}\Phi;\boldsymbol{gram}'\rightarrow\boldsymbol{gram}$ is a natural transformation.
- $\beta{:}\boldsymbol{mod}\rightarrow\Phi;\boldsymbol{mod}'$ is a natural transformation.

such that the following property (the invariance condition) holds for any signature $\Sigma\in|\boldsymbol{SIGN}|$, $m\in|\boldsymbol{mod}(\Sigma)|$ and $\phi'\in\boldsymbol{gram}'(\Phi(\Sigma))$: $m\models_\Sigma\alpha_\Sigma(\phi')$ iff $\beta_\Sigma(m)\models'_{\Phi(\Sigma)}\phi'$. ♦

7.4.11 Definition – Institution Maps

Let $\iota=<\boldsymbol{SIGN},\boldsymbol{gram},\boldsymbol{mod},\models>$ and $\iota'=<\boldsymbol{SIGN}',\boldsymbol{gram}',\boldsymbol{mod}',\models'>$ be institutions. An institution map $\rho{:}\iota\rightarrow\iota'$ is a triple $<\Phi,\alpha,\beta>$ where:

- $\Phi{:}\boldsymbol{SIGN}\rightarrow\boldsymbol{SIGN}'$ is a functor.
- $\alpha{:}\boldsymbol{gram}\rightarrow\Phi;\boldsymbol{gram}'$ is a natural transformation.
- $\beta{:}\Phi;\boldsymbol{mod}'\rightarrow\boldsymbol{mod}$ is a natural transformation.

such that the following property (the invariance condition) holds for any signature $\Sigma\in|\boldsymbol{SIGN}|$, $m'\in|\boldsymbol{mod}(\Phi(\Sigma))|$ and $\phi\in\boldsymbol{gram}(\Sigma)$: $\beta_\Sigma(m')\models_\Sigma\phi$ iff $m\models'_{\Phi(\Sigma)}\alpha_\Sigma(\phi)$. ♦

7.4.12 Proposition

Through Pars. 7.4.4, 7.4.5 and 7.4.6 we have defined both a map *LTL→CTL** and a morphism *CTL*→LTL*. ♦

The fact that the relationship between the two institutions is based on the identity functor between their categories of signatures blurs the difference between the concepts of morphism and map. The existence of the two functors $\mathcal{T}$ and $\mathcal{U}$ between $\boldsymbol{THEO}_{LTL}$ and $\boldsymbol{THEO}_{CTL*}$ is also a consequence of the existence of a map and a morphism between the institutions.

7.4.13 Proposition

Let $\rho=<\Phi,\alpha,\beta>:\iota\rightarrow\iota'$ be an institution map. The functor Φ extends to $\boldsymbol{THEO}_{\iota}\rightarrow\boldsymbol{THEO}_{\iota'}$ by establishing $\Phi(<\Sigma,\Gamma>)=<\Phi(\Sigma),c(\alpha_{\Sigma}(\Gamma))>$. ♦

7.4.14 Proposition

Let $\rho=<\Psi,\alpha',\beta'>:\iota\rightarrow\iota'$ be an institution morphism. The functor Ψ extends to $\boldsymbol{THEO}_{\iota'}\rightarrow\boldsymbol{THEO}_{\iota}$ through $\Psi(<\Sigma',\Gamma'>)=<\Psi(\Sigma'),\alpha'_{\Sigma'}{}^{-1}(\Gamma')>$. ♦

We can now generalise the results on the adjunction between the categories of theories of two institutions.

7.4.15 Proposition

Let $\iota=<\boldsymbol{SIGN},\boldsymbol{gram},\boldsymbol{mod},\models>$ and $\iota'=<\boldsymbol{SIGN}',\boldsymbol{gram}',\boldsymbol{mod}',\models'>$ be institutions, $\rho=<\Phi,\alpha,\beta>:\iota\rightarrow\iota'$ an institution map and $<\Psi,\alpha',\beta'>:\iota'\rightarrow\iota$ a morphism such that Ψ is a right adjoint of Φ, and, for every $\Sigma\in|\boldsymbol{SIGN}|$, $\alpha_{\Sigma}=\boldsymbol{gram}(\eta_{\Sigma});\alpha'_{\Phi(\Sigma)}$ where η is the unit of the adjunction. Then,

1. The functor $\mathcal{U}:\boldsymbol{THEO}_{\iota'}\rightarrow\boldsymbol{THEO}_{\iota}$ induced by the institution morphism $<\Psi,\alpha',\beta'>$, is a right adjoint of the functor $\boldsymbol{THEO}_{\iota}\rightarrow\boldsymbol{THEO}_{\iota'}$ induced by the institution map $<\Phi,\alpha,\beta>$.
2. If each component of β' is surjective, i.e. if the institution morphism is sound in the sense of [62], then the units η_{Σ} are conservative.

Proof:

This is a direct generalisation of the proof of Proposition 7.4.9. ♦

In other words, adjunctions on signatures can be lifted to adjunctions of theories provided that the left adjoint is associated with a map and the right adjoint with a morphism of institutions. A compatibility result is required, $\alpha_{\Sigma}=\boldsymbol{gram}(\eta_{\Sigma});\alpha'_{\Phi(\Sigma)}$, to make sure that both the map and the morphism make essentially the same translations. Notice that the invariance condition relating α and β, automatically generates a similar property for β. The result on "conservative" representations of one formalism into another is also important: basically, it says that no new theorems arise when a theory is translated from one formalism to another.

This result shows that there is a very strong relationship between institution morphisms and maps, as suggested by the fact that they make use of essentially the same transformations between languages and models. The difference between them, which is evident in the directions taken by the transformations vis-à-vis the functor between the categories of signatures, can be explained more easily when we see that they correspond to the two directions of an adjunction. Note that the map takes the direction of the left adjoint, while the morphism takes the direction of the right adjoint. These directions are consistent with the accepted view of maps as providing representations and morphisms projections of one institution into another.

In fact, provided that there is an adjunction between the categories of signatures of two institutions, maps and morphisms between them can be defined, interchangeably, that provide adjunctions for the functor between the corresponding categories of theories.

7.4.16 Proposition

Let $\iota = \langle \mathbf{SIGN}, \mathbf{gram}, \mathbf{mod}, \models \rangle$ and $\iota' = \langle \mathbf{SIGN}', \mathbf{gram}', \mathbf{mod}', \models' \rangle$ be institutions.

1. If $\rho = \langle \Phi, \alpha, \beta \rangle : \iota \to \iota'$ is a map such that Φ has a right adjoint Ψ, then
   a. The triple $\langle \Psi, \alpha', \beta' \rangle$, where α' is the natural transformation defined by $\alpha'_{\Sigma'} = \alpha_{\Psi(\Sigma')} ; \mathbf{gram}'(\varepsilon_{\Sigma'})$ and β' is the natural transformation defined by $\beta'_{\Sigma'} = \mathbf{mod}'(\varepsilon_{\Sigma'}) ; \beta_{\Psi(\Sigma')}$, is an institution morphism $\iota' \to \iota$.
   b. The functor $\mathcal{T}: \mathbf{THEO}_{\iota} \to \mathbf{THEO}_{\iota'}$ induced by the map has a right adjoint: the functor $\mathcal{U}: \mathbf{THEO}_{\Gamma'} \to \mathbf{THEO}_{I}$ induced by the morphism $\mathcal{U}(\langle \Sigma', \Gamma' \rangle) = \langle \Psi(\Sigma'), \alpha_{\Psi(\Sigma')}^{-1}(\varepsilon_{\Sigma'}^{-1}(\Gamma')) \rangle$.
2. If $\langle \Psi, \alpha', \beta' \rangle : \iota' \to \iota$ is a morphism such that Ψ has a left adjoint Φ, then
   a. The triple $\langle \Phi, \alpha, \beta \rangle$, where α is the natural transformation defined by $\alpha_{\Sigma} = \mathbf{gram}(\eta_{\Sigma}) ; \alpha'_{\Phi(\Sigma)}$ and β is the natural transformation defined by $\beta_{\Sigma} = \beta'_{\Phi(\Sigma)} ; \mathbf{mod}(\eta_{\Sigma})$, is an institution map from $\iota \to \iota'$.
   b. The functor $\mathcal{U}: \mathbf{THEO}_{\iota'} \to \mathbf{THEO}_{\iota}$ induced by the morphism has a left adjoint: the functor $\mathcal{T}: \mathbf{THEO}_{\iota} \to \mathbf{THEO}_{\iota'}$ induced by the map $\mathcal{T}(\langle \Sigma, \Gamma \rangle) = \langle \Phi(\Sigma), \alpha'_{\Psi(\Sigma)} c(\eta_{\Sigma}(\Gamma))) \rangle$. ♦

7.5 Coordinated Categories

In this last section of the last chapter of Part II, we address one of the topics at the heart of the research programme on CommUnity, i.e. the subject of Part III: the formalisation of the separation of concerns that is known as "coordination". This provides a good justification for stopping our introduction to category theory here, because the reader will not need any more categorical "ammunition" to attack Part III.

An introduction to coordination was given in Sect. 5.2 as part of the motivation for studying the behaviour of functors in relation to universal constructions. The reader is invited to read it (once again) as well as, if possible, what I consider to be the best introduction to "coordination": Arbab's gem "What Do You Mean, Coordination?" [3]. The central idea of this research area is to investigate the extent up to which a given formalism can separate between the mechanisms that coordinate the interactions that are responsible for emergent behaviour from the description of what in systems is responsible for the computations that ensure the functionalities of the services that individual system components provide.

For instance, object-oriented systems do not go a long way in supporting that separation. Because interactions in object-oriented approaches are based on *identities* [73], in the sense that, through clientship, objects interact by invoking specific methods of specific objects (instances) to get something specific done, the resulting systems are too rigid to support the levels of agility required by the "just-in-time" binding mechanisms of (Web) services. Any change on the collaborations that an object maintains with other objects needs to be performed at the level of the code that implements that object and, possibly, of the objects with which the new collaborations are established. That is, as put in [98], feature calling is for interconnections what assembly language represents for computations.

On the contrary, interactions in a service-oriented approach should be based only on the description of what is required, thus decoupling the "what one wants to be done" from the "who does it". In the context of the societal metaphor that we used in the book, it is interesting to note that this shift from "object"-oriented to "service"-oriented interactions mirrors what has been happening in human society: more and more, business relationships are being established in terms of acquisition of services (e.g. 1000 Watt of lighting for your office) instead of products (10 lamps of 100 Watt each for the office).

Our introduction to Sect. 5.2 discloses most of the "secrets" of the mathematical characterisation that we started to develop in [35] as a systematic study of the nature and properties of the separation between "computation" and "coordination". Now that the reader has more categorical background, we can revisit the motivation that has been already delivered. Note that we shall systematically work with colimits just to fix a direction of the "component-of" relationship and use it consistently. However, those that are more accustomed to limits can simply switch the direction of the arrow, i.e. work in the opposite category. Summarising:

- We model this separation by a forgetful functor $\boldsymbol{int}$:$\boldsymbol{SYS} \rightarrow \boldsymbol{INT}$, where the category $\boldsymbol{SYS}$ stands for the representations (models, behaviours, specifications, programs, etc) of the components out of which systems can be put together. The category $\boldsymbol{INT}$ captures the "interfaces" through which interconnections between system components can be established.
- The functor $\boldsymbol{int}$ should be faithful (as in Par. 5.1.7) so that morphisms in $\boldsymbol{SYS}$ (the "component-of" relationship) do not induce more relationships

between components than those that can be captured through their underlying interfaces. That is, by taking into consideration the computational part, we should not get additional observational power over the external behaviour of systems. Using the terminology that we introduced in Chap. 6, ***SYS*** is concrete over ***INT***.

- Because we use diagrams for modelling configurations of complex systems and colimits to obtain emergent behaviour, ***int*** should lift colimits as defined in Par. 5.2.1. That is, when we interconnect system components in a (configuration) diagram, any colimit of the underlying diagram of interfaces establishes an interface for which a computational part exists that captures the joint behaviour of the interconnected components as given by the colimit of the original diagram. We have already mentioned that this property expresses (non)interference between computation and coordination. On the one hand, the computations assigned to the components cannot interfere with the viability (in the sense of the existence of a colimit) of the underlying configuration of interfaces. On the other hand, the computations assigned to the components cannot interfere in the calculation of the interface of the resulting system. For instance, we saw in Par. 6.1.22 that split fibre-(co)complete (co)fibrations lift limits.
- It is also clear that ***int*** should preserve colimits in the sense of Par. 5.2.1. That is, every interconnection of system components should be an interconnection of the underlying interfaces. In other words, computations should not make a configuration of system components "viable", in the sense that it admits a colimit, when the underlying configuration of interfaces is not. This is another form of the required "noninterference". Given that ***int*** is faithful, this means that all colimits in ***SYS*** are concrete as defined in Par. 6.1.9.

Lifting and preservation of colimits imply that any colimit in ***SYS*** can be computed by first translating the diagram to ***INT***, then computing the colimit in ***INT***, and finally lifting the result back to ***SYS***. We have already encountered this situation for ***PROC*** through the functor ***alph*** and for the category of theories (or presentations) ***THEO*** of any (π-)institutions through the functor ***sign***. In the case of processes, this means that the set of behaviours does not interfere with the interconnections; and in the case of theories, that interconnections are established just by name bindings.

Both examples allow us to illustrate another intuitive property of the separation that is not captured by those mentioned so far. Consider, for instance, processes. Taking pullbacks as the most basic form of interconnection, we can notice that the "middle" process through which we express the interconnection is "always" idle, i.e. has all possible behaviours. Indeed, the set of behaviours that is present in the middle process does not interfere either in the interconnection, which is expressed at the level of the alphabets, or in the calculation of the set of behaviours of the resulting process, which is defined through the intersection of the inverse images of the sets of behaviours of the other two component processes. Hence, there is a sort of "canonical" middle processes: those that are idle. Notice that their duals, the empty processes, do not make good middle processes because they do not admit any incoming morphisms.

The same happens with theories and theory presentations: the middle object in a pushout is always empty (or the closure of the empty set of axioms) because the theorems that result from the pushout are computed from the pushout of the signatures and the theorems of the other two components. This seems to be saying that the middle objects that we use for interconnecting components, be it for pullbacks or pushouts, are essentially interfaces, which makes all the sense from the point of view of the separation of coordination from computation. How can we express this property in categorical terms?

Basically, and taking the colimit approach as exemplified by, for instance, theories, what we want is to be able to assign to every interface C:***INT*** a component $s(C)$:***SYS*** such that, for every morphism $f:C \rightarrow \boldsymbol{int}(S)$, there is a morphism $g:s(C) \rightarrow S$ such that $\boldsymbol{int}(g)=f$. That is, we want every interface C to have a "realisation" as a system component $s(C)$ in the sense that, using C to interconnect a component S, which is achieved through a morphism $f:C \rightarrow \boldsymbol{int}(S)$, is tantamount to using $s(C)$ through any $g:s(C) \rightarrow S$ such that $\boldsymbol{int}(g)=f$.

Notice that, because ***int*** is faithful, there is only one such g, which means that f and g are essentially the same. That is, sources of morphisms in diagrams in ***SYS*** are essentially interfaces. We would use the dual property to characterise what happens with processes. Such a realisation is called a *discrete lift* in [1]. A functor ***int*** for which every object C:***INT*** admits a discrete lift is said to have *discrete structures*.

7.5.1 Definition – Discrete Lifts/Structures

Given a concrete category $\varphi:\boldsymbol{D} \rightarrow \boldsymbol{C}$, a *discrete lift* for c:***C*** is a ***D***-object d such that $\varphi(d)=c$ and, for every morphism $f:c \rightarrow \varphi(d')$, there is a morphism $g:d \rightarrow d'$ such that $\varphi(g)=f$. The functor (concrete category) is said to have *discrete structures* whenever every ***C***-object admits a discrete lift. ◆

The dual notion is called *indiscrete lift* and the functor (concrete category) is said to have *indiscrete structures*.

When ***int*** lifts and preserves colimits, this property allows us to replace every middle object in a configuration diagram by the discrete lift of the underlying interface: both diagrams will have the same colimits. For all practical purposes, this means that we can use more economical representations for configuration diagrams by showing only the interfaces of the middle objects that interconnect components.

It is easy to see that the indiscrete lift of a process alphabet $A_\perp$ is $<A_\perp,\boldsymbol{tra}(A_\perp)>$, and the discrete lift of a signature Σ is the theory $<\Sigma,c_\Sigma(\emptyset)>$.

They also admit their dual versions, i.e. signatures have indiscrete lifts (inconsistent theories) and alphabets have discrete lifts (empty processes), but these are not the ones that interest us for system configuration: they disable rather than enable interaction! Notice that the (in)discrete lifts are the objects involved in the (co)reflections that define the corresponding forgetful functors as (co)reflectors (see Pars. 7.3.9 and 7.3.10).

7.5.2 Proposition

Every concrete category φ:$\boldsymbol{D}$→$\boldsymbol{C}$ that has discrete structures is reflective, the reflections (i.e. the components of the unit) being identities. ♦

The proof of this result is immediate once one transcribes the definition of discrete lifts to diagrams:

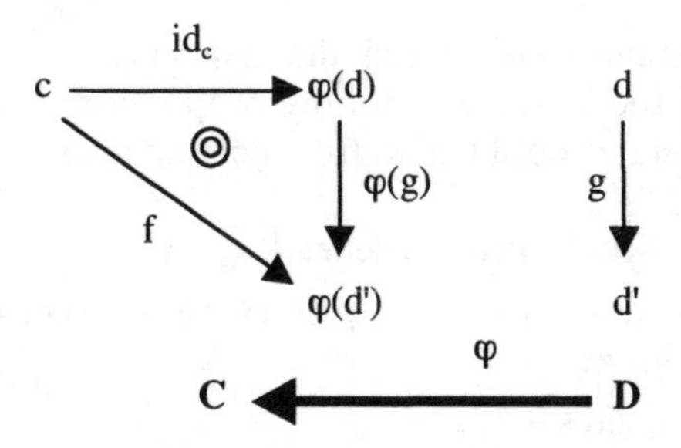

Notice that, φ being faithful, the coreflections are epis as shown in Par. 7.2.11. Actually, this is the property that allows us to replace the middle objects that perform interconnections by the discrete lifts of their underlying interfaces.

Indeed, denoting by ***sys*** the reflector of ***int***, every diagram

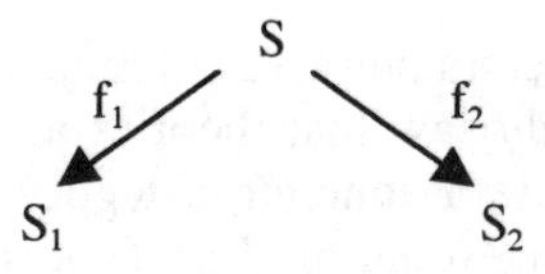

extends to

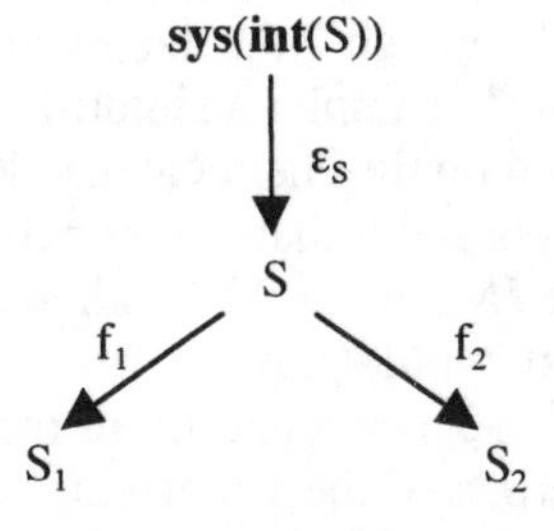

Both diagrams admit the same pushouts because, ε_S being epi, we have that $\varepsilon_S;f_1;g_1=\varepsilon_S;f_2;g_2$ implies $f_1;g_1=f_2;g_2$.

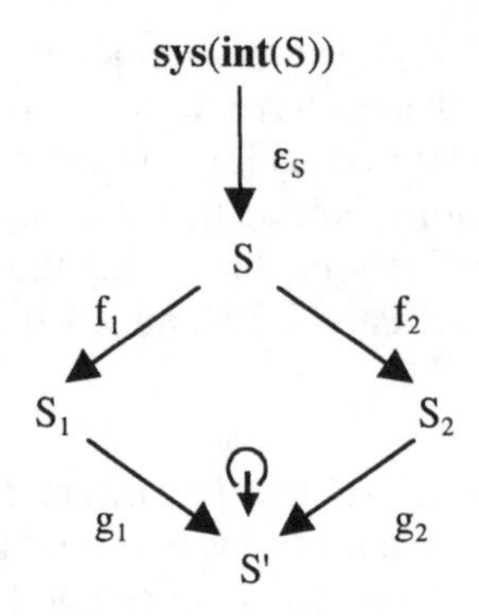

7.5.3 Exercise

Prove that both diagrams have, indeed, the same pushouts. ♦

We now have all the ingredients for our proposed characterisation of the formalisms that separate coordination from computation:

7.5.4 Definition – Coordinated Categories

A concrete category (faithful functor) φ:***D***→***C*** is said to be *coordinated* if:

- φ lifts colimits.
- φ has discrete structures.

In these circumstances, we also say that ***D*** is coordinated over ***C*** (via the functor φ). ♦

We have omitted the requirement on the preservation of colimits because it can be inferred from the two properties.

7.5.5 Exercise

Prove that coordinated functors preserve colimits. ♦

As examples, we already saw that theories and theory presentations of any (π)institution constitute a concrete category that is coordinated over their signatures, and that the (dual of) ***PROC*** is coordinated over (the dual of) the category of alphabets. In Part III, we study an example related to architectural description languages, the language CommUnity. We end this section with a "genuine" example: a simplified version of the language Gamma [11], which is based on the chemical reaction paradigm [16].

First, we would like to point out that the properties that characterise ***SYS*** as being coordinated over ***INT*** make ***SYS*** "almost" topological over ***INT***. To be topological [1], ***int*** would have to lift colimits uniquely, which would make the concrete category amnestic in the sense of Par. 6.1.4. As far as the algebraic properties of the underlying formalism are concerned, this is not a problem because every concrete category can be modified to produce an amnestic, concretely equivalent version. However, and although ***PROC*** is indeed amnestic, ***PRES***, for instance, is not, and neither is CommUnity.

This is the "closest" characterisation we have to a "classical" mathematical structure: topological categories abound in mathematics and other areas of computer science. In the areas related to software engineering, namely those in which one welcomes, or cannot avoid, "user intervention", one tends to work "up to isomorphism" more than "up to equality". In the case of the lifting of colimits, this means that there can be room for choosing between different, but isomorphic, system representations, for instance, alternative presentations of the same theory. One tends not to care whether a given conjunction ends up represented as $a \wedge b$ or $b \wedge a$.

7.5.6 Definition – Gamma Programs

A Gamma program P consists of:

- A signature Σ=<S,Ω,Π>, where S is a set of sorts, Ω is a set of operation symbols and Π is a set of relation symbols, representing the data types that the program uses.
- A set of reactions, each of which is of the form:

$$R \equiv X,\ t_1,\ ...,\ t_n \rightarrow t'_1,\ ...,\ t'_m \Leftarrow c$$

  where

  1. X is a set (of variables); each variable is typed by a data sort in S.
  2. $t_1,\ ...,\ t_n \rightarrow t'_1,\ ...,\ t'_m$ is the action of the reaction – a pair of sets of terms over X.
  3. c is the reaction condition – a proposition over X. ♦

An example of a Gamma program is the following producer of burgers and salads from, respectively, meat and vegetables.

```
PROD: sorts       meat, veg, burger, salad
      ops         vprod: veg→salad, mprod: meat→burger
      reactions   m:meat, m → mprod(m)
                  v:veg, v → vprod(v)
```

The parallel composition of Gamma programs, as defined in [11], is a program consisting of all the reactions of the component programs. Its behaviour is obtained by executing the reactions of the component programs in any order, possibly in parallel. This leads us to the following notion of morphism.

7.5.7 Definition – Morphisms of Gamma Programs

A morphism σ between Gamma programs P_1 and P_2 is a morphism between the underlying data signatures such that $\sigma(P_1) \subseteq P_2$, i.e. P_2 has more reactions than P_1. ♦

In order to illustrate system configuration in Gamma, let us consider that we want to interconnect the producer with the following consumer:

```
CONS: sorts        food, waste
      ops          cons: food →waste
      reactions    f:food, f → cons(f)
```

The interconnection of the two programs is based on the identification of the food the consumer consumes, that is, the interconnection is established between their data types. For instance, the coordination of the producer and the consumer based on meat is given as follows:

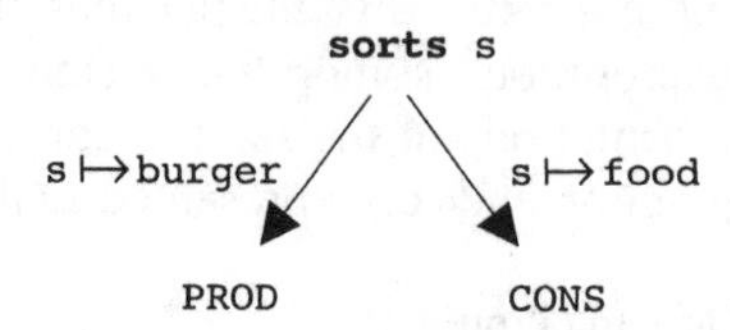

Gamma is, indeed, coordinated over the category of data types:

- The forgetful functor *dt* from Gamma programs to data types is faithful.
- Given any diagram in Gamma, a colimit $\sigma_i:(dt(P_i)\rightarrow\Sigma)i{:}\mathbf{I}$ of the corresponding diagram in the category of data types is lifted to the following colimit of programs $\sigma_i:(P_i\rightarrow<\Sigma, \cup\sigma_j(R_j)>)i{:}\mathbf{I}$.
- The discrete lift of a data type is the program with the empty set of reactions.

7.5.8 Exercise

Work out the full characterisation of the category of Gamma programs and prove that it is indeed coordinated over the data types. ♦

Part III

Applications

8 CommUnity

8.1 A Language for Program Design

CommUnity is a language similar to Unity [19] and Interacting Processes [48]. It was initially developed [44] to show how "programs" could fit into Goguen's categorical approach to general systems theory. Since then, the language and the design framework have been extended to provide a formal platform for testing ideas and experimenting techniques for the architectural design of open, reactive and reconfigurable systems.

One of the extensions that we have made to CommUnity since its original definition concerns the support for higher levels of design. At its earlier stages, the architecture of the system is normally given in terms of components that are not necessarily programs but abstractions of programs – called *designs* – that can be *refined* into programs in later steps of the development process. Designs may also account for components of the real world with which the software components are interconnected. Typically, such abstractions derive from requirements specified in some logic or other mathematical models of the behaviour of real-world components.

The goal of supporting abstraction is not only to address a typical stepwise approach to software *construction*, but to also address the definition of an architectural design layer that is close enough to the application domain for the evolution of the system to be driven directly as a reflection of the changes that occur in the domain. An important part of this evolution may consist of changes in the nature of components, with real-world components being replaced or controlled by software components, or software components being reprogrammed in another language.

The support for abstraction in CommUnity is twofold. On the one hand, designs account for what is usually called *underspecification*, i.e. they are structures that do not denote unique programs but collections of programs. On the other hand, designs can be defined over a collection of data types that do not correspond necessarily to those that are available in the final implementation platform. Therefore, there are two refinement procedures that have to be accounted for in CommUnity. First, the removal of underspecification from designs in order to define programs over the layer of abstraction defined by the data types that have been used. A second form of refinement consists of the reification of the data types in order to bring programs into the target implementation environment.

The choice of data types determines, essentially, the nature of the elementary computations that can be performed locally by the components, which are abstracted as operations on data elements. Such elementary computations also determine the granularity of the services that components can provide and, hence, the granularity of the interconnections that can be established at a given layer of abstraction. Nevertheless, data refinement is more concerned with the computational aspects of systems than with the coordination mechanisms that are responsible for interactions among system components. Because the support that category theory can provide to the specification of abstract data types is already well established and available in the literature [8, 28, 29, 76, 92], we do not address this aspect of CommUnity in depth but, rather, concentrate on the broader architectural aspects. We give more emphasis to the refinement of designs for a fixed choice of data types and omit any discussion on data refinement. An approach similar to the refinement calculus for actions systems [9] can also be followed to support data refinement in the context of reactive systems.

Given this, we assume a fixed collection of data types. In order to remain independent of any specific language for the definition of these data types, we take them in the form of a first-order algebraic specification. That is, we assume a data signature $<S,\Omega>$, where S is a set (of sorts) and Ω is a $S{*}{\times}S$-indexed family of sets (of operations), to be given together with a collection Φ of first-order sentences specifying the functionality of the operations.

A CommUnity design for a component over such a data type specification is of the form

```
design P is
out    out(V)
in     in(V)
prv    prv(V)
do
[]g∈sh(Γ)          g[D(g)]: L(g), U(g) → R(g)
[]g∈prv(Γ) prv     g[D(g)]: L(g), U(g) → R(g)
```

where

- V is a set (of *communication channels*). A communication channel (or, simply, channel) can be declared as *input*, *output* or *private*. Each channel v is typed with a sort $sort(v){\in}S$ that reflects the nature of the data that is exchanged through it. Input channels are used for reading data from the environment of the component. The component has no control of the values that are made available in such channels. Moreover, reading a value from an input channel does not "consume" it; the value remains available until the environment decides to replace it.

Output and private channels are controlled locally by the component, i.e. the values that, at any given moment, are available on these channels cannot be modified by the environment. Output channels allow the environment to read data produced by the component. Private channels support internal activity that does not involve the environment in any way. We use *loc(V)* to denote the union $prv(V) \cup out(V)$, i.e. the set of local channels.

In some earlier papers on CommUnity, we named the elements of *V* *variables* or *attributes*. The change to *channels* reinforces the idea that they are means that components have to communicate rather than store data. This is consistent with the black-box view of components that we intend to model, which should hide the representation of the state of components and provide only means for it to be observed.

Channels cater for asynchronous communication between components in the sense that reading and writing into a channel are independent operations. A value that is written on a channel will remain there, regardless of how many times it is read, until it is overwritten.

- Γ is a set (of *action names*). Actions can be either *private* or *shared* (for simplicity, we only declare those that are private). Private actions represent internal computations in the sense that their execution is uniquely under the control of the component. Shared actions are used for synchronous interactions with the environment, meaning that their execution is also under the control of the environment.

  The significance of naming actions will become obvious below. The idea is to provide points of rendez vous at which components can synchronise, for instance, as a means of ensuring that the right values are being exchanged through the channels.
- For each action name *g*, the following attributes are defined:
  - *D(g)* is a subset of *loc(V)* consisting of the local channels into which executions of the action can write. This is sometimes called the *write frame* of *g*. For simplicity, we will omit the explicit reference to the write frame when *R(g)* is a conditional multiple assignment (see below), in which case *D(g)* can be inferred from the assignments. Given a local channel *v*, we will denote by *D(v)* the set of actions *g* such that $v \in D(g)$, i.e. the actions that write into *v*.
  - *L(g)* and *U(g)* are two conditions such that $U(g) \supset L(g)$. These conditions establish an interval in which the enabling condition of any guarded command that implements *g* must lie. The condition *L(g)* is a lower bound for enabledness in the sense that it is implied by the enabling condition. Therefore, its negation establishes a *blocking* condition. On the other hand, *U(g)* is an upper bound in the sense that it implies the enabling condition, therefore establishing a *progress* condition. Hence, the enabling condition is fully determined only if *L(g)* and *U(g)* are equivalent, in which case we write only one.

- *R(g)* is a condition on *V* and *D(g)'*, where by *D(g)'* we denote the set of primed local channels from the write frame of *g*. As usual, primed channels account for references to the values that the channels display after the execution of the action. These conditions are usually a conjunction of implications of the form *pre⊃pos* where *pre* does not involve primed channels. They correspond to pre/postcondition specifications in the sense of Hoare. When *R(g)* is such that the primed version of each local channel in the write frame of *g* is fully determined, we obtain a conditional multiple assignment, in which case we use the notation that is normally found in programming languages. When *D(g)* is empty, *R(g)* is tautological, which we denote by *skip*.

CommUnity supports several mechanisms for underspecification. Actions may be underspecified in the sense that their enabling conditions may not be fully determined (subject to refinement by reducing the interval established by *L* and *U*), and their effects on the channels may also be undetermined. When, for every $g \in \Gamma$, *L(g)* and *U(g)* coincide, and the relation *R(g)* defines a conditional multiple assignment, then the design is called a *program* and the traditional notation for guarded commands is used.

Notice that a program with a nonempty set of input channels is *open* in the sense that its execution is only meaningful in the context of a configuration in which these inputs have been connected with local outputs of other components. The notion of configuration, and the execution of an open program in a given configuration, are discussed after Par. 8.2.4.

The behaviour of a closed program is as follows. At each execution step, one of the actions whose enabling condition holds of the current state is selected, and its assignments are executed atomically. Furthermore, private actions that are infinitely often enabled are guaranteed to be selected infinitely often. See [77] for a model-theoretic semantics of CommUnity.

Designs can be parameterised by data elements (sorts and operations) indicated after the name of the component (see an example below). These parameters are instantiated at configuration time, i.e. when a specific component needs to be included in the configuration of the system being built, or as part of the reconfiguration of an existing system.

As an example, consider the following parameterised design:

```
design buffer [t:sort, bound:nat] is
in     i:t
out    o:t
prv    rd: bool, b: list(t)
do     put: |b|<bound → b:=b.i
[] prv next: |b|>0∧¬rd → o:=head(b) || b:=tail(b) || rd:=true
[]     get: rd → rd:=false
```

The parameters of this design consist of the sort *t* of data elements that the buffer can handle and the capacity *bound* of the buffer. The buffer itself is defined over a list with elements of *t*. As already discussed, we are assuming that the data type *list* is available through an algebraic specification that includes the traditional operations such as |_| returning the current size of the list, *head(_)* returning the first element of the list, *tail(_)* returning the list after the first element, and _._ for appending an element to the end of the list.

This design is actually a (parameterised) program, and the traditional notation of guarded commands was used accordingly. Notice in particular that the reference to the write frame of the actions was omitted; it can be inferred from the multiple assignments that they perform. As already mentioned, because we are dealing with multiple assignments, the traditional notation involving the symbol := was used instead of the logical language over channels and their primed versions. In the case above, this corresponds to:

$R(put)$: $b'=b.i$
$R(next)$: $o'=head(b) \wedge b'=tail(b) \wedge rd'$
$R(get)$: $\neg rd'$

This program models a buffer with a limited capacity and a FIFO discipline. It can store, through the action *put*, messages of sort *t* received from the environment through the input channel *i*, as long as there is space for them. The buffer can also discard stored messages, making them available to the environment through the output channel *o* and the action *next*. Naturally, this activity is possible only when there are messages in store and the current message in *o* has already been read by the environment (which is modelled by the action *get* and the private channel *rd*).

In order to illustrate the ability of CommUnity to support higher-level component design, consider the design of a typical sender of messages.

```
design sender[t:sort] is
out    o:t
prv    rd: bool
do     prod[o,rd]: ¬rd,false→rd'
[]     send[rd]: rd,false→¬rd'
```

In this design, we are primarily concerned with the interaction between the sender and its environment, ignoring details of internal computations such as the production of messages. This is why the output channel *o* is included in the write frame of *prod*, but *R(prod)* does not place any constraint on how it is updated. Notice that the component *sender* cannot produce another message before the previous one has been processed: after producing a message, the sender expects an acknowledgement (modelled through the execution of *send*) to produce a new message.

In order to leave unspecified when and how many messages the *sender* will send and in which situations it will produce a new message, the progress conditions of *prod* and *send* are false. Furthermore, the discipline of production is also left completely unspecified: the action *prod* includes the output channel o in its write frame, but the design does not commit to any specific way of updating the values in this channel.

From a mathematical point of view, (instantiated) CommUnity designs are structures defined as follows.

8.1.1 Definition – Signatures and Designs

A *signature* in CommUnity is a tuple $<V,\Gamma,tv,ta,D>$, where

- V is an S-indexed family of mutually disjoint finite sets,
- Γ is a finite set,
- $tv: V \rightarrow \{out,in,prv\}$ is a total function,
- $ta: \Gamma \rightarrow \{sh,prv\}$ is a total function,
- $D: \Gamma \rightarrow 2^{loc(V)}$ is a total function.

A *design* in CommUnity is a pair $<\theta,\Delta>$, where $\theta=<V,\Gamma,tv,ta,D>$ is a signature and Δ, the body of the design, is a tuple $<R,L,U>$, where:

- R assigns to every action $g \in \Gamma$, a proposition over $V \cup D(g)'$,
- L and U assign a proposition over V to every action $g \in \Gamma$. ♦

The reader who is familiar with parallel program design languages or earlier versions of CommUnity may have noticed the absence of initialisation conditions. They are not included in CommUnity designs because they are part of the configuration language of CommUnity, not the parallel program design language. That is, we take initialisation conditions as part of the mechanisms that relate to the building and management of configurations out of designs, not of the construction of designs themselves.

8.2 Interconnecting Designs

So far, we have presented the primitives for the design of individual components, which are another variation on guarded commands, albeit with some "twists" of originality such as the use of an interval as a specification for the enabling conditions of commands. The main distinguishing features of CommUnity are those that concern design "in the large", i.e. the ability to design large systems from simpler components.

The model of interaction between components in CommUnity is based on action synchronisation and the interconnection of input channels of a component with output channels of other components. These are standard means of interconnecting software components. What distinguishes CommUnity from other parallel program design languages is the fact that such interactions between components have to be made explicit by providing the corresponding name bindings. Indeed, parallel program design languages normally leave such interactions implicit by relying on the use of the same names in different components. In CommUnity, names are local to designs. This means that the use of the same name in different designs is treated as being purely accidental and, hence, expresses no relationship between the components.

In CommUnity, name bindings are established as relationships between the signatures of the corresponding components, matching channels and actions of these components. These bindings are made explicit in configurations. A configuration determines a diagram containing nodes labelled with the components that are part of the configuration. Name bindings are represented as additional nodes representing the actual interactions, and edges labelled with the projections that map each interaction to the signatures of the corresponding components.

For instance, a configuration in which the messages from a *sender* component are sent through a bounded buffer is defined through the following diagram:

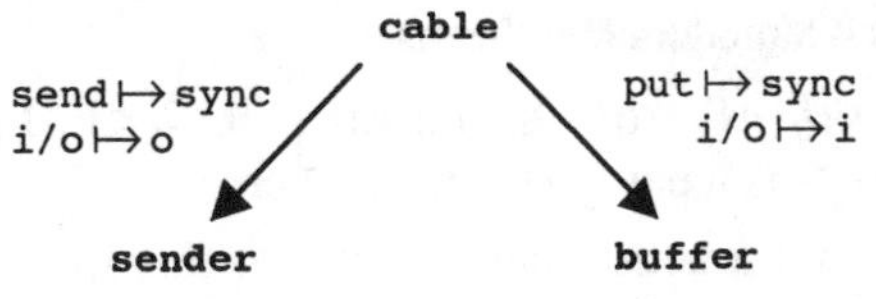

The node labelled *cable* is the representation of the set of bindings. It stands for the following design:

```
design cable[t:sort] is
in     i/o:t
do     sync: true,false → skip
```

By using the word "cable" we mean to suggest analogies with the use of physical cables for interconnecting mechanical or electrical components. Because, as we have seen, channels and action names are typed and classified in different categories, not every pair of names is a valid binding. To express the rules that determine valid bindings, it is convenient to define *cable* as a component itself (just like electric cables are made of coloured wires so that we know what should be connected to what). Hence, in the case above, *cable* consists of an input channel *i/o* to model the medium through which data is to be transmitted between the sender and the buffer, and a shared action *sync* for the two components to synchronise in order to transmit the data. Because, as we have already mentioned, names in CommUnity are local, the identities of the shared input channel and the shared action in *cable* are not relevant; they are just placeholders for the projections to define the relevant bindings.

The bindings themselves are established through the labels of the edges of the diagram. In the case above, the input channel of *cable* is mapped to the output channel *o* of *sender* and to the input channel *i* of *buffer*. This establishes an i/o-interconnection between *sender* and *buffer*. On the other hand, the actions *send* of *sender* and *put* of *buffer* are mapped to the shared action of *cable*. This defines that *sender* and *buffer* must synchronise each time either of them wants to perform the corresponding action. The fact that the mappings on action names and on channels go in opposite directions is discussed after Par. 8.2.1.

Notice that *sync* does not perform any activity: it just provides the place for the rendez vous between the sender and the buffer to take place. This is in analogy with cables that are completely neutral, i.e. they do not interfere with the computations that are going on in the components. Hence, cables are, essentially, signatures. This observation is formalised in Pars. 8.2.4 and 8.2.5.

The arrows that we are using to define interconnections between components are also mathematical objects: they are examples of signature morphisms.

8.2.1 Definition – Signature Morphisms

A morphism $\sigma:\theta_1\rightarrow\theta_2$ of signatures $\theta_1=<V_1,\Gamma_1,tv_1,ta_1,D_1>$ and $\theta_2=<V_2,\Gamma_2,tv_2,ta_2,D_2>$ is a pair $<\sigma_{ch},\sigma_{ac}>$, where

- σ_{ch}: $V_1\rightarrow V_2$ is a total function satisfying:
  1. $sort_2(\sigma_{ch}(v))=sort_1(v)$ for every $v\in V_1$.
  2. $\sigma_{ch}(o)\in out(V_2)$ for every $o\in out(V_1)$.
  3. $\sigma_{ch}(i)\in out(V_2)\cup in(V_2)$ for every $i\in in(V_1)$.
  4. $\sigma_{ch}(p)\in prv(V_2)$ for every $p\in prv(V_1)$.
- σ_{ac}: $\Gamma_2\rightarrow\Gamma_1$ is a partial mapping satisfying for every $g\in\Gamma_2$ s.t. $\sigma_{ac}(g)$ is defined:
  5. If $g\in sh(\Gamma_2)$, then $\sigma_{ac}(g)\in sh(\Gamma_1)$.
  6. If $g\in prv(\Gamma_2)$, then $\sigma_{ac}(g)\in prv(\Gamma_1)$.
  7. $\sigma_{ch}(D_1(\sigma_{ac}(g)))\subseteq D_2(g)$.
  8. σ_{ac} is total on $D_2(\sigma_{ch}(v))$ and $\sigma_{ac}(D_2(\sigma_{ch}(v)))\subseteq D_1(v)$, $v\in loc(V_1)$. ♦

Signature morphisms represent more than the projections that arise from name bindings as illustrated above. A morphism $\sigma$from θ_1 to θ_2 is intended to support the identification of a way in which a component with signature θ_1 is embedded in a larger system with signature θ_2. This justifies the various constructions and constraints in the definition.

The function σ_{ch} identifies for each channel of the component the corresponding channel of the system. The partial mapping σ_{ac} identifies the action of the component that is involved in each action of the system, if ever. The fact that the two mappings go in opposite directions is justified as follows. Actions of the system constitute synchronisation sets of actions of the components. Because not every component is necessarily involved in every action of the system, the action mapping is partial. On the other hand, because each action of the component may participate in more than one synchronisation set, but each synchronisation set cannot induce internal synchronisations within the components, the relationship between the actions of the system and the actions of every component is functional from the former to the latter. Hence, actions are dealt with in the category ***PAR*** of partial functions. As seen in Par. 7.1.12, this category is equivalent to the category that we used for modelling alphabets of processes, meaning that the intuitions that we developed on the way universal constructions capture composition can be used for CommUnity as well.

Input/output communication within the system is not modelled in the same way as action synchronisation. Synchronisation sets reflect parallel composition, whereas with i/o-interconnections we wish to identify communication channels declared in the components. This means that, in the system, channels should be identified rather than paired. This is why mappings on channels and mappings on actions go in opposite directions. As a result, the mathematical semantics of configuration diagrams as defined after Par. 8.2.3 induces fibred products of actions (synchronisation sets) and amalgamated sums of channels (equivalence classes of connected channels).

The constraints are concerned with typing. Sorts associated with channels have to be preserved, but, in terms of their classification, input channels of a component may become output channels of the system in the sense that, as a default, they should remain open for communication with other components. In most languages for parallel design, the default is to hide the communication, which in CommUnity would correspond to classify the resulting channel as being private. In our opinion, closing/hiding the channel should not be a default but a design decision that should be performed explicitly. Hence, in CommUnity, mechanisms for internalising communication can be applied but they are not the default in a configuration. The last two conditions on write frames (7 and 8) imply that actions of the system in which a component is not involved cannot have local channels of the component in their write frame. That is, change within a component is completely encapsulated in the structure of actions defined for the component.

Given the ingredients out of which signatures are assembled, the proof of the following result is purely routine and is left as an exercise.

8.2.2 Proposition – Category of Signatures

Signatures in CommUnity together with their morphisms constitute a category that we shall denote by ***c-SIGN***. ♦

The notation can be simplified and made more friendly by adopting features that are typical of languages for configurable distributed systems like [87]. For instance, the interconnection defined before can be described as follows:

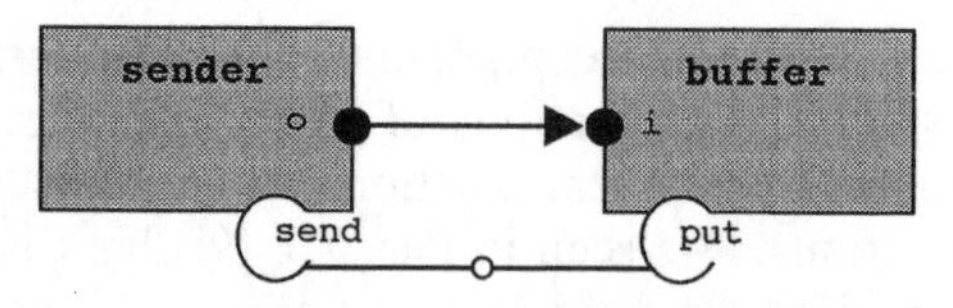

The notation should be self-explanatory. Components are represented through boxes, their channels through bullets and their actions through circles. Normally we only depict the actions and channels involved in the configuration. Hence, as discussed below, private actions and channels in particular do not figure up.

Interconnections, i.e. name bindings, are still represented explicitly, but, instead of being depicted as a component, the cable is now represented, perhaps more intuitively, in terms of arcs that connect channels and actions directly. The direction of the arcs is from output to input channels. Configurations in this notation are easily translated into categorical diagrams by transforming the interconnections into channels and morphisms, something that, again, we abstain from formalising in this book.

So far, we explained how interconnections between components can be established at the level of the signatures of their designs. It remains to explain how the corresponding designs are interconnected, i.e. what is the semantics of the configuration diagram once designs are taken into account. For that purpose, we need to extend the notion of morphism from signatures to designs.

8.2.3 Definition/Proposition – Design Morphisms

A morphism $\sigma{:}P_1 \rightarrow P_2$ of designs $P_1{=}{<}\theta_1,\Delta_1{>}$ and $P_2{=}{<}\theta_2,\Delta_2{>}$ consists of a signature morphism $\sigma{:}\theta_1 \rightarrow \theta_2$ such that, for every $g \in \Gamma_2$ for which $\sigma_{ac}(g)$ is defined:

1. $\Phi \vdash (R_2(g) \supset \underline{\sigma}(R_1(\sigma_{ac}(g))))$.
2. $\Phi \vdash (L_2(g) \supset \underline{\sigma}(L_1(\sigma_{ac}(g))))$.
3. $\Phi \vdash (U_2(g) \supset \underline{\sigma}(U_1(\sigma_{ac}(g))))$.

where Φ is the axiomatisation of the data type specification, $\vdash$ denotes validity in the first-order sense, and $\underline{\sigma}$ is the extension of σ to the language of expressions and conditions as discussed in Par. 6.5.3 for institutions in general. We normally simplify the notation and overload the use of σ in place of $\underline{\sigma}$ as mentioned in Par. 6.5.4. Designs and their morphisms constitute a category ***c-DSGN***. This category is concrete over ***c-SIGN*** through the obvious forgetful functor. ◆

A morphism $\sigma{:}P_1 \rightarrow P_2$ identifies a way in which P_1 is "augmented" to become P_2 so that P_2 can be considered as having been obtained from P_1 through the superposition of additional behaviour, namely the interconnection of one or more components. The conditions on the actions require that the computations performed by the system reflect the interconnections established between its components. Condition 1 reflects the fact that the effects of the actions of the components can only be preserved or made more deterministic in the system. This is because the other components in the system cannot interfere with the transformations that the actions of a given component make on its state, except possibly by removing some of the underspecification present in the component design.

Conditions 2 and 3 allow the bounds that the component design specifies for the enabling of the action to be strengthened but not weakened. Strengthening of the lower bound reflects the fact that all the components that participate in the execution of a joint action have to give their permission for the action to occur. On the other hand, it is clear that progress for a joint action can only be guaranteed when all the designs of the components involved can locally guarantee so.

This notion of morphism captures what in the literature on parallel program design is called "superposition" or "superimposition" [10, 19, 48, 72]. See [44] for a categorical formalisation of different notions of superposition and their algebraic properties.

The semantics of configurations is given by a categorical construction: the colimit of the underlying diagrams. As explained in Chap. 4, taking the colimit of a diagram collapses the configuration into an object by internalising all the interconnections, thus delivering a design for the system as a whole. Furthermore, the colimit provides a morphism σ_i from each component design P_i in the configuration into the new design (that of the system) – the edge of the cocone at P_i as seen in Par. 4.4.1. Each such morphism is essential for identifying the corresponding component within the system because the construction of the new design typically requires that the features of the components be renamed in order to account for the interconnections.

Again, given the nature of the "ingredients", it is not difficult to understand how colimits of designs work: because channels are handled through total functions, colimits amalgamate channels as seen in Par. 4.3.2. Because actions are handled as partial functions in the opposite direction, i.e. in the dual category, colimits operate on actions as limits and compute fibred products as explained in Par. 4.3.8. For instance, in the case of actions, the colimit represents every synchronisation set $\{g_1,\ldots,g_n\}$ of actions of the components, as defined through the interconnections, by a single action $g_1\|\ldots\|g_n$ whose occurrence captures the joint execution of the actions in the set (recall the discussions around Pars. 4.3.8 and 4.4.11). Because, as mentioned after Par. 4.2.6, limits perform conjunctions of logical conditions, the transformations performed by a joint action are specified by the conjunction of the specifications of the local effects of each of the synchronised actions:

$$R(g_1\|\ldots\|g_n)=\sigma_1(R(g_1))\wedge\ldots\wedge\sigma_n(R(g_n))$$

where the σ_i are the morphisms that connect the components to the system (the edges of the cocone). The bounds on the guards of joint actions are also obtained through the conjunctions of the bounds specified by the components, i.e.

$$L(g_1\|\ldots\|g_n)=\sigma_1(L(g_1))\wedge\ldots\wedge\sigma_n(L(g_n))$$
$$U(g_1\|\ldots\|g_n)=\sigma_1(U(g_1))\wedge\ldots\wedge\sigma_n(U(g_n))$$

Finally, because morphisms require inclusions of write frames, the write frame of a joint action is given by the union of the (translations of) the write frames of the component actions. This way of computing colimits derives from the strong algebraic properties of the category of designs.

8.2.4 Proposition

The forgetful functor ***c-sign*** that maps CommUnity designs to the corresponding signatures defines ***c-DSGN*** as a category coordinated over ***c-SIGN*** as defined in Par. 7.5.4. We shall call a *cable* the discrete lift of a signature: given a signature θ, the corresponding cable ***dsgn***(θ) has θ for signature and, for every action g, $R(g)$, $L(g)$ and $U(g)$ are all *true*, which we normally denote by *skip*. ◆

Summarising, colimits in CommUnity capture a generalised notion of parallel composition in which the designer makes explicit what interconnections are used between components. Because the category of designs is coordinated over signatures, all interconnections can be performed through cables, i.e. they do not involve the computational part of components, only their "interfaces" – i/o communication through channels and rendez vous through action synchronisation. We can see this operation as a generalisation of the notion of superimposition as defined in [48].

The colimit of the configuration, when it returns a closed program, can also be used for providing an operational semantics for the system thus configured. As explained in Sect. 8.1, at each execution step, any action whose guard is true can be executed, with the guarantee that private actions that are infinitely often enabled are selected infinitely often. Because actions of the system are synchronisation sets of actions of the components, the evaluation of the guard of the chosen action can be performed in a distributed way by evaluating the guards of the component actions in the synchronisation set. According to the semantics that we have just given, the joint action will be executed iff all the local guards evaluate to *true*. The execution of the multiple assignment associated with the joint action can also be performed in a distributed way by executing each of the local assignments. What is important is that the atomicity of the execution is guaranteed, i.e. the next system step should only start when all local executions have completed, and the i/o-communications should be implemented so that every local input channel is instantiated with the correct value – that which holds before any execution starts (synchronicity).

Hence, the colimit of the configuration diagram should be seen as an abstraction of the actual distributed execution that is obtained by coordinating the local executions according to the interconnections, rather than the program that is going to be executed as a monolithic unit. The fact that the computational part, i.e. the one concerned with the execution of the actions on the state, can be separated from the coordination aspects is therefore an essential property for guaranteeing that the operational semantics is compositional on the structure of the system as given through its configuration.

Not every diagram of designs reflects a meaningful configuration. For instance, it does not make sense to interconnect components by connecting two output channels because it creates conflicts among the actions that output to these channels. Such constraints are not "structural": they cannot be captured by morphisms between designs because they concern general interconnections, not just the component-of relationship. In other words, the constraints are not technical but, rather, methodological.

8.2.5 Definition/Proposition – Well-Formed Configurations

Let ***chan*** be the forgetful functor from ***c-DSGN*** to ***SET*** that maps designs to their underlying sets of channels. A *configuration* is a finite diagram **dia**:$\boldsymbol{I}\rightarrow\boldsymbol{c\text{-}DSGN}$ together with a subset J of $|\boldsymbol{I}|$ (the nodes that represent the components being interconnected) such that:

1. For any $f{:}i\rightarrow j$ in $\boldsymbol{I}$, either $i=j$ and $f=id_i$, or $j\in J$, $i\notin J$ and $\boldsymbol{dia(i)}$ is a cable.
2. For every $i\in|\boldsymbol{I}|\backslash J$ s.t. $\boldsymbol{dia(i)}$ is a cable, there exist distinct nodes $j,k\in|\boldsymbol{I}|$ with morphisms $f{:}i\rightarrow j$ and $g{:}i\rightarrow k$.
3. If $\{\mu_i{:}\boldsymbol{chan(dia(i))}\rightarrow V{:}\ i\in|\boldsymbol{I}|\}$ is a colimit of $\boldsymbol{dia;chan}$ then, for every $v\in V$, there exists at most one $i\in|\boldsymbol{I}|$ s.t. $\mu_i^{-1}(v)\cap out(V_{\boldsymbol{dia(i)}})\neq\emptyset$, and for such i, $\mu_i^{-1}(v)\cap out(V_{\boldsymbol{dia(i)}})$ is a singleton.

A configuration ***dia*** is *well-formed* if for every $i\in|\boldsymbol{I}|\backslash J$ s.t. $\boldsymbol{dia(i)}$ is a cable, $\boldsymbol{dia(i)}$ has neither private actions nor private channels. ♦

Condition 1 states that the elementary interconnections are established through cables. Condition 2 ensures that a configuration diagram does not include cables that are not used. Finally, condition 3 prevents the identification of output channels. The explicit reference to the subset J of components is necessary because the distinction between nodes that are being used as channels and as components is a pragmatic, not formal one. It is possible that, in a given configuration, a node is intended to represent a component, but, because it is still totally underspecified, it is the discrete lift of a signature, i.e. what we have called a cable.

Well-formed configurations are such that private actions and channels are not involved in the interconnections, i.e. they support the intuitive semantics we gave in Sect. 8.1 according to which private channels cannot be read by the environment and that the execution of shared actions is uniquely under the control of the component.

An example of a more complex configuration is given below. It models the interconnection between a user and a printer via a buffer.

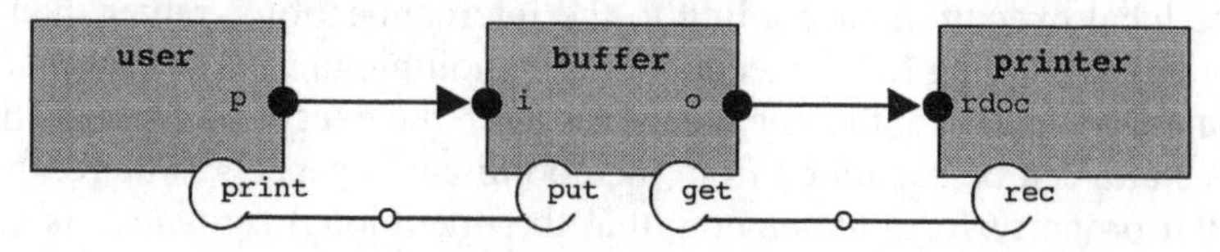

The user produces files that it stores in the private channel *w*. It can then convert them either to Postscript or Pdf formats, after which it makes them available for printing in the output channel *p*.

```
design user is
out     p: ps+pdf
prv     s,t: bool, w: Lowtex
do      work[w,s,t]: ¬t,false → t'
[]      pr_ps: ¬s∧t,false → p:=ps(w) || s:=true
[]      pr_pdf: ¬s∧t,false → p:=pdf(w) || s:=true
[]      print: s → s:=false || t:=false
```

The printer copies the files it downloads from the input channel *rdoc* into the private channel *pdoc*, after which it prints them.

```
design printer is
in      rdoc: ps+pdf
prv     busy: bool, pdoc: ps+pdf
do      rec: ¬busy → pdoc:=rdoc || busy:=true
[] prv  end_print: busy → busy:=false
```

The configuration connects the user to the printer via a buffer as expected. The user "prints" by placing the file in the buffer: this is achieved through the synchronisation set *{print,put}* and the i/o-interconnection *{p,i}*. The printer downloads from the buffer the files that it prints: this is achieved through the synchronisation set *{get,rec}* and the i/o-interconnection *{o,rdoc}*.

The design of the system that results from the colimit of the configuration diagram contains two channels that account for the two i/o-interconnections *{p,i}* and *{o,rdoc}*, together with the private channels of the components. At the level of its actions, it generates the following shared actions (synchronisation sets):

print|put, get|rec

as required by the interconnections, and

work, pr_ps, pr_pdf, work|get|rec, pr_ps|get|rec, pr_pdf|get|rec

which reflect the concurrent executions that respect the interconnections. No other shared actions are possible because of the synchronisation requirements imposed on the components.

8.3 Refining Designs

The notion of morphism defined in the previous section does not capture a refinement relation in the sense that it does not ensure that any implementation of the target provides an implementation for the source. For instance, it is easy to see that morphisms do not preserve the interval assigned to the guard of each action. Given that the aim of the defined morphisms was to capture the relationship that exists between systems and their components, this is hardly surprising.

The same holds in languages such as CSP [69]: in the failure or ready semantics, parallel composition does not induce refinement. On the one hand, $P||Q$ is not necessarily a refinement of P. On the other hand, refinement cannot always be expressed as the result of a parallel composition; P may refine Q and, yet, there may not exist a Q' such that P is $Q||Q'$.

Because refinement is an important dimension in structuring development, it is natural that we investigate ways of supporting it in a categorical setting. This would be especially useful for analsying the way refinement and composition can work together. A notion of morphism can indeed be defined that captures a refinement relation for CommUnity designs.

8.3.1 Definition/Proposition – Refinement Morphisms

A refinement morphism $\sigma:P_1 \rightarrow P_2$ between two designs $P_1=<\theta_1,\Delta_1>$ and $P_2=<\theta_2,\Delta_2>$ is a pair $<\sigma_{ch},\sigma_{ac}>$ satisfying:

- $\sigma_{ch}:V_1 \rightarrow V_2$ is a total function satisfying, for every $v \in V_1$, $o \in out(V_1)$, $i \in in(V_1)$, $p \in prv(V_1)$:
  1. $sort_2(\sigma_{ch}(v))=sort_1(v)$.
  2. $\sigma_{ch}(o) \in out(V_2)$.
  3. $\sigma_{ch}(i) \in in(V_2)$.
  4. $\sigma_{ch}(p) \in prv(V_2)$.
  5. $\sigma_{ch}\downarrow(out(V_1) \cup in(V_1))$ is injective.
- $\sigma_{ac}:\Gamma_2 \rightarrow \Gamma_1$ is a partial mapping satisfying, for every $v \in loc(V_1)$:
  6. σ_{ac} is total on $D_2(\sigma_{ch}(v))$.
  7. $\sigma_{ac}(D_2(\sigma_{ch}(v))) \subseteq D_1(v)$.
- For every $g \in \Gamma_2$ s.t. $\sigma_{ac}(g)$ is defined:
  8. If $g \in sh(\Gamma_2)$ then $\sigma_{ac}(g) \in sh(\Gamma_1)$.
  9. If $g \in prv(\Gamma_2)$ then $\sigma_{ac}(g) \in prv(\Gamma_1)$.
  10. If $g \in sh(\Gamma_1)$ then $\sigma_{ac}^{-1}(g) \neq \emptyset$.
  11. $\sigma_{ch}(D_1(\sigma_{ac}(g))) \subseteq D_2(g)$.
  12. $\Phi \vdash (R_2(g) \supset \underline{\sigma}(R_1(\sigma_{ac}(g))))$.
  13. $\Phi \vdash (L_2(g) \supset \underline{\sigma}(L_1(\sigma_{ac}(g))))$.

- For every $g_1 \in \Gamma_1$,

  14. $\Phi \vdash (\underline{\sigma}(U_1(g_1)) \supset \bigvee_{\sigma_{ac}(g_2)=g_1} U_2(g_2))$.

As in Par. 8.2.3, we denote by Φ the axiomatisation of the data type specification, and by $\vdash$ the validity relation of first-order logic; $\underline{\sigma}$ is, again, the extension of σ to the language of expressions and conditions.

Designs and their refinement morphisms constitute a category ***r-DSGN***. This category is concrete over ***c-SIGN*** through the functor ***r-sign*** that, like ***c-sign***, projects designs to their signatures. ♦

A refinement morphism identifies a way in which a design P_1 (its source) is refined by a more concrete design P_2 (its target). The function σ_{ch} identifies, for each channel of P_1, the corresponding channel of P_2. Notice that, contrary to what happens with design morphisms (recall Par. 8.2.3), refinement does not change the border between the system and its environment, and hence, input channels can no longer be mapped to output channels (3). This is also why the mapping is required to be injective on input and output channels (5): identifying channels is a configuration operation to be achieved through interconnections, not a refinement step.

As for design morphisms, refinement morphisms are required to preserve the sorts of channels (1). As discussed at the beginning of this chapter, data refinement is a dimension that, for simplicity, we are deliberately ignoring in the book. A more general notion of refinement can be given by mapping channels to terms defined over the language of the data types enriched by channels as constants. See [37] for details.

The mapping σ_{ac} identifies for each action g of P_1 the set $\sigma_{ac}^{-1}(g)$ of actions of P_2 that implements g. This set is a menu of refinements that is made available for implementing action g; different choices can be made at different states to take advantage of the structures available at the more concrete design level. This menu can be empty for private actions, i.e. one may choose not to implement the private actions of the more abstract design. Because private actions do not intervene in interconnections, what is important is that the overall behaviour of the component as made observable through shared actions and output channels be implemented. This is also why every shared action has to be implemented (10); again, such actions model interaction between the component and its environment, and refinement should not interfere with the border between them.

The actions for which σ_{ac} is left undefined (the new actions) and the channels which are not involved in $\sigma_{ch}(V_1)$ (the new channels) introduce more detail in the description of the component. As for the "old actions", the interval defined by their blocking and progress conditions (in which the enabling condition of any implementation must lie) must be preserved or reduced (14 and 15). This is intuitive because refinement, pointing in the direction of implementations, should reduce underspecification. Hence, the lower bound cannot be weakened (14), and contrary to design morphisms, the upper bound cannot be strengthened (15). This is also the reason why the effects of the actions of the more abstract design are required to be preserved or made more deterministic (13).

Notice that the forgetful functors ***r-sign*** and ***c-sign*** are essentially the same; they are only formally different because their sources are not the same category. Indeed, the only difference between design and refinement morphisms at the level of signatures is on the additional properties that refinement morphisms need to satisfy: 3, 5 and 10.

As an example, it is easy to see that *sender* is refined by *user* via the refinement morphism η:*sender*→*user* defined by

$$\eta_{ch}(o)=p,\ \eta_{ch}(rd)=(s)$$
$$\eta_{ac}(pr\_ps)=\eta_{ac}(pr\_pdf)=prod,\ \eta_{ac}(print)=send$$

In *user*, the production of messages (to be sent) is modelled by any of the actions *pr_ps* and *pr_pdf*; the messages are made available in the output channel *p*. Notice that the production of messages, which was left unspecified in *sender,* is completely defined in *user*: it corresponds to the conversion of the files stored in *w* to ps or pdf formats.

In the simplified graphical notation that we have been using, refinement is represented through patterned arrows. Notice that, for depicting refinement, all the actions and channels of the source should be represented, including private ones.

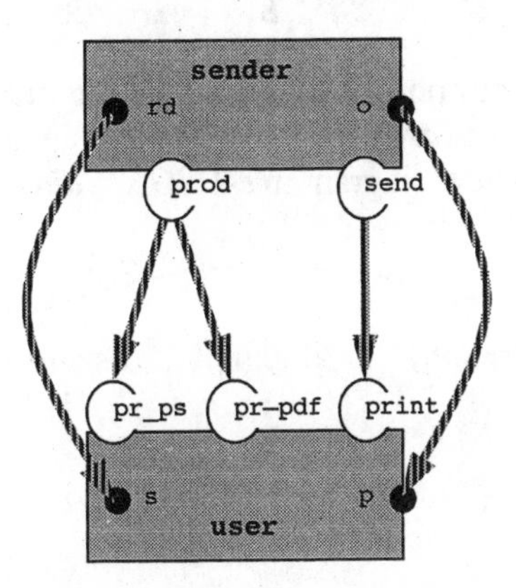

Summarising what we have built so far, we have two categories ***c-DSGN*** (defined in Par. 8.2.3) and ***r-DSGN*** (defined in Par. 8.3.1), both over the same notion of object – CommUnity design – but with different notions of morphism. That is, they capture different aspects of their social lives: one tells us about their ability to relate with other designs at the same level of abstraction, and the other about the way they can be made more "concrete" by reducing underspecification. Furthermore, ***c-DSGN*** is coordinated over ***c-SIGN*** through the functor ***c-sign*** as seen in Par. 8.2.4. We are now interested in the way interconnection relates to refinement.

The first important property relates to the requirement that refinement should not be based on the specificities of each particular design as far its ability to be interconnected to other designs is concerned. In other words, refinement morphisms should be such that designs that are isomorphic in ***c-DSGN*** refine, and are refined exactly by, the same designs.

8.3.2 Proposition

Every isomorphism in ***c-DSGN*** defines an isomorphism in ***r-DSGN***. ◆

Another crucial property is in the ability to refine a complex system from refinements of its individual components. Consider a well-formed configuration ***dia*** of a system with components $S_1,...,S_n$ and refinement morphisms $\eta_i:S_i \to S'_i$: $i \in 1..n$,

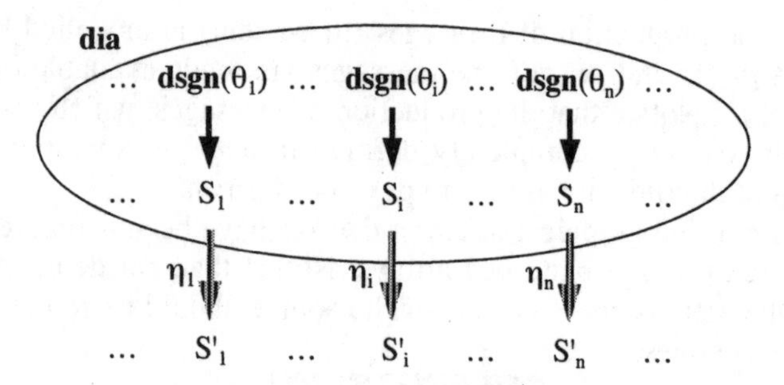

By composing the morphisms η_i with those in ***dia*** that originate in cables (designs of the form $\boldsymbol{dsgn(\theta)}$, where θ is a signature) and have the S_i as targets, we obtain a new diagram in ***c-DSGN*** – $\boldsymbol{dia+(\eta_i)}$:

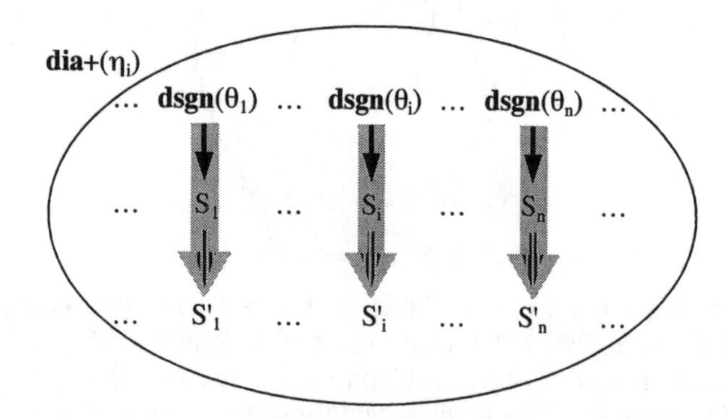

This composition is possible because, ***c-DSGN*** being coordinated over ***c-SIGN***, any morphism $\sigma_i:\boldsymbol{dsgn(\theta_i)} \to S_i$ is also $\boldsymbol{c\text{-}sign(\sigma_i)}:\theta_i \to \boldsymbol{c\text{-}sign(S_i)}$. Given now a refinement morphism $\eta_i:S_i \to S'_i$, we can compose $\boldsymbol{c\text{-}sign(\sigma_i)}$ with $\boldsymbol{r\text{-}sign(\eta_i)}$ to obtain a signature morphism $\theta_i \to \boldsymbol{c\text{-}sign(S'_i)}$ that can be lifted back to ***c-DSGN*** as a morphism $\boldsymbol{dsgn(\theta_i)} \to S'_i$.

The two diagrams satisfy the following important property.

8.3.3 Proposition

In the circumstances laid out above, if $p{:}dia \to S$ and $p'{:}\mathbf{dia}+(\eta_i) \to S'$ are colimits, there is a unique refinement morphism $S \to S'$ such that, for every $f{:}i \to j$ in $\mathbf{I}$, $\mathbf{c\text{-}sign}(\mathbf{dia}(f));\mathbf{r\text{-}sign}(\eta_j)=\mathbf{r\text{-}sign}(\eta_i);\mathbf{c\text{-}sign}(\mathbf{dia}'(f))$.

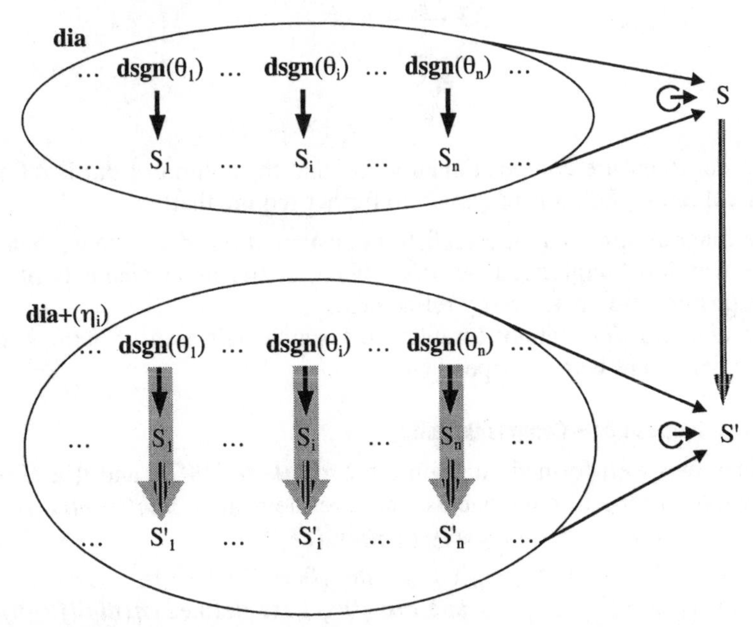

◆

This property is another form of *compositionality*: it states that refinement of the whole can be obtained from refinements of the parts. Compositionality, as discussed in Par. 6.2.4, is a key issue in the design of complex systems because it makes it possible to reason about a system using the descriptions of their components at any level of abstraction, without having to know how these descriptions are refined in the lower levels (which includes their implementation).

This form of compositionality can be formulated more precisely in CommUnity by extending the notion of refinement to configurations much in the same way as we extended the notion of realisation to configurations of specifications in Par. 6.2.4.

8.3.4 Definition – Refinement of Configurations

Given two configurations $\mathbf{dia}{:}\mathbf{I} \to \mathbf{c\text{-}DSGN}$ and $\mathbf{dia}'{:}\mathbf{I} \to \mathbf{c\text{-}DSGN}$, a *refinement of* **dia** *over* **dia'** is an $|\mathbf{I}|$-indexed family $(\eta_i{:}\mathbf{dia}(i) \to \mathbf{dia}'(i))_{i \in |\mathbf{I}|}$ of morphisms in $\mathbf{r\text{-}DSGN}$ s.t., for every $f{:}i \to j$ in $\mathbf{I}$, $\mathbf{c\text{-}sign}(\mathbf{dia}(f));\mathbf{r\text{-}sign}(\eta_j)=\mathbf{r\text{-}sign}(\eta_i);\mathbf{c\text{-}sign}(\mathbf{dia}'(f))$.

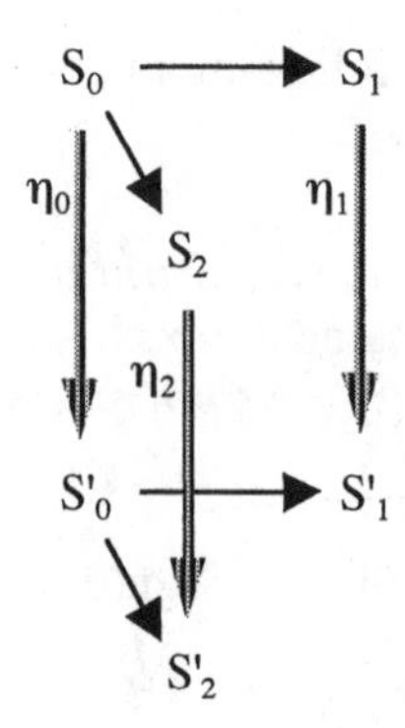

◆

In order to ensure compositionality, i.e. that the colimit of ***dia*** is refined by the colimit *of* ***dia'***, it is necessary to further require that:

- The diagram ***dia'*** cannot establish the instantiation of any input channel that was left "unplugged" in ***dia***. That is, the input channels of the composition are preserved by refinement.
- The diagram ***dia'*** cannot establish the synchronisation of actions that were defined as being independent in ***dia***.

8.3.5 Proposition – Compositionality

Consider two well-formed configurations **dia**:$I \rightarrow$***c-DSGN*** and **dia'**:$I \rightarrow$***c-DSGN*** over a set J of components, and a refinement $(\eta_i$:***dia***$(i) \rightarrow$***dia'***$(i))_{i \in |I|}$ of ***dia*** over ***dia'*** such that, for every $f{:}i \rightarrow j$ in I,

1. For every $v' \in in(V'_i)$, if $v' \notin \eta_i(V_i)$ then ***dia'***$(f)(v') \notin \eta_j(in(V_j))$.
2. For every $g' \in \Gamma'_j$, if $\eta_j(g)$ and ***dia'***$(f)(g')$ are defined $\eta_i($***dia'***$(f)(g'))$ is also defined.
3. For every $i \in |I| \backslash J$, $\eta_{i_{ac}}$ is injective.

Then, there is a unique morphism $\eta{:}S \rightarrow S'$ in ***r-DSGN*** s.t., for every $i \in |I|$, ***c-sign***(μ_i);***r-sign***(η)=***r-sign***(η_i);***c-sign***(μ'_i), where $(\mu_i$: ***dia***$(i) \rightarrow S)_{i \in |I|}$ and $(\mu'_i$: ***dia'***$(i) \rightarrow S')_{i \in |I'|}$ are colimits of **dia** and **dia'**, respectively.

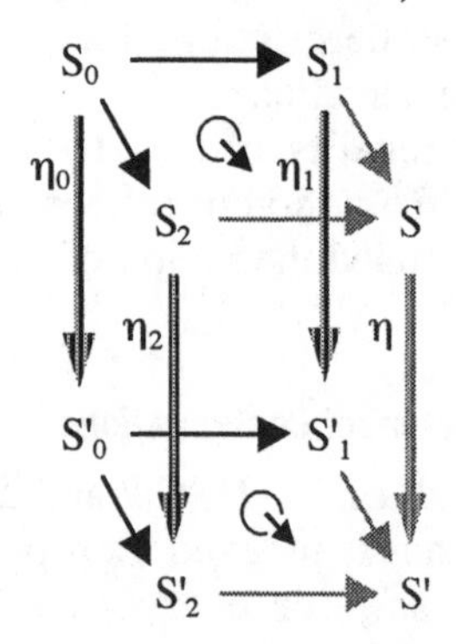

◆

An outline of the proof of this result can be found in [78].

9 Architectural Description

9.1 Motivation

Although components have always been considered to be the fundamental building blocks of software systems, it is in the way that the components of a system interact that the emergence of global properties of the system resides. With no interaction there is no emergence of new behaviour and, therefore, no value to the system as a whole that is not already provided through its components in isolation.

We can safely say that most of the complexity of system construction lies in the definition of the interconnections that should regulate how components interact. Designing small, encapsulated components that, through the computations that they perform locally, provide services with certain functionalities is something that can be mastered, without much difficulty, with existing methods and development techniques.

Knowing how to interconnect components so that, from the interactions, the global properties that are required of the system can emerge is a totally different matter. Most of the time, it is an error-prone process. What in the literature is known as the "feature interaction problem" [111] is just a symptom of this difficulty: the emergence of "strange", "unexpected" or "undesired" behaviour from feature composition is intrinsic to the use of methods for putting together systems from individual features as basic units of functionality. While we compose features having in mind the emergence of certain properties that constitute requirements on the behaviour of the system, it is difficult to predict which other forms of behaviour will also emerge, namely ones that are not of interest and whose "negation" is normally omitted from the requirements specification because one never thought of them being possible.

Situations like feature interaction are not problems that need to be solved. They are phenomena that are intrinsic to the way we build systems and that "just" need to be controlled. For that purpose, we need first-class representations of the interconnections.

This level of complexity is aggravated by the need to evolve systems. As the world of business in general becomes more and more aggressive and competitive, for instance, as a consequence of the impact of the Internet and wireless technologies, companies need their information systems to be easily adaptable to changes in the business rules with which they operate, most of the time in a way that does not imply interruptions to the services that they provide. Quoting directly from [47], "... the ability to change is now more important than the ability to create e-commerce systems in the first place. Change becomes a first-class design goal and requires business and technology architecture whose components can be added, modified, replaced and reconfigured". All this means that the "complexity" of software has definitely shifted from *construction* to *evolution*, and that methods and technologies are required that address this new level of complexity and adaptability.

Software architecture [50, 88] is a "recent" topic in software engineering aimed at addressing the gross decomposition and organisation of systems in which, through so-called connectors, component interactions are recognised as being first-class design entities [98]. According to [2], a connector (type) can be defined by a set of *roles* and a *glue* specification. For instance, a typical client–server architecture can be captured by a connector type with two roles – client and server – that describe the expected behaviour of clients and servers, and a glue that describes how the activities of the roles are coordinated (e.g. asynchronous communication between the client and the server). The roles of a connector type can be *instantiated* with specific components of the system under construction, which leads to an overall system structure consisting of components and connector instances establishing the interactions between the components.

The similarities between architectural constructions as informally described above and parameterised programming [56] are rather striking and have been developed in [59] in the context of the emerging interest in software architecture. The view of architecture that is captured by the principles and formalisms of parameterised programming is reminiscent of module interconnection languages and interface definition languages [55]. This perspective is somewhat different from the one we discussed above in the sense that, whereas they capture functional dependencies between the modules that need to be linked to constitute a given program, we focus instead on the organisation of the *behaviour* of systems as compositions of components ruled by protocols for communication and synchronisation.

In this chapter, we show that the mathematical "technology" of parameterised programming can also be used for the formalisation of architectural connectors in the interaction sense. The mathematical framework that we propose for formalising architectural principles is not specific to any particular Architecture Description Language (ADL). In fact, it will emerge from the examples that we shall provide that, contrary to most other formalisations of software architectural concepts that we have seen, category theory is not another semantic domain for the formalisation of the description of components and connectors (like, say, the use of CSP in [2] or first-order logic in [87]). Instead, it provides for the very semantics of "interconnection", "configuration", "instantiation" and "composition", i.e. the principles and design mechanisms that are related to the gross modularisation of complex systems. Category theory does this at a very abstract level because what it proposes is a toolbox that can be applied to whatever formalism is chosen for modelling the behaviour of systems as long as that formalism satisfies some structural properties. It is precisely the structural properties that make a formalism suitable for supporting architectural design that we make our primary focus. However, we need some concrete language in which to illustrate and motivate our approach. Not surprisingly, we use CommUnity for that purpose.

9.2 Connectors in CommUnity

According to [2], a connector (type) can be defined by a set of *roles* that can be instantiated with specific components of the system under construction, and a *glue* specification that describes how the activities of the role instances are to be coordinated. Using the mechanisms that we introduced in the previous chapter for configuration design in CommUnity, it is not difficult to come up with a formal notion of connector that has the same properties as those given in [2] for the language WRIGHT.

9.2.1 Definition – Architectural Connectors

A *connection* consists of

- Two designs G and R, the glue and the role, respectively.
- A signature θ and two morphisms σ:*dsgn*(θ)→G,μ:*dsgn*(θ)→R connecting the glue and the role.

A *connector* is a finite set of connections with the same glue that, together, constitute a well-formed configuration (see Par. 8.2.5). Its *semantics* is the colimit of the diagram formed by its connections.

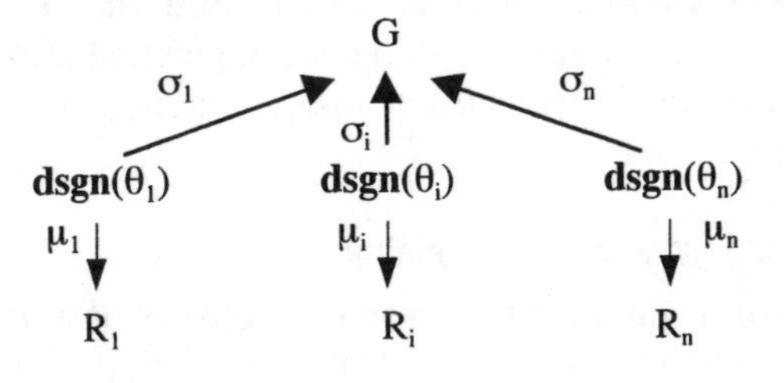

◆

Because we are working in a coordinated category, we know already that we can adopt a simplified notation for diagrams such as these by using signatures directly in place of their discrete lifts (cables):

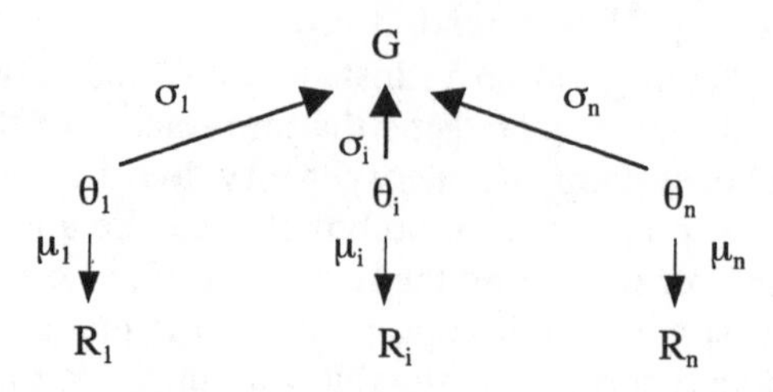

For instance, asynchronous communication through a bounded channel can be modelled by a connector *ASYNC* with two connections, as depicted below using the graphical notation that we introduced for configurations:

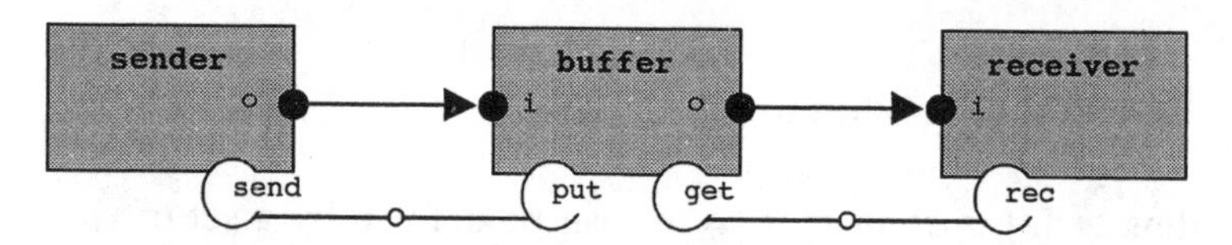

The glue of *ASYNC* is the bounded buffer with FIFO discipline presented in Sect. 8.1. It prevents the *sender* from sending a new message when there is no space, and prevents the *receiver* from reading a new message when there are no messages. The two roles, *sender* and *receiver*, define the behaviour required of the components to which the connector can be applied. For the *sender*, we require that no message be produced before the previous one has been processed. Its design is the one given already in Sect. 8.1. For the *receiver*, we simply require that it have an action that models the reception of a message.

```
design receiver [t:sort] is
in     i: t
do     rec: true,false → skip
```

What we described are connector *types* in the sense that their roles can be instantiated with specific designs. In WRIGHT [2], role instantiation has to obey a compatibility requirement expressed via the refinement relation of CSP [69]. In CommUnity, the refinement relation is formalised through the morphisms defined in Par. 8.3.1, leading to the following notion of instantiation:

9.2.2 Definition – Connector Instantiation

An *instantiation* of a connection with role *R* consists of a design *P* together with a refinement morphism $\phi{:}R\to P$. An instantiation of a connector consists of an instantiation for each of its connections. ♦

In order to define the semantics of such an instantiation, notice that, as discussed in Par. 8.3.2, each instantiation $\phi:R\rightarrow P$ of a connection can be composed with μ:***dsgn***$(\theta)\rightarrow R$ to define $\mu;\phi:\theta\rightarrow$***c-sign***$(P)$. Because, as seen in Par. 8.2.4, the category of designs is coordinated over signatures, every such signature morphism can be lifted to a design morphism $\mu;\phi$:***dsgn***$(\theta)\rightarrow P$. Hence, an instantiation of a connector defines a diagram in ***c-DSGN*** that connects the role instances to the glue.

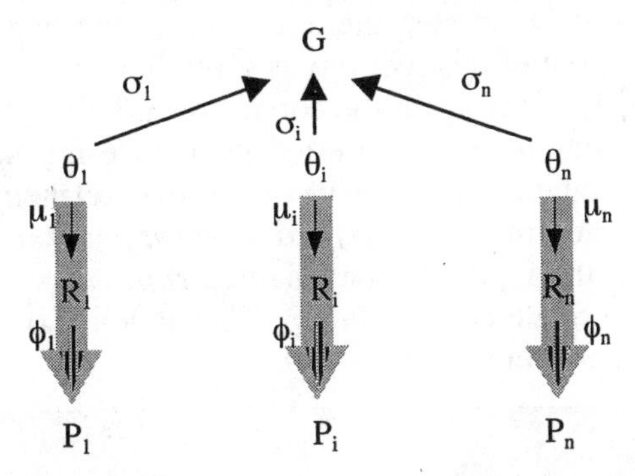

Because each connection is according to the rules set for well-formed configurations as detailed in Par. 8.2.5, the diagram defined by the instantiation is, indeed, a configuration and, hence, has a colimit.

9.2.3 Definition – Semantics of Connector Instantiation

The interconnection (configuration) defined by a connector instantiation is the diagram in ***c-DSGN*** formed as described above by composing the role morphism of each connection with its instantiation. The semantics of an instantiation is the colimit of the interconnection that it defines. ♦

Because, as already argued, colimits in ***c-DSGN*** express parallel composition, this semantics agrees with the one provided in [2] for the language WRIGHT. In the next section, we take this analogy with WRIGHT one step further. Moreover, through compositionality as formalised in Par. 8.3.5, the categorical formalisation makes it possible to prove that the design that results from the semantics of the instantiation is a refinement of the semantics of the connector itself.

As a simple example, consider a connector with just one role.

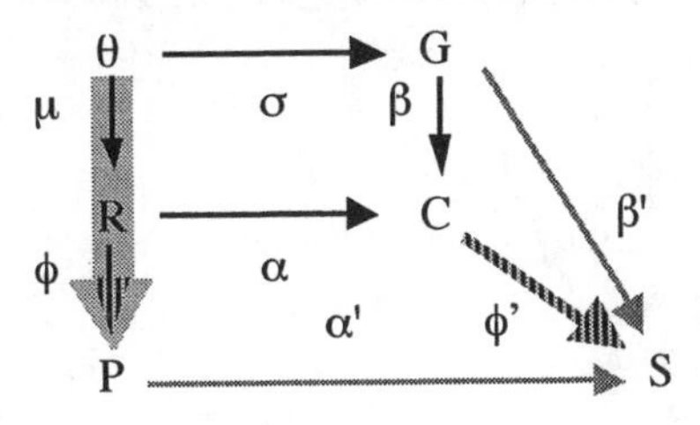

The semantics of the connector is given by the colimit of the pair $<\mu,\sigma>$ – $<\alpha:R\rightarrow C,\beta:G\rightarrow C>$. The instantiation of the role with the component P through the refinement morphism ϕ is given by the colimit of $<\mu;\phi,\sigma>$ – $<\alpha':P\rightarrow S,\beta':G\rightarrow S>$. We can easily prove that there exists a refinement morphism $\phi':C\rightarrow S$, which establishes the "correctness" of the instantiation mechanism. This is because all the different objects and morphisms involved can be brought into a more general category in which the universal properties of colimits guarantee the existence of the required refinement morphism. A full proof of this property is given in [78].

As an example, consider again the connector *ASYNC*. We have already seen in Sect. 8.3 that *sender* is refined by the design *user*. Likewise, *printer* is a refinement of *receiver* via the refinement morphism $\kappa:receiver\rightarrow printer$ defined by $\kappa_{ch}(i)=rdoc$ and $\kappa_{ac}(rec)=rec$: the reception of a message from the input channel (named *rdoc)* corresponds to downloading it into the private channel *pdoc*. This action is only enabled if the previous message has been printed.

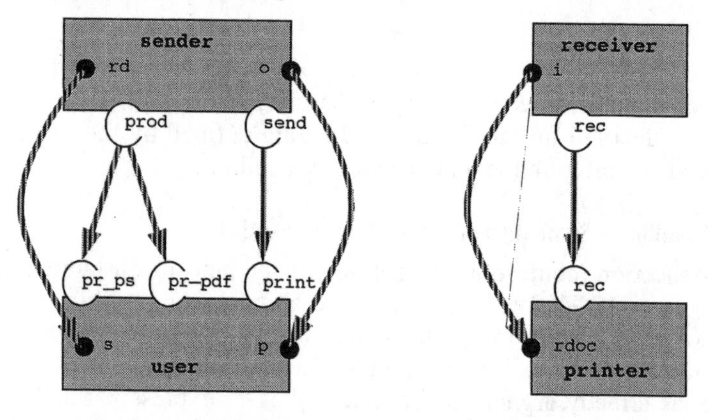

Using these refinement morphisms, we can instantiate the connector *ASYNC* to connect the user to the printer:

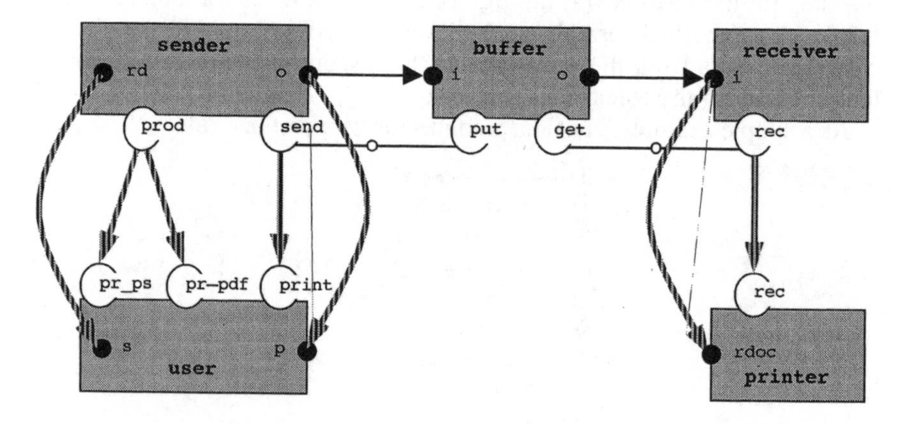

The final configuration is obtained by calculating the composition of the signature morphisms that define the two connections of *ASYNC* with the refinement morphisms η,κ. For instance, the channel p of *user* gets connected to the input channel i of *buffer* because $\eta(o)=p$ and o is connected to i of *buffer*. The resulting configuration is the one presented in Sect. 8.2.

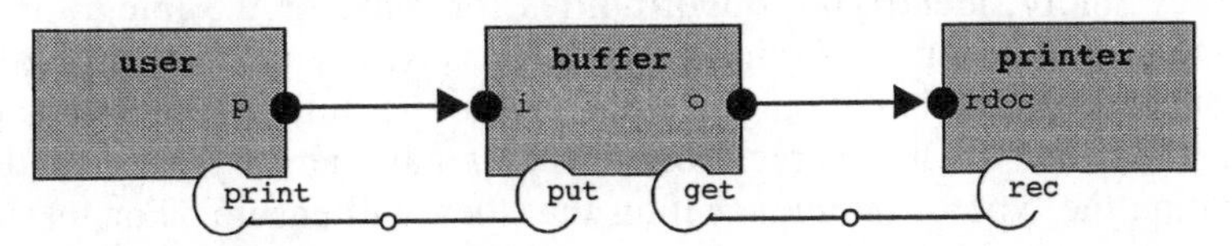

In order to simplify the notation when making use of connectors, we normally hide the glue and its connections to the roles, leaving just the roles visible to suggest that they provide the "interface" of the connector:

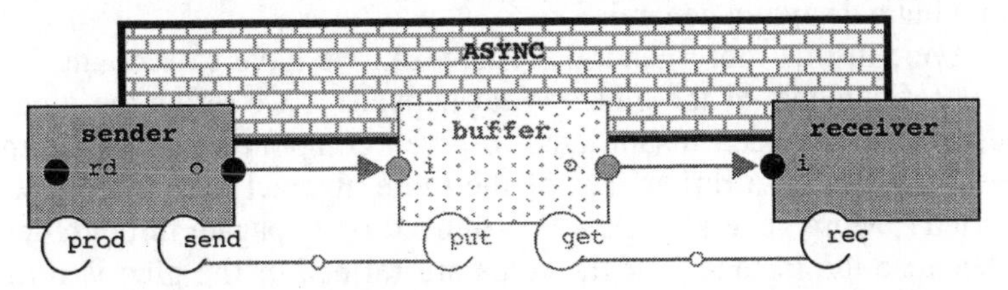

Instantiation is denoted as follows:

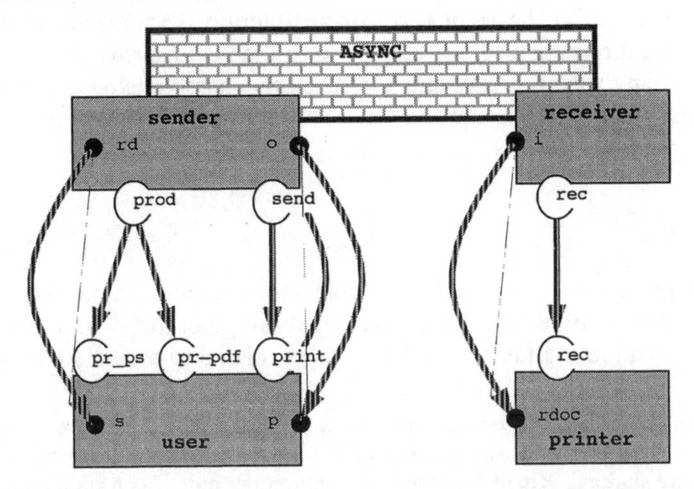

Architectural connectors are used for systematising software development by offering standard means for interconnecting components that can be reused from one application to another. In this sense, the typical glue is a program that implements a well-established pattern of behaviour (e.g. a communication protocol) that can be superposed to existing components of a system through the instantiation of the roles of the connector.

However, architectures also fulfil an important role in supporting a high-level description of the organisation of a system by identifying its main components and the way these components are interconnected. An early identification of the architectural elements intended for a system will help to manage the subsequent design phases according to the organisation that they imply, identifying opportunities for reuse or the integration of third-party components. From this point of view, it seems useful to allow for connectors to be based on glues that are not yet fully developed as programs, but for which concrete commitments have already been made to determine the type of interconnection that they will ensure. For instance, at an early stage of development, one may decide to adopt a client–server architecture without committing to a specific protocol of communication between the client and the server. This is why, in the definition of connector in CommUnity, we left open the possibility for the glue not to be a program but a design in general.

However, in this more general framework, we have to account for the possible refinements of the glue. What happens if we refine the glue of a connector that has been instantiated to given components of a system? Is the resulting design a refinement of the more abstract design from which we started? More generally, how do connectors propagate through design, be it because the instances of the roles are refined or the glue is refined? One of the advantages of using category theory as a mathematical framework for formalising architectures is that answers to questions like these can be discussed at the right level of abstraction. Another advantage is that the questions themselves can be formulated in terms that are independent of any specific ADL and answered by characterising the classes of ADLs that satisfy the given properties. This is what we do further on.

9.3 Examples

We now present more examples of connectors, namely some that we will need in later sections for illustrating algebraic operations on connectors. Their application is illustrated on examples related to a case study on mobility [93]:

> One or more carts move continuously in the same direction on a U-units long circular track. A cart advances one unit at each step. Along the track there are stations. There is at most one station per unit. Each station corresponds to a check-in counter or to a gate. Carts take bags from check-in stations to gate stations. All bags from a given check-in go to the same gate. A cart transports at most one bag at a time. When it is empty, the cart picks a bag up from the nearest check-in. Carts must not bump into each other. Carts also keep a count of how many laps they have done, starting at some initial location.

The program that controls a cart is

```
design cart is
in     idest: 0..U-1, ibag:int
out    obag, laps : int
prv    loc: 0..U-1, dest: -1..U-1, initloc: int
do     move: loc≠dest →
             loc:=loc+U1 || laps:=if(loc=initloc,laps+1,laps)
[]     get:  dest=-1 → obag:=ibag || dest:=idest
[]     put:  loc=dest → obag:=0 || dest:=-1
```

where $+_U$ is addition modulo U.

Locations are represented by integers from zero to the track length minus one. Bags are represented by integers, the absence of a bag being denoted by zero. Whenever the cart is empty, its destination is an unreachable location (–1), so that the cart keeps moving until it gets a bag and a valid gate location through action *get*. When it reaches its destination, the cart unloads the bag through action *put*. Notice that, because input channels may be changed arbitrarily by the environment, the cart must copy their values to output/private channels to make sure the correct bag is unloaded at the correct gate.

A check-in counter manages a queue of bags that it loads one by one onto passing carts.

```
design check-in is
out    bag: int, dest: 0..U-1,
prv    loc: 0..U-1, next: bool, q: list(int)
do     new: q≠[]∧next → bag:=head(q)||q:=tail(q)||next:=false
[]     put: ¬next → next:=true
```

Channel *next* is used to impose sequentiality among the actions. In a configuration in which a cart is loading at a gate, the *put* action must be synchronised with a cart's *get* action and channels *bag* and *dest* must be shared with *ibag* and *idest*, respectively.

A gate keeps a queue of bags and adds each new bag to the tail.

```
design gate is
in     bag: int
prv    loc: 0..U-1, q: list(int)
do     get: true → q:=q.bag
```

In a configuration in which a cart is unloading at a gate, action *get* of the gate must be synchronised with the cart's *put* action, and channel *bag* must be shared with *obag*.

9.3.1 Synchronisation

We begin with the connector that allows us to synchronise two actions of different components. A plain cable would suffice for this purpose, but it is not able to capture the general case of transient synchronisation [93]. Having already a connector for the simpler case makes the presentation more uniform. The glue and roles of the synchronisation connector are the same.

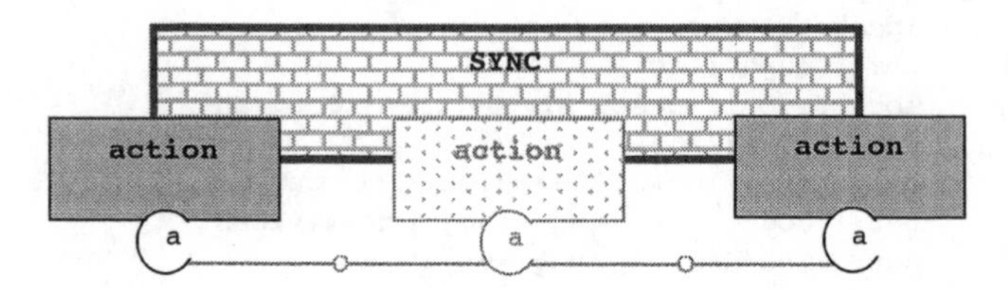

where

```
design action is
do     a: true,false → skip
```

Notice that the action has the least deterministic specification possible: its guard is given the widest possible interval and no commitments are made on its effects. Hence, it can be refined by any action.

According to the colimit semantics of connectors, when the two roles are instantiated with particular actions a_1 and a_2 of particular components, the components have to synchronise with each other every time one of them wants to execute the corresponding action. That is, either both execute the joint action, or none executes.

As an example of using this connector, if we wish to count how often a cart unloads, we can monitor its *put* action with a counter:

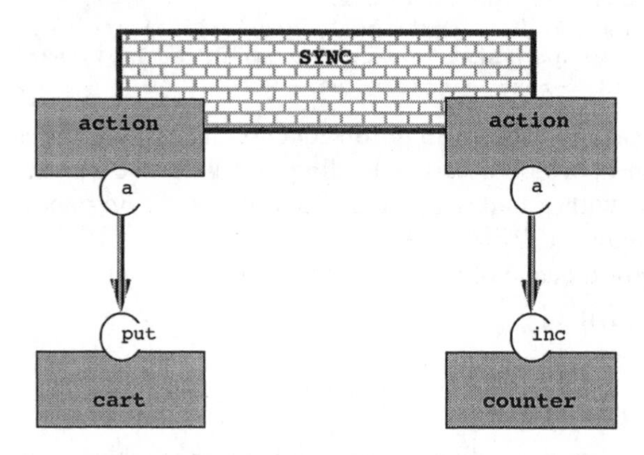

where

```
design counter is
out    c:int
do     inc: true → c:=c+1
[]     reset: true → c:=0
```

According to what is defined in Par. 9.2.2, the interconnection defined by this instantiation is the following configuration:

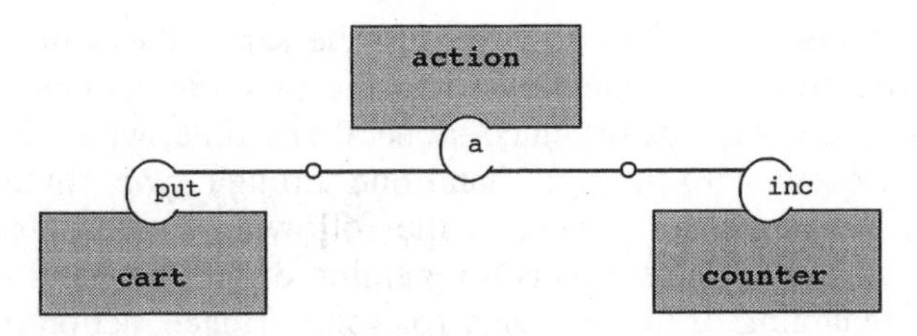

The resulting semantics is the synchronisation of *put* and *inc*. The following program captures the joint behaviour of the interconnected components:

```
design monitored_cart is
in     idest: 0..U-1, ibag: int
out    obag, laps, unloads : int
prv    loc: 0..U-1, dest: -1..U-1, initloc : int
do     move: loc≠dest
              → loc:=loc+U1 || laps:=if(loc=initloc,laps+1,laps)
[]     get:  dest=-1 → obag:=ibag || dest:=idest
[]     put|inc: loc=dest
              → obag:=0 || dest:=-1 || unloads:=unloads+1
```

9.3.2 Subsumption

Intuitively, synchronisation corresponds to an equivalence between the occurrence of two actions: the occurrence of each of the actions implies the occurrence of the other. In many circumstances, we are interested in one of the implications, like in remote method invocation: a call requests the execution of the method, but that method may be invoked by other calls.

For instance, to avoid a cart colliding with a cart that is right in front of it, we only need one implication: if the first one moves, so must the one in front. The other implication is not necessary. The analogy with implication also extends to the counter-positive: if the front car cannot move, for instance, because it is (un)loading a bag, then neither can the rear one.

We call this one-way synchronisation *action subsumption*. The subsumption connector is given by the following configuration:

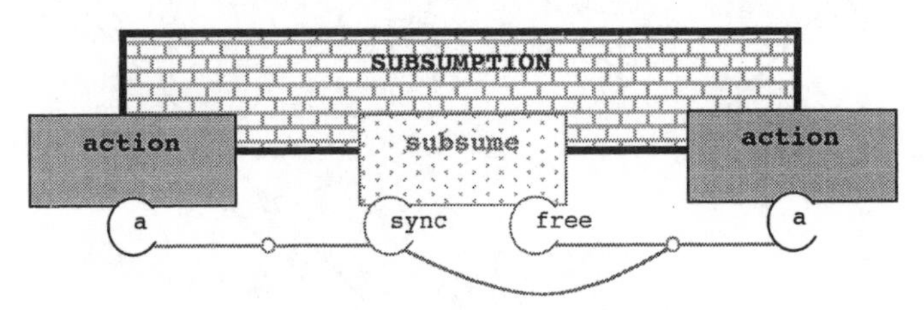

where the glue is now given by

```
design subsume is
do     sync: true,false → skip
[]     free: true,false → skip
```

Notice that although the two roles are the same, the connector is not symmetric because the connections treat the two role actions differently: the right-hand one may be executed alone at any time, while the left-hand one must co-occur with the right-hand one through *sync*. Indeed, the semantics of the connector generates the following synchronisation sets: $a_1|sync|a_2$ and $free|a_2$, where a_1 is a renaming of the left-hand role action and a_2 is a renaming of the left-hand role one. Hence, action a_1 can only occur together with a_2, but a_2 can occur without a_1.

In order to understand how subsumption works, let us detail the construction of the semantics of the connector, starting with the corresponding diagram of signatures. Because only actions are involved, we take the diagram directly over pointed sets:

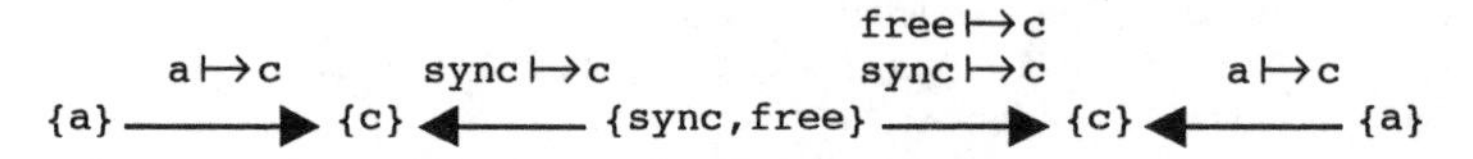

The limit can be computed by first taking the two obvious pullbacks:

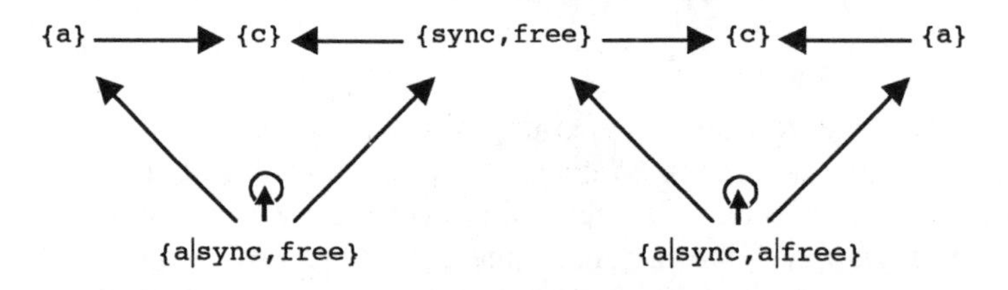

Notice that, because the interconnecting morphisms are different, different synchronisations are generated even if the nodes are the same. Finally, the middle pullback relates the pairwise synchronisations to make up the global interconnection:

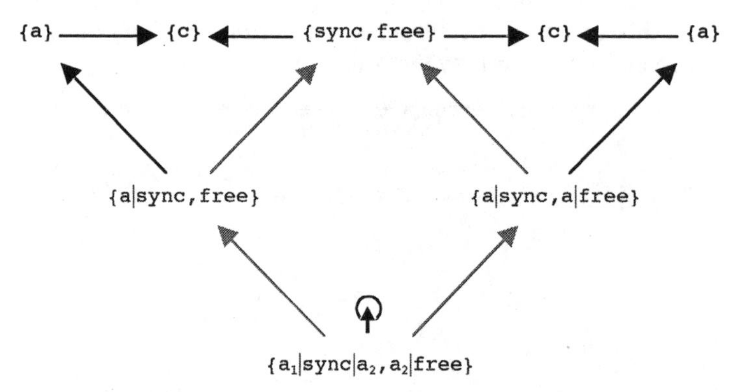

As an example of the application of this connector, consider its instantiation with the *move* actions of two carts: any movement of the cart on the left implies a movement of the cart on the right. Hence, this instantiation can be used to prevent collision when the left cart is too close behind the right cart.

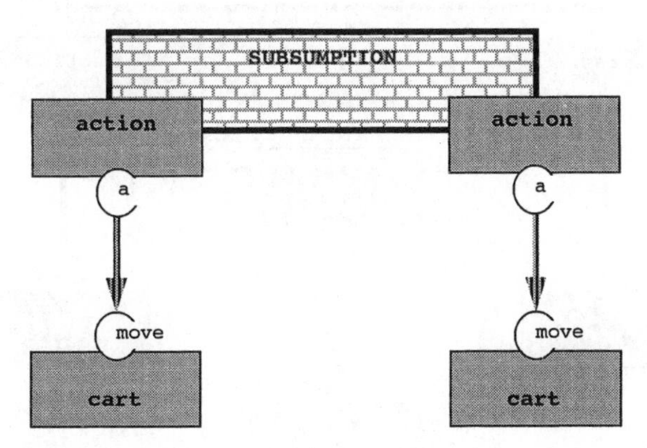

9.3.3 Extension Cord

A generalisation of the subsumption and synchronisation connectors is to allow an action to synchronise, independently, with actions of two different components, achieving an effect similar to an extension cord that one uses to connect two devices to the same power supply.

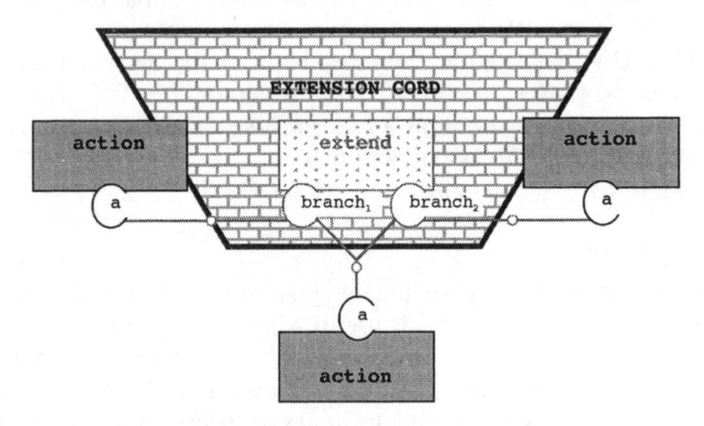

where

```
design extend
do     branch1: true, false → skip
[]     branch2: true, false → skip
```

If we instantiate the left-hand role with an action a_1, the right-hand role with an action a_2 and the middle role with an action b, the semantics of the interconnection, as obtained through the colimit, is given by two synchronisation sets: $a_1|branch_1|b$ and $a_2|branch_2|b$. Notice that action b will always occur simultaneously with either a_1 or a_2 but not with both. For instance, a gate that can handle two carts downloading simultaneously can configured as follows:

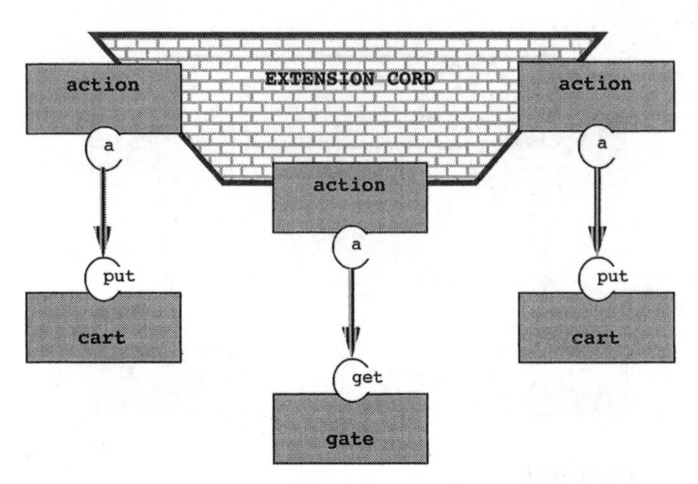

This connector can be generalised to any finite number n of ramifications (branch actions), giving rise to an *n-EXTENSION CORD* through which a server can be connected simultaneously, but independently, to a fixed maximum number of clients.

The more attentative reader may also have noticed that the glue *extend* of *EXTENSION CORD* is the same as (isomorphic to) *subsume* of *SUBSUMPTION*. Indeed, the difference between the two connectors is in the roles that they offer; *SUBSUMPTION* does not offer the "free" ramification of the subsumed action as a role for interconnection. We can also say that *SUBSUMPTION* is no more than a *1-EXTENSION CORD*. This relationship is discussed in more detail in Par. 10.1.2.

9.3.4 Inhibition

Another basic connector type is the one that allows us to inhibit an action by making its guard false. This is useful when, for some reason, we need to prevent an action from occurring but without having to reprogram the component. Indeed, the mechanism of superposition that we use as a semantics for the application of architectural connectors allows us to disable an action without changing the guard directly but by just inducing this effect: it suffices to synchronise the action with one that has a false guard.

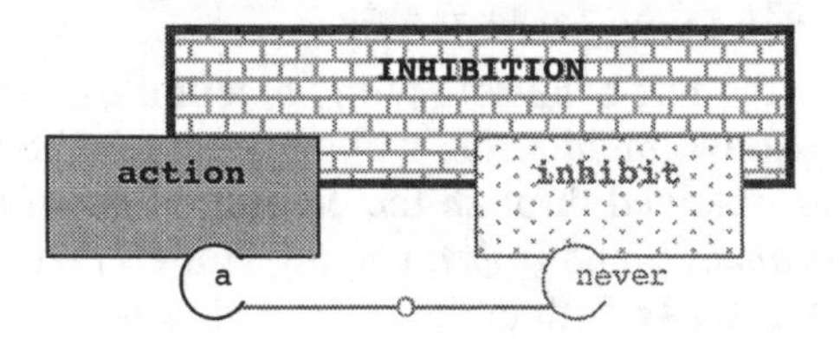

The glue is:

```
design inhibit is
do     never: false → skip
```

When the role is instantiated with an action with guard B, the result of the interconnection is the same action guarded by $B \wedge \textit{false}$.

This connector can be generalised to arbitrary conditions with which one may strengthen the guards of given actions. The inhibitor just has to be provided with the data that is necessary to compute the condition C that will strengthen the guard. This can be done, for instance, through the use of input channels through which we can select the sources of the information that will disable the action.

```
design inhibit(C) is
in     ...
do     never: C → skip
```

The result of instantiating the role with an action with guard B is the same action guarded by $B \wedge C$.

9.4 An ADL-Independent Notion of Connector

The notion of connector presented in Sect. 9.2 can be generalised to design formalisms other than CommUnity. In this section, we discuss the properties that such formalisms need to satisfy for supporting the architectural concepts and mechanisms that we have illustrated for CommUnity.

Before embarking on this discussion, we need to fix a framework in which designs, configurations and relationships between designs, such as refinement, can be formally described.

9.4.1 Definition – Design Formalisms

A formalism supporting system design consists of:

- A category ***c-DESC*** of component descriptions in which systems of interconnected components are modelled through diagrams.
- For every set *CD* of component descriptions, a set *Conf(CD)* consisting of all well-formed configurations that can be built from the components in *CD*. Each such configuration is a diagram in ***c-DESC*** that is guaranteed to have a colimit. Typically, *Conf* is given through a set of rules that govern the interconnection of components in the formalism.
- A category ***r-DESC*** with the same objects as ***c-DESC***, but in which morphisms model refinement, i.e. a morphism $\eta{:}S \rightarrow S'$ in ***r-DESC*** expresses that S' refines S, identifying the design decisions that lead from S to S'. Because the description of a composite system is given by a colimit of a diagram in ***c-DESC*** and, hence, is defined up to an isomor-

phism in ***c-DESC***, refinement morphisms must be such that descriptions that are isomorphic in ***c-DESC*** refine, and are refined exactly by, the same descriptions. ♦

Summarising, all we require is a notion of system description, a relationship between descriptions that captures components of systems, another relationship that captures refinement and criteria for determining when a diagram is a well-formed configuration.

In the context of this categorical framework, we shall now discuss the properties that are necessary for supporting software architecture. A key property for supporting architectural design is a clear separation between the description of individual components and their interaction in the overall system organisation. In other words, the formalism must support the separation between what, in the description of a system, is responsible for its computational aspects and what is concerned with coordinating the interaction between its different components.

In the case of CommUnity, as we have seen, only signatures are involved in interconnections. The body of a component design describes its functionality and, hence, corresponds to the computational part of the design. At the more general level that we are discussing, we shall take the separation between coordination and computation to be materialised through a functor ***sign:c-DESC→SIGN*** mapping descriptions to signatures, forgetting their computational aspects. The fact that the computational side does not play any role in the interconnection of systems can be captured by requiring ***sign*** to be coordinated in the sense of Par. 7.5.4.

Another crucial property for supporting architectural design is in the interplay between structuring systems in architectural terms and refinement. We have already pointed out that one of the goals of software architectures is to support a view of the gross organisation of systems in terms of components and their interconnections that can be carried through the refinement steps that eventually lead to the implementation of all its components. Hence, it is necessary that the application of architectural connectors to abstract designs, as a means of making early decisions on the way certain components need to be coordinated, will not be jeopardised by subsequent refinements of the component designs towards their final implementations. Likewise, it is desirable that the application of a connector may be made on the basis of an abstract design of its glue as a means of determining main aspects of the required coordination without committing to the final mechanisms that will bring about that coordination.

One of the advantages of the categorical framework that we propose is that it makes the formulation of these properties relatively easy, leading to a characterisation of the design formalisms that support them in terms of the structural properties that we have been discussing. For instance, we show in Sect. 8.3 that, in the situations in which refinement morphisms map directly to signature morphisms, we can simply put together, in a diagram of signatures, the morphisms that define the interactions and the morphisms that establish the refinement of the component descriptions.

More precisely, in the situations in which there exists a forgetful functor ***r-sign:r-DESC→SIGN*** that agrees with the coordination functor ***sign*** on signatures, i.e. ***r-sign(S)=sign(S)*** for every *S:**c-DESC***, and given a well-formed configuration diagram ***dia*** of a system with components $S_1,\ldots,S_n$ and refinement morphisms $\eta_i:S_i \rightarrow S'_i$, $i \in 1..j$,

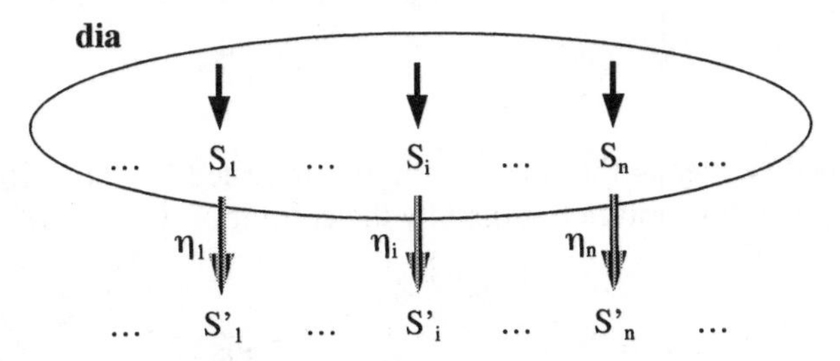

we can obtain a new diagram in ***c-DESC*** and, hence, a new configuration, by composing the morphisms ***r-sign(***η_i***)*** with those in ***dia*** that originate in channels (signatures) and have the S_i as targets.

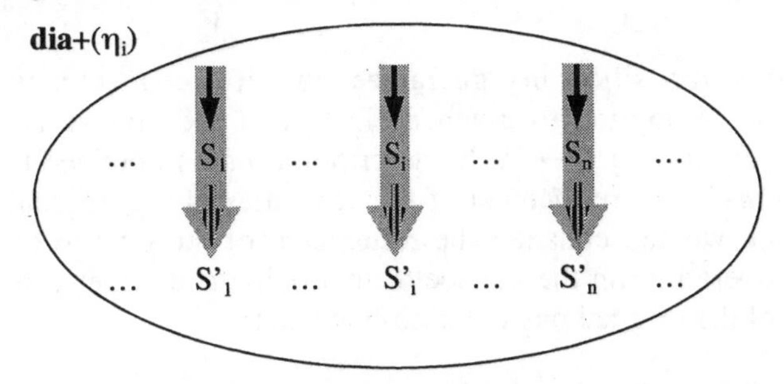

In general, it may be possible to propagate the interactions between the components of a system when their descriptions are replaced by more concrete ones, even when refinement morphisms do not map to signature morphisms. This more general situation can be characterised as follows.

9.4.2 Definition – Compositional Design Formalisms

We say that a design formalism is *compositional* whenever, for every well-formed configuration ***dia*** involving descriptions $\{S_1,\ldots,S_n\}$ and refinements morphisms $\{\eta_i:S_i \rightarrow S'_i: i \in 1..n\}$, there is a well-formed configuration diagram ***dia***+(η_i) that characterises the system obtained by replacing the S_i by their refinements and satisfies the following correctness criterion: the colimit of ***dia***+(η_i) provides a refinement for the colimit of ***dia***. ♦

When we consider the specific case of the configurations obtained by direct instantiation of an architectural connector, this property reflects the compositionality of the connector as an operation on configurations.

Compositionality ensures that the semantics of the connector is preserved (refined) by any system that results from its instantiation. For instance, given instantiations $\eta_1:R_1 \rightarrow P_1$ and $\eta_2:R_2 \rightarrow P_2$ of a binary connector

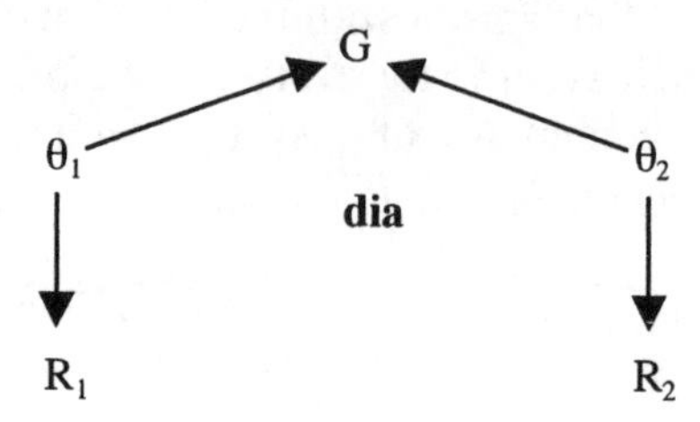

compositionality means that the description returned by the colimit of ***dia*** is refined by the description returned by the colimit of ***dia***$+(\eta_1,\eta_2)$.

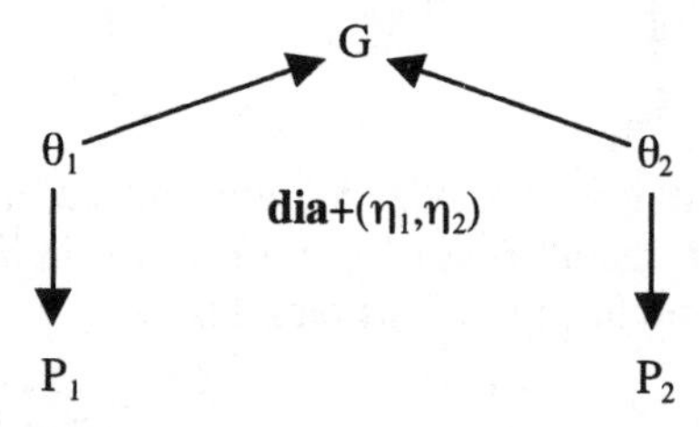

Likewise, compositionality guarantees that if a connector with an abstract glue G is applied to given designs, and the glue is later refined through a morphism $\eta:G \rightarrow G'$, the description that is obtained through the colimit of ***dia***$+\eta$ is a refinement of the semantics of the original instantiation. In fact, we can consider the refinement of the glue to be a special case of an operation on the connector that delivers another connector – a refinement of the original one in the case at hand.

9.4.3 Definition – Architectural Schools

A design formalism ***F=<c-DESC,Conf,r-DESC>*** supports architectural design, and is called an *architectural school,* iff

- ***c-DESC*** is coordinated over ***SIGN*** through ***sign:c-DESC→SIGN***.
- *F* is compositional. ◆

9.5 Adding Abstraction to Connectors

The mathematical framework that we presented in the previous sections provides not only an ADL-independent semantics for the principles and techniques that can be found in existing approaches to software architectures, but also a basis for extending the capabilities of existing ADLs. In the remainder of this chapter, we present and explore some of the avenues that this mathematical characterisation has opened, hoping that the reader will want to explore them even further, or find new ones.

As already mentioned, the purpose of the roles in a connector is to impose restrictions on the local behaviour of the components that are admissible as instances. In the approach to architectural design outlined in the previous sections, this is achieved through the notion of correct instantiation via refinement morphisms. As also seen above, roles do not play any part in the calculation of the resulting system. They are used only for defining what a correct instantiation is. This separation of concerns justifies the adoption of a more declarative formalism for the specification of roles, namely one in which it is easier to formulate the properties required of components to be admissible instances.

In this section, we place ourselves in the situation in which the glues are designs, the roles are specifications, and the instantiations of the roles are, again, designs. We are going to consider that specifications are given as a category ***SPEC***, e.g. the category of theories of a logic formalised as an institution (see Par. 6.5.11). We take the relationship between specifications and designs to be captured through the following elements:

- A functor ***spec***$:$***SIGN***$\rightarrow$***SPEC*** mapping signatures and their morphisms to specifications. The idea behind the functor ***spec*** is that like through ***c-desc*** (the left adjoint of ***sign***) signatures provide the means for interconnecting designs, they should also provide the means for interconnecting specifications. Hence, every signature generates a canonical specification – the specification of a cable. However, it is not necessary for ***spec*** to satisfy as many structural properties as ***c-desc*** because, for the purposes of this section, we are limiting the use of specifications to the definition of connector roles. Naturally, if we wish to address architecture building at the specification level, then we will have to require ***SPEC*** to satisfy the properties that we discussed in Sect 9.4.
- A satisfaction relation $\models$ between design morphisms and specification morphisms satisfying the following properties:
  1. If $\pi:P\rightarrow P' \models \sigma:S\rightarrow S'$, then $id_P \models id_S$ and $id_{P'} \models id_{S'}$.
  2. $\pi_1;\pi_2 \models \sigma_1;\sigma_2$ if $\pi_1:P_1\rightarrow P_2 \models \sigma_1:S_1\rightarrow S_2$ and $\pi_2:P_2\rightarrow P_3 \models \sigma_2:S_2\rightarrow S_3$.
  3. Let $\boldsymbol{s}:\boldsymbol{I}\rightarrow$***SPEC*** be a diagram of specifications and $\boldsymbol{p}:\boldsymbol{I}\rightarrow$***c-DESC*** a diagram of designs with the same shape such that, for every edge $f:i\rightarrow j$ in I, $\boldsymbol{p}_f:\boldsymbol{p}_i\rightarrow \boldsymbol{p}_j \models \boldsymbol{s}_f:\boldsymbol{s}_i\rightarrow \boldsymbol{s}_j$. We require that, if $\boldsymbol{p}$ admits a colimit $\pi_i:\boldsymbol{p}_i\rightarrow P$, then $\boldsymbol{s}$ admits a colimit $\sigma_i:\boldsymbol{s}_i\rightarrow S$ such that, for every node $i:I$, $\pi_i \models \sigma$.
  4. If $id_P \models id_S$ and $\rho:P\rightarrow P'$ is a refinement morphism, then $id_{P'} \models id_S$
  5. For every signature θ, $id_{\textbf{\textit{c-desc}}(\theta)} \models id_{\textbf{\textit{spec}}(\theta)}$.

The satisfaction relation is defined directly on morphisms because our ultimate goal is to address interconnections, not just components. Satisfaction of component specifications by designs is given through the identity morphisms. The properties required of the satisfaction relation address its compatibility with the categorical constructions that we use, namely composition of morphisms and colimits. The last two properties mean that refinement of component designs leaves the satisfaction relation invariant, and that the design (cable) generated by every signature satisfies the specification (cable) generated by the same signature.

We saw in Par. 5.1.3 how a satisfaction relation between specifications and programs is defined by a functor ***r-DESC→SPEC*** that maps every design to the set of properties that it "satisfies". Intuitively, this functor is an extension of ***spec*** in the sense that, for every θ, ***spec(θ)*** is isomorphic to ***spec(c-desc(θ))***. Therefore, we call it ***spec***. In this case, it makes sense to define $\pi{:}P{\rightarrow}P' \models \sigma{:}S{\rightarrow}S'$ iff there exist $\rho{:}S{\rightarrow}\mathbf{spec}(P)$, $\rho'{:}S'{\rightarrow}\mathbf{spec}(P')$ s.t. $\rho;\mathbf{spec}(\pi)=\sigma;\rho'$.

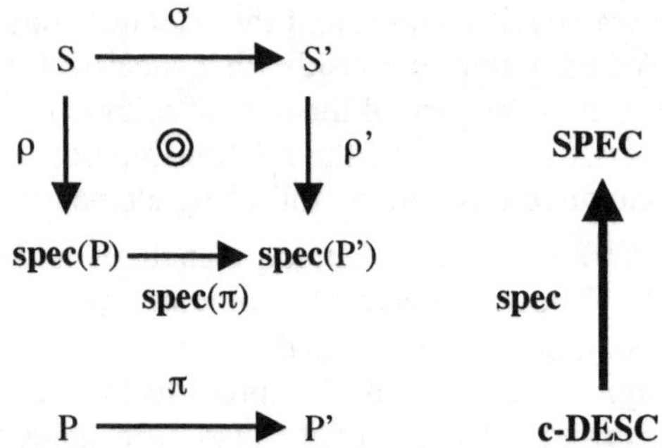

Notice that we get $id_P \models id_S$ iff there exists $\rho{:}S{\rightarrow}\mathbf{spec}(P)$, which is exactly how satisfaction was defined in Par. 5.1.3. Moreover, properties 1–4 are obtained free from the fact that ***spec*** is a functor. Condition 5 is satisfied because ***spec(θ)*** was required to be isomorphic to ***spec(c-desc(θ))***.

In such a setting, we generalise the notion of connector as follows.

9.5.1 Definition – (Generalised) Architectural Connectors

A (generalised) connection consists of

- A design G and a specification R, its glue and role, respectively.
- A signature θ and two morphisms $\mu{:}\mathbf{c\text{-}desc}(\theta){\rightarrow}G$, $\sigma{:}\mathbf{spec}(\theta){\rightarrow}R$ in ***c-DESC*** and ***SPEC***, respectively, connecting the glue and the role via the signature (cable).

A *(generalised) connector* is a finite set of connections with the same glue. An instantiation of a connection with signature θ and role morphism σ consists of a design P and a design morphism $\pi{:}\mathbf{c\text{-}desc}(\theta){\rightarrow}P$ such that $\pi \models \sigma$. An instantiation of a connector consists of an instantiation for each of its connections. An instantiation is correct if the diagram defined by the instantiation morphisms and the glue morphisms is a well-formed configuration. The colimit of this configuration defines the semantics of the instantiation, and is guaranteed to exist if the instantiation is correct. ♦

Although the generalisation seems to be quite straightforward, we do not have an immediate generalisation for the semantics of connectors. This is because the glue is a design and the role is a specification, which means that a connector does not provide us with a diagram like in the homogeneous case that we studied in Sect. 9.2. However, if we are provided with a specification for the glue, we can provide semantics for the connector at the specification level.

9.5.2 Definition – Complete Architectural Connector

A complete connection consists of

- A design G and a specification R, called the glue and the role of the connection, respectively.
- A signature θ and two morphisms μ:***c-desc***$(\theta)\to G$,σ:***spec***$(\theta)\to R$ in ***c-DESC*** and ***SPEC***, respectively, connecting the glue and the role via the signature (cable).
- A specification S and a morphism τ:***spec***$(\theta)\to S$ such that $\mu \models \tau$. Notice that this means that the design G satisfies the specification S.

A *complete connector* is a finite set of complete connections with the same glue design and specification. Its semantics is given by the colimit, if it exists, of the ***SPEC***-diagram defined by the σ_i and the τ_i. ♦

9.5.3 Proposition

The semantics of the instantiation of a complete connector satisfies the semantics of the connector. ♦

An illustration of an abstract architectural school can be given in terms of a first-order extension[1] of the linear temporal logic that we studied in Sect. 3.5. As an example, we present below the specifications of a typical sender and receiver of messages through a pipe.

```
specification   pipe_sender is
signature       eof, send
axioms          eof ⊃ G(¬send∧eof)

specification   pipe_receiver is
signature       cl, eof, rec
axioms          cl ⊃ G(¬rec ∧ cl)
                (Geof ∧ ¬cl) ⊃ (¬recUcl)
```

The specification *pipe_sender* accounts, through *send*, for the transmission of data. The end of data transmission is signalled through channel *eof*: the axiom requires that *eof* be stable (remains true once it becomes true) and transmission of messages to cease once *eof* becomes true.

[1] The extension is straightforward: the reader is encouraged to formalise it as an exercise, for which [38, 39, 67] can be consulted.

The specification *pipe_receiver* accounts, through *rec*, for the reception of data. The other means of interaction with the environment is concerned with the closure of communication. Through channel *eof*, a Boolean can be received that indicates if transmission has ceased. Closure of communication is signalled in the channel *cl*. The first axiom requires that *cl* be stable and the reception of messages to cease once *cl* becomes true. The second axiom expresses that, if the information received through *eof* is stable, the receiver is obliged to close the communication as soon as it is informed that there will be no more data. However, the receiver may decide to close the communication before that.

It remains to capture the relationship between specifications and designs in CommUnity. On the one hand, every design signature θ can be mapped into a temporal signature by forgetting the different classes of channels and actions, as well as the write frames of action. On the other hand, part of the semantics of designs can be encoded in LTL:

- For every action g, the negation of $L(g)$ is a blocking condition for its occurrence: $(g \supset L(g))$.
- For every local channel v, $D(v)$ consists of the set of actions that can modify it: $\bigvee_{g \in D(v)} g \vee (Xv{=}v)$.
- For every action g, the condition $R(g)$ holds in every state in which g is executed: $(g \supset \tau(R(g)))$, where τ is a translation that replaces any primed variable v' by the term (Xv).
- Private actions that are infinitely often enabled are guaranteed to be selected infinitely often: $(\mathbf{GF}\, U(g) \supset \mathbf{GF}\, g)$

This encoding extends to refinement morphisms, establishing a functor ***r-DESC→THEO***$_{\boldsymbol{FOLTL}}$, from which we can define a satisfaction relation. Notice that the fact that refinement of CommUnity designs is contravariant on the upper bound of actions is crucial: refinement morphisms are liveness preserving but interconnection morphisms are not.

In order to illustrate the generalised notion of connector in this setting, we present below the connector *cpipe* whose glue is given by

```
design pipe [t:sort, bound:t] is
in      i:t, scl:bool
out     o:t, eof:bool
prv     rd: bool, b: list(t)
do      put: true → b:=b.i
[] prv  next: |b|>0∧¬rd → o:=head(b) || b:=tail(b) || rd:=true
[]      get: rd → rd:=false
[] prv  sig: scl∧|b|=0 → eof:=true
```

This design models a buffer with unlimited capacity and a FIFO discipline. It signals the end of data to the consumer of messages as soon as the buffer gets empty and the sender of messages has already informed, through the input channel *scl*, that it will not send anymore messages.

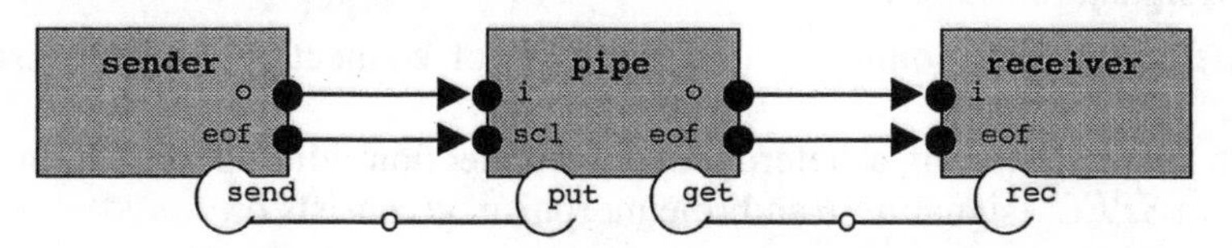

The two roles – the specifications *sender* and *receiver* introduced before – define the behaviour required of the components to which the connector *cpipe* can be applied. It is interesting to notice that, due to the fact that LTL is more abstract than CommUnity, we were able to abstract away completely the production of messages in the role *sender*. In the design of *sender* presented in Sect. 8.1 we had to consider an action modelling the production of messages.

We can now generalise these constructions even further by letting the connections use different specification formalisms and the instantiations to be performed over components designed in different design formalisms. In this way, it is possible to support the reuse of third-party components, namely legacy systems, as well as the integration of non-software components. We can thus highlight the role of architectures in promoting a structured and incremental approach to system construction and evolution.

However, to make sense of the interconnections, we have to admit that all the design formalisms are coordinated over the same category of signatures. That is, we assume that the integration of heterogeneous components is made at the level of the coordination mechanisms, independently of the way each component brings about its computations. Hence, we will assume given

- A family $\{\mathbf{DSGN}_d\}_{d:D}$ of categories of designs, all of which are coordinated over the same category $\mathbf{SIGN}$ via a family of functors $\{\mathbf{dsgn}_d\}_{d:D}$.
- A family $\{\mathbf{SPEC}_c\}_{c:C}$ of categories of specifications together with a family $\{\mathbf{spec}_c\!:\!\mathbf{SPEC}_c\!\rightarrow\!\mathbf{SIGN}\}_{c:C}$ of functors.
- A family $\{\models_s\}_{s:S}$ of satisfaction relations, each of which relates a design formalism $d(s)$ and a specification formalism $c(s)$. We do not require S to be the Cartesian product $D\times C$, i.e. there may be pairs of design and specification formalisms for which no satisfaction relation is provided.

Given such a setting, we can generalise the notion of connector.

9.5.4 Definition – (Heterogeneous) Architectural Connectors

A *heterogeneous* connection consists of

- A design formalism $\mathbf{DSGN}_d$ and a specification formalism $\mathbf{SPEC}_c$.
- A design $G\!:\!\mathbf{DSGN}_d$ and a specification $R\!:\!\mathbf{SPEC}_c$, called the glue and the role of the connection, respectively.

- A signature θ:***SIGN*** and two morphisms μ:***dsgn***$_d(\theta)\rightarrow G$, σ:***spec***$_c(\theta)\rightarrow R$ in ***DSGN***$_d$ and ***SPEC***, respectively, connecting the glue and the role via the signature (cable).

An heterogeneous connector is a finite set of connections with the same glue.

An instantiation of a heterogeneous connection with specification formalism ***SPEC***$_c$, signature θ and role morphism σ consists of

- A design formalism ***DSGN'***$_{d'}$ and a satisfaction relation $\models_{<d',c>}$ between ***DSGN'***$_{d'}$ and ***SPEC***$_c$ such that $<d',c>\in S$.
- A design P and a design morphism π:***dsgn***$_d'(\theta)\rightarrow P$ such that $\pi\models_s\sigma$.

An instantiation of a connector consists of an instantiation for each of its connections. ♦

Given that in the context of heterogeneous connectors we have to deal with several design formalisms, providing a semantics for the resulting configurations requires a homogeneous formalism to which all the design formalisms can be mapped. Clearly, because we want the heterogeneity of formalisms to be carried through to the implementations in order to be able to support the integration of legacy code, third-party components and even non-software components, this common formalism cannot be at the same level as that of designs, and the mapping cannot be a simple translation. What seems to make more sense is to choose a behaviour model that can be used to provide a common semantics to all the design formalisms so that the integration is not performed at the level of the descriptions, but of the behaviours that are generated from the descriptions. Indeed, when we talk about the integration of heterogeneous components, our goal is to coordinate their individual behaviours. An architecture should provide precisely the mechanisms through which this coordination is effected.

10 An Algebra of Connectors

As argued in [84], the level of support that Architectural Description Languages (ADLs) provide for connector building is still far from the one provided for components. For instance, although considerable amounts of work can be found on several aspects of connectors [2, 14, 84], further steps are still necessary to achieve a systematic way to construct new connectors from existing ones. Yet, the ability to manipulate connectors in a systematic and controlled way is essential for promoting reuse and incremental development, and to make it easier to address complex interactions.

At an architecture level of design, component interactions can be very simple (for instance, a shared variable), but they can be very complex as well (for example, database-accessing and networking protocols). Hence, it is very important that we have mechanisms for designing connectors in an incremental and compositional way, as well as principled ways of extending existing ones, promoting reuse. This is especially important for connectors that are used at lower levels of design because it is well known that the implementation of complex protocols is a very difficult and error-prone part of system development.

It is not always possible to adapt components to work with existing connectors. Even in those cases where it is feasible, a better alternative may be to modify the connectors because, typically, there are fewer connector types than component types. Moreover, most ADLs either provide a fixed set of connectors or only allow the creation of new ones from scratch, hence requiring from the designer a deep knowledge of the particular formalism and tools at hand. Conceptually, operations on connectors allow one to factor out common properties for reuse and to better understand the relationships between different connector types.

The notation and semantics of such connector operators are, of course, among the main issues to be dealt with. Our purpose in this chapter is to show how typical operators can be given an ADL-independent semantics by formalising them in the categorical framework that we presented in the previous chapter.

For instance, given a connector expressed through a configuration diagram ***dia***

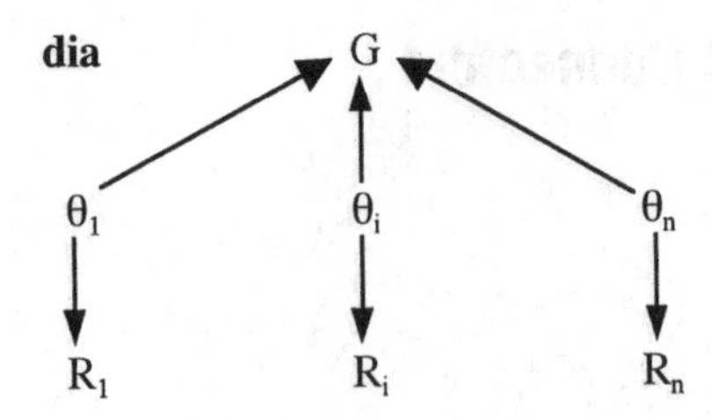

and a refinement $\eta:G \rightarrow G'$ of its glue, we can construct through ***dia+η*** another connector that has the same roles as the original one, but whose glue is now G'.

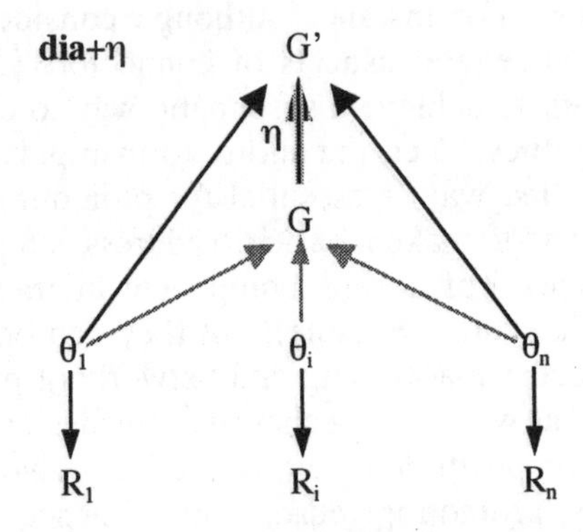

A fundamental property of this construction, given by compositionality, is that the semantics of the original connector, as expressed by the colimit of its diagram, is preserved in the sense that it is refined by the semantics of the new connector. This means that all instantiations of the new connector are refinements of instantiations of the old one. This operation supports the definition of connectors at higher levels of abstraction by delaying decisions on concrete representations of the coordination mechanisms that they offer, thus providing for the definition of specialisation hierarchies of connector types.

In this chapter we present three connector transformations that operate on the roles rather than the glue. Transformations that, like above, operate at the level of the glue are more sensitive in that they interfere more directly with the semantics of the connector to which they are applied. Hence, they should be restricted to engineers who have the power, and ensuing responsabilities, to change the way connectors are implemented. Operations on the roles are less critical and can be performed more liberally by users who have no access to the implementation (glue). In the last section, we present higher-order mechanisms that can be applied to connectors to obtain other connectors.

Throughout the chapter, we will be working over a fixed architectural school ***F=<c-DESC,Conf,r-DESC>*** (see Par. 9.4.3).

10.1 Three Operations on Connectors

10.1.1 Role Refinement

To tailor general-purpose connectors for a specific application, it is necessary to replace the generic roles by specialised ones that can effectively act as "formal parameters" for the application at hand. Role replacement is done in the same way as applying a connector to components: there must be a refinement morphism from the generic role to the specialised one. The old role is cancelled, and the new role morphism is the composition of the old one with the refinement morphism as discussed in Sect. 8.3.

Given an n-ary connector

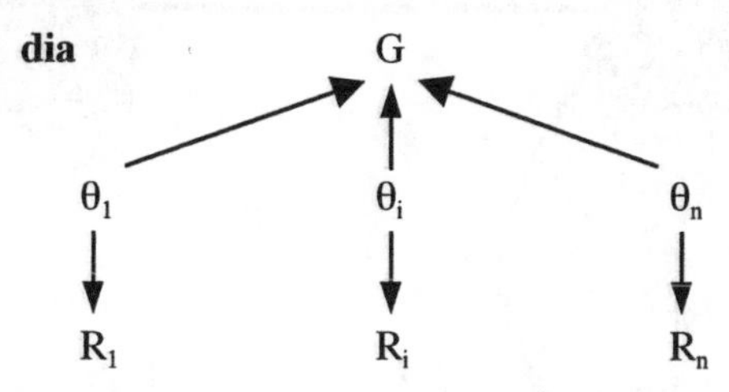

and a refinement morphism η_i:R_i→R'_i for some $1 \le i \le n$, the role refinement operation yields the connector

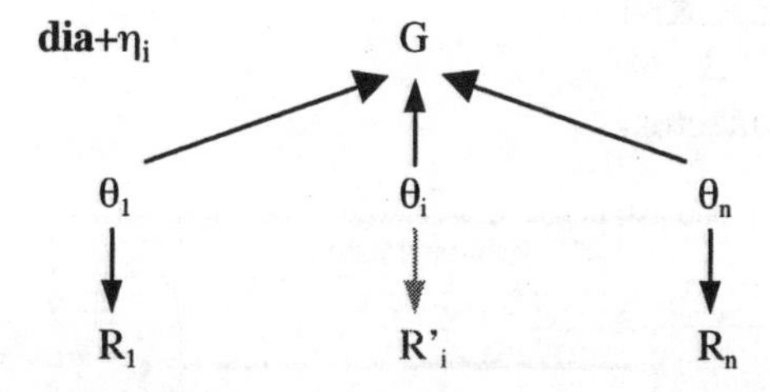

This operation can be applied to both abstract and heterogeneous connectors. It preserves the semantics of the connectors to which it is applied in the sense that any instantiation of the new connector is also an instantiation of the original one. This is because the refinement morphism that instantiates R'_i can be composed with η_i to yield an instantiation of R_i.

As an example of role refinement, consider the asynchronous connector shown previously. This connector is too general for our luggage distribution service because the sender and receiver roles do not impose any constraints on the admissible instances. We would like to refine these roles in order to prevent meaningless applications to our example, like sending the location of the check-in as a bag to the cart. This can be done through the refinement of *sender* with

```
design cart-role is
out    obag: int
prv    dest: -1..U-1
do     get[obag]: dest=-1 → dest'>-1
[]     put: dest>-1,false → obag:=0 || dest:=-1
```

where the channel *rd* of *sender* is refined by the term *dest>–1*. Notice that this is a generalised refinement as observed after Par. 8.3.1 in the sense that we are using a term of the target language to refine a channel. See [37] for details on this extension.

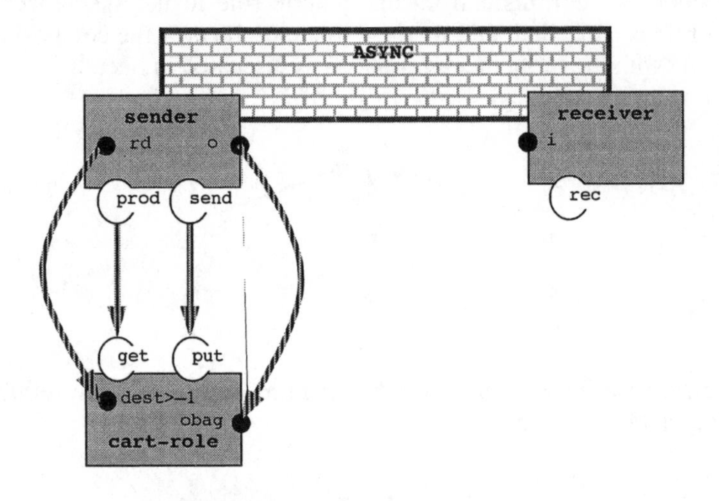

The resulting connector is

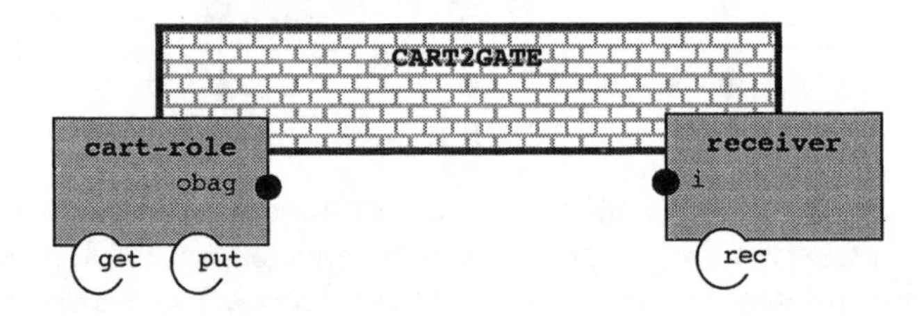

Notice that the invalid combinations are not possible because *cart-role* cannot be refined with a gate or a check-in. Moreover, the *obag* channel of *cart-role* cannot be refined by channel *laps* of cart.

10.1.2 Role Encapsulation

To prevent a role from being further refined, the second operation we consider, when executed repeatedly, decreases the arity of a connector by encapsulating some of its roles, making the result part of the glue.

Given an n-ary connector

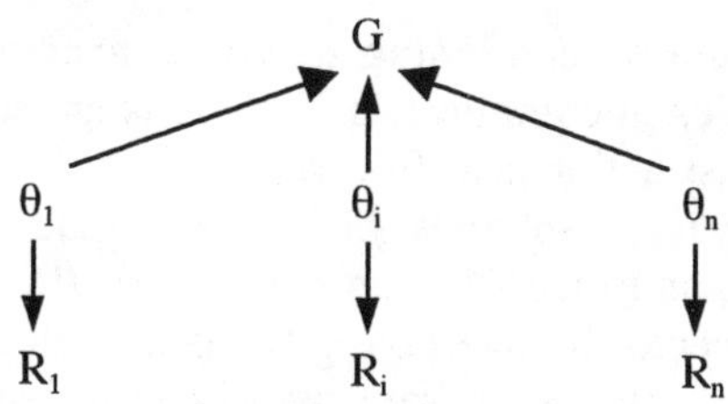

the encapsulation of the ith role is performed as follows: the pushout of the ith connection is calculated, and the other connections are changed by composing the morphisms that connect the channels to the glue with the morphism that connects the glue with the apex of the pushout, yielding a connector of arity *n–1*.

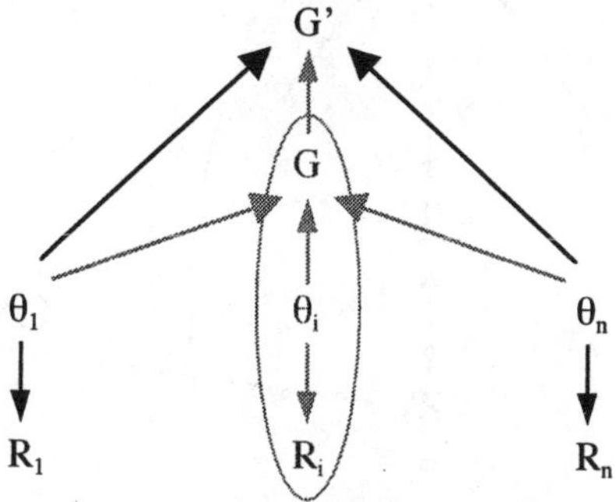

For instance, we can obtain *SUBSUMPTION* as defined in Par. 9.3.2, from *EXTENSION CORD* as defined in Par. 9.3.3, through encapsulation of the right-hand side *action* role. Indeed, as observed in Par. 9.3.3, both connectors have isomorphic glues: the difference between them is in the roles that they offer. On the other hand, the pushout of the glue–role connection that is involved in the encapsulation returns, up to isomorphism, that same glue because the role does not require any properties of the instances; it just performs name bindings.

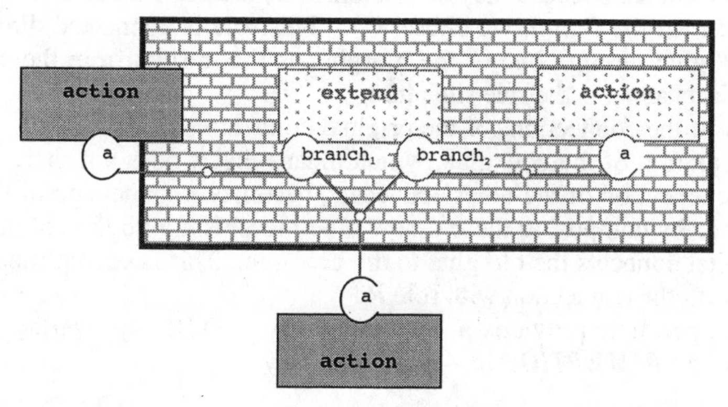

10.1.3 Role Overlay

The third operation allows combining several connectors into a single one if they have some roles in common, i.e. if there is an isomorphism between those roles. The construction is as follows.

Consider a connector with roles $\{R_i\}_{1\leq i\leq n}$ and glue G, and a connector with roles $\{R'_k\}_{1\leq k\leq m}$ and glue G' such that R_j and R'_l are isomorphic. We construct a new connector by overlaying the isomorphic roles. The glue of the new connector is calculated from the diagram that consists of all the connections that have isomorphic roles together with the isomorphisms, and calculate its colimit. The apex of the colimit is the new glue.

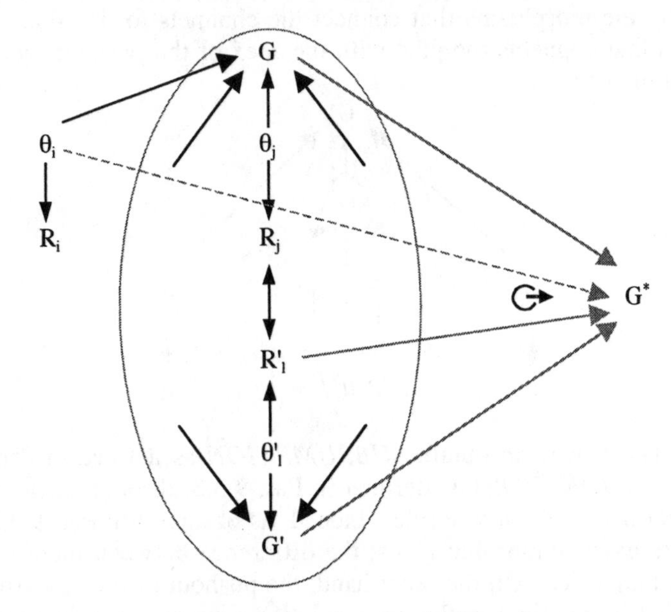

Each of the pairs of connections involved gives rise to a connection of the new connector. The role of this connection is one of the roles of the old connections; because they are isomorphic, it does not matter which one is chosen (in the figure, we chose R'_l). This role is connected directly to the new glue through one of the morphisms that results from the colimit; hence, its channel is the role signature.

Each of the connections of the original connectors that is not involved in the calculation of the new glue, which means that it does not share its role with a connection of the other connector, becomes a connection of the new connector by composing the old new glue morphism with the colimit morphism that connects the old glue to the new one. This is exemplified in the figure with the connection with role R_i.

This operation provides a way of building *SYNC* by overlaying two copies of *SUBSUMPTION* in a symmetric way.

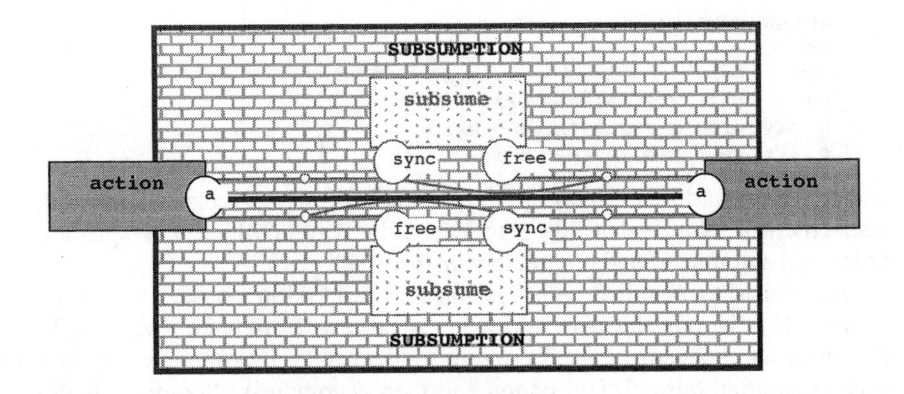

Notice that the colimit of the underlying diagram makes all actions collapse into a single one. Hence the glue of the resulting connector is isomorphic to the "sync" component. Moreover, each pair of overlaid roles results into a single one, and therefore there will be only two "action" roles. In summary, the resulting connector is *SYNC*.

10.2 Higher-Order Connectors

As explained before, it is important to have principled forms of adapting connectors to new situations, for instance, in order to incorporate compression, fault-tolerance, security, monitoring, etc. Let us consider *compression* as an example. The goal is to be able to adapt any connector that represents a communication protocol in order to compress data for transmission in a transparent way. In order to give a first-class description of this form of adaptation, the kind of communication protocol modelled by the adapted connector needs to be made more precise. We shall describe the *compression* adaptation mechanism only for connectors that model unidirectional communication protocols.

A generic unidirectional communication protocol can be modelled by the binary connector *Uni-comm[s]*:

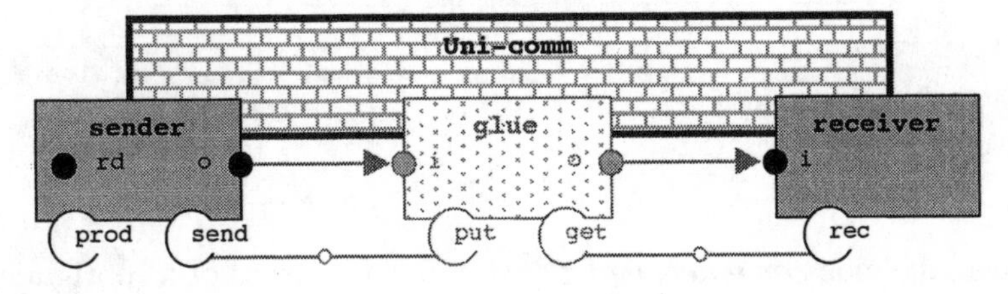

where

```
design glue[s] is
in     i:s
out    o:s
do     put: true,false → skip
[] prv prod[o]: true,false → skip
[]     get: true,false → skip
```

and *sender[s]* and *receiver[s]* are defined as before (see Sects. 8.1 and 9.2, respectively). Notice that this glue leaves the way in which messages are processed and transmitted completely unspecified.

Our aim is to install a compression/decompression service over *Uni-comm*. That is, our aim is to apply an operator to *Uni-comm* such that, in the resulting connector, a message sent by the sender is compressed before it is transmitted through *Uni-comm* and then decompressed before it is delivered to the receiver. We describe such an operator by a higher-order connector, where the compression and decompression algorithms are taken as parameters as captured by the following algebraic specification:

```
spec   Ξcd is Ξnat +
sorts  s,t
ops    comp:t->s
       decomp:s->t
       size_s:s->nat
       size_t:t->nat
axioms decomp(comp(x))=x, for any x:t
       size_s(comp(x)) ≤ size_t(x), for any x:t
```

Sorts t and s represent the types of original and compressed messages, respectively. The operation *comp* represents the process of compression of a single message, and *decomp* the inverse process of decompression. The size of the compressed message is required not to be greater than the size of the original message. At configuration time, these data elements must be instantiated with specific sorts and operations.

The higher-order connector itself, which we name *Compression(Uni-comm)[*Ξ^{cd}*]*, is defined by:

- The binary connector *COMPRESSION*

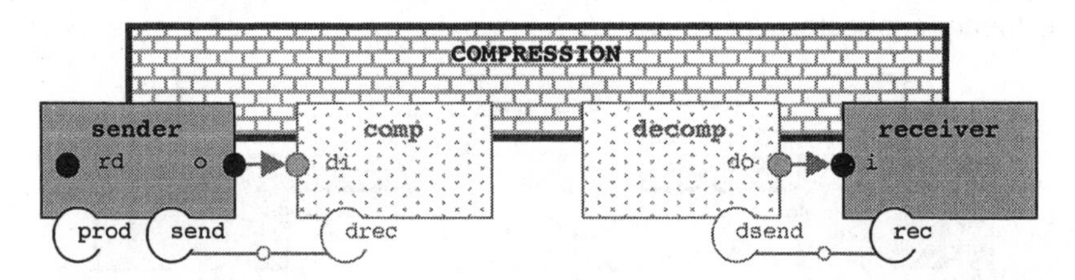

where the glue *comp-decomp[*Ξ^{cd}*]* is defined in terms of a configuration with the following two components:

```
design comp[Ξcd] is
in      di:t
out     co:s
prv     v:t; rd,msg:bool
do      drec: ¬msg → v:=di ‖ msg:=true
[] prv  comp:¬rd∧msg → co:=comp(v) ‖ rd:=true
[]      csend:rd → rd:=false ‖ msg:=false

design decomp[Ξcd] is
in      ci:s
out     do:t
prv     v:s; rd,msg:bool
do      crec: ¬msg → v:=ci ‖ msg:=true
[]      dec:¬rd∧ msg → do:=decomp(v) ‖ rd:=true
[]      dsend: rd → rd:=false ‖ msg:=false
```

Design $comp[\Xi^{cd}]$ models the compression of messages of type *t* received through *di* into messages of type *s* that are then transmitted through *co*. Design $decomp[\Xi^{cd}]$ models the decompression of messages of type *s* received through *ci* into messages of type *t* that are then transmitted through *do*.

- The connector *Uni-comm[s]* — the formal parameter.
- The refinement morphisms $\eta_s{:}sender[s] \rightarrow comp\text{-}decomp[\Xi^{cd}]$ and $\eta_r{:}receiver[s] \rightarrow comp\text{-}decomp[\Xi^{cd}]$ induced, respectively, by

$$\eta^*_s(o)=co,\ \eta^*_s(rd)=rd,\ \eta^*_s(comp)=prod,\ \eta^*_s(csend)=send$$
$$\eta^*_r(i)=ci,\ \eta^*_r(crec)=rec$$

Because components *comp* and *decomp* do not interact, any component refined by one of them is also refined by their composition $comp\text{-}decomp[\Xi^{cd}]$. The corresponding induced morphisms have only to take into account the renaming of variables and actions.

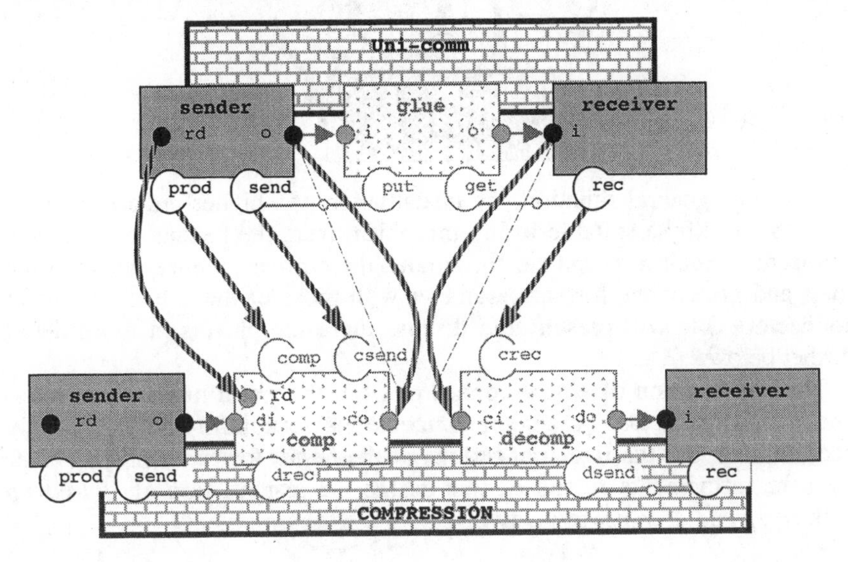

In summary, *Compression(Uni_comm)[Ξ^{cd}]* has the formal parameter *Uni-comm[s]*, which restricts the actual connectors to which the service of compression/decompression can be applied — it requires that any actual parameter (connector) models a unidirectional communication protocol. The connector *COMPRESSION* describes, on the one hand, that messages sent by the actual sender are transmitted to *comp*, which compresses them, and, on the other hand, that *decomp* decompresses the messages it receives and delivers the result to the actual receiver. Finally, the two refinement morphisms establish the instantiation of *Uni-comm[s]* with *comp[s]* in the role of *sender*, and *decomp[s]* in the role of *receiver*. In this way, the formal parameter *Uni-comm[s]* is the connector used to transmit compressed messages.

It remains to explain the procedure of parameter passing, i.e. how the service just described can be installed over a specific connector and how the resulting connector is obtained. As an example, consider again *ASYNC*. It is not difficult to realize that we may replace *Uni-comm[s]*, as the formal parameter of *Compression(Uni-comm)[Ξ^{cd}]*, by *ASYNC* because this connector does model a unidirectional communication protocol. More concretely, *ASYNC* has exactly the same roles as *Uni-comm* and its glue – *buffer[s]* – is a refinement of *Uni-comm*'s glue.

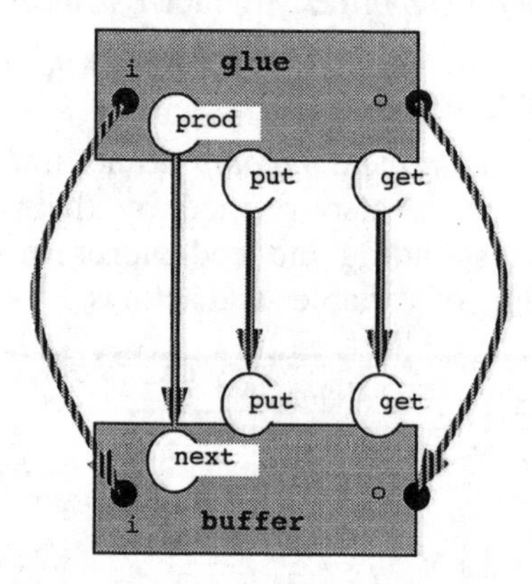

In a more general situation, the instantiation of a higher-order connector is established by a suitable fitting morphism from the formal to the actual connector. Such a morphism formulates the correspondence between the roles and glue of the formal parameter with those of the actual parameter connector. We will present and discuss these morphisms in more detail further below.

The construction of a new connector from the given higher-order connector and the actual parameter connector is straightforward. We only need to compose the interconnections of the *buffer* to *sender* and *receiver* with the refinements η_s and η_r that define the instantiation of *Uni-comm* with *comp* and *decomp*, respectively.

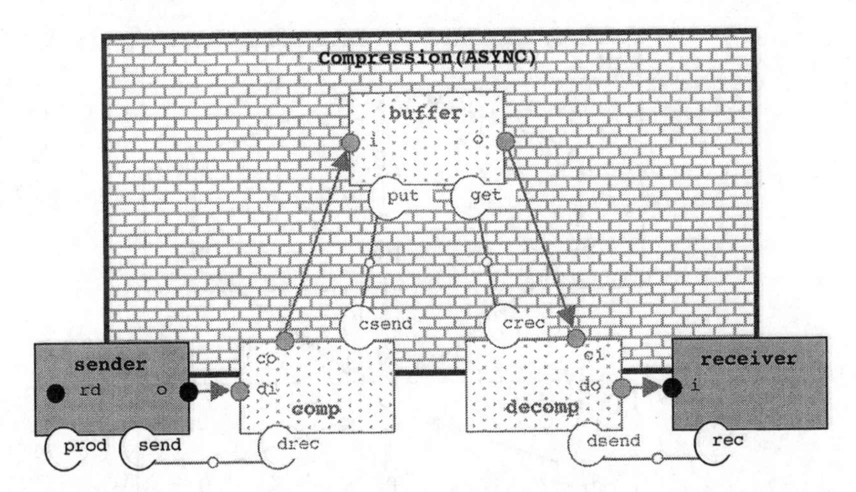

For example, channel *co* of *comp* becomes connected to the input chanel *i* of *buffer* because *co* corresponds to the chanel *o* of *sender*, which in turn is, in *ASYNC*, connected to *i*. The resulting configuration fully defines the connector *Compression(ASYNC)*$[\Xi^{cd}+K]$. Its roles are *sender* and *receiver*, and its glue is defined in terms of a configuration involving *comp*, *decomp* and *buffer*, as shown above.

Summarising, we described the installation of a compression-decompression service over a unidirectional communication protocol as a parameterised entity that has connectors as parameters and delivers a connector as a result of its instantiation. This is why it is called a higher-order connector. Then we explained how the higher-order connector can be instantiated with a specific connector, and finally, we showed how the resulting connector is obtained.

10.2.1 Definition – Higher-Order Connectors

A *higher-order connector* (hoc) consists of:

- A connector *pC*, called the formal parameter of the hoc; its roles, glue and connections are called, respectively, the parametric roles, the parametric glue and the parametric connections of the hoc.
- A connector *C* – its roles and glue are also called the roles and the glue of the hoc.
- An instantiation of the formal parameter connector with the glue of the hoc, i.e. a refinement morphism η_i from each of the parametric roles to the glue, such that the diagram in ***c-DESC*** obtained by composing the role morphism of each parametric connection with its instantiation constitutes a well-formed configuration.

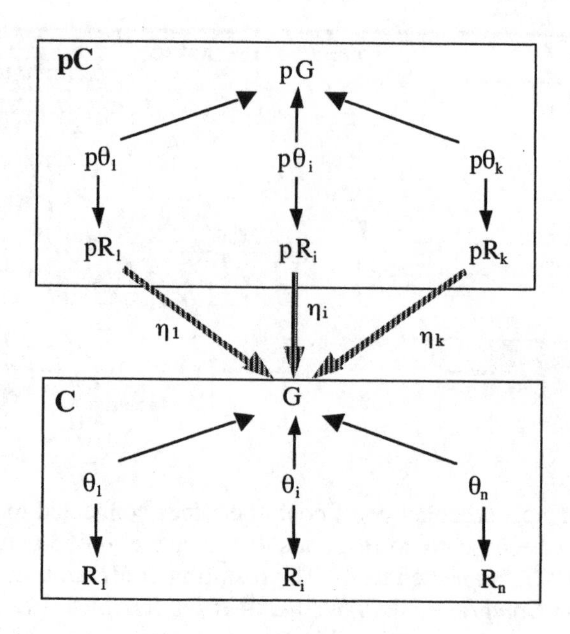

◆

10.2.2 Definition – Semantics of Higher-Order Connectors

The semantics of a higher-order connector is the connector depicted below. Its roles are the roles of C and its glue is G', the design returned by the colimit of the configuration $pC+(\eta_i)$. ◆

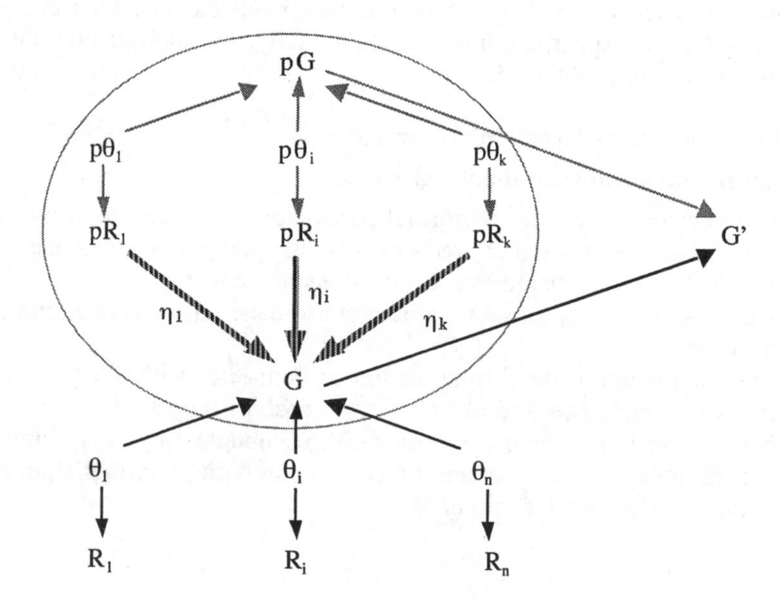

For simplicity, we have assumed one single parameter. However, the definition can be extended to the case of several parameters in a straightforward way. Intuitively, instantiation of can be regarded as the replacement of a connector (the formal parameter *pC*) that was instantiated to given components of a system (the glue of the hoc) by another connector (the actual parameter). In addition, the type of interconnection that *pC* ensures must be preserved. In other words, the design that results from the replacement must be a refinement of the design from which we started.

Like for connectors, the instantiation of the formal parameter of a higher-order connector is established via a fitting morphism from the formal to the actual parameter. These morphisms, on the one hand, formulate the correspondence between roles and glue of the formal parameter with those of the actual parameter and, on the other hand, capture conditions under which the "functionality" of the formal parameter is preserved.

In order to be able to use, in the design of a given system, a connector *C* in place of a connector *C*', it is obvious that the two connectors must have the same number of roles. Furthermore, *C*' must be able to be instantiated with the same components as *C*. In other words, every restriction on the components to which *C*' can be applied must also be a restriction imposed by *C*. In this way, fitting morphisms must require that each of the roles of *C*' is refined by the corresponding role of *C*.

As shown through the connector *Uni-comm*, connectors may be based on glues that are not fully developed as designs (that is, may be underspecified). Nevertheless, the concrete commitments that they make may determine, to some extent, the type of interconnection that the connector will ensure. The type of interconnection is clearly preserved if we simply consider a less unspecified glue, i.e. if we refine the glue. Hence, fitting morphisms must allow for arbitrary refinements of the glue. Having this in mind, we arrive at the following notion of fitting morphism.

10.2.3 Definition – Fitting Morphisms

A morphism ϕ from a connection $<\sigma_1:\mathbf{desc}(\theta_1)\to G_1,\mu_1:\mathbf{desc}(\theta_1)\to R_1>$ to a connection $<\sigma_2:\mathbf{desc}(\theta_2)\to G_2,\mu_2:\mathbf{desc}(\theta_2)\to R_2>$, also called a *fitting morphism*, consists of a pair $<\phi_G:G_1\to G_2,\phi_R:R_2\to R_1>$ of refinement morphisms in ***r-DESC*** s.t. the interconnection $<\sigma_1,\mu_1>+\phi_G$ of R_1 with G_2 is refined by the interconnection $<\sigma_2,\mu_2>+\phi_R$. A fitting morphism ϕ from a connector C_1 to a connector C_2 with the same number of connections consists of a fitting morphism ϕ from each of C_1's connections to each of C_2's connections, all with the same glue refinement ϕ_G.

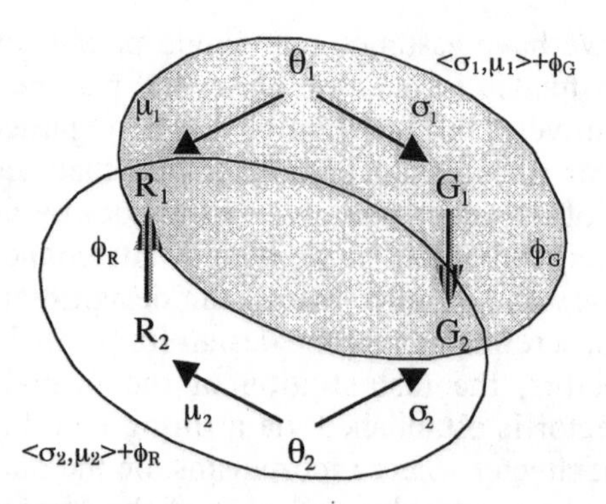

◆

If there exists a fitting morphism from a connector C_1 to a connector C_2, then we may replace each occurrence of C_1 by an occurrence of C_2. Compositionality ensures that every coordination decision is preserved.

10.2.4 Definition – Instantiation of Higher-Order Connectors

An instantiation of a higher-order connector with formal parameter pC consists of a connector C^A (the actual parameter) together with a fitting morphism $\phi:pC \rightarrow C^A$ such that a well-formed configuration is obtained by first composing the role morphisms of each actual connection with the corresponding fitting component, and then with the role instantiation.

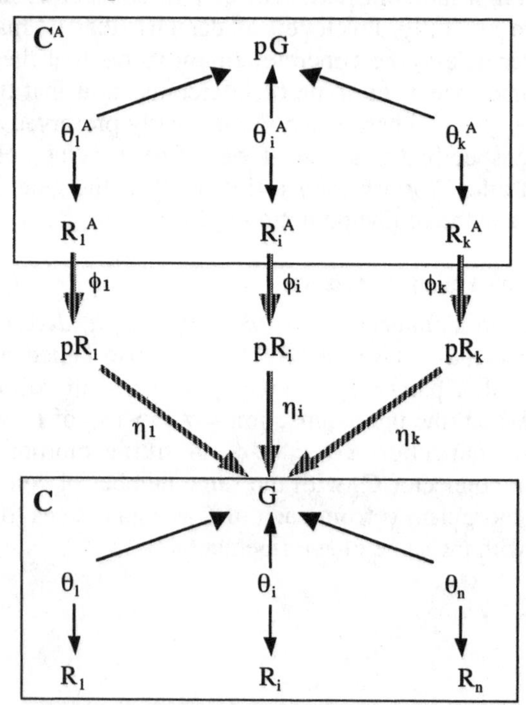

The semantics of a higher-order connector instantiation is the connector with the same roles as C and whose glue is a design returned by the colimit of the configuration $C^A+(\phi_i;\eta_i)$. ♦

Higher-order connectors facilitate the separation of concerns in the development of complex connectors and their compositional construction. An important feature of our notion of higher-order connector is that different kinds of functionality, modelled separately by different higher-order connectors, can be combined, giving rise also to a higher-order connector. In this way, it is possible to analyse the properties that such compositions exhibit, namely to investigate whether undesirable properties emerge and desirable properties are preserved.

The key idea for composition of hocs is the instantiation of a hoc with a hoc – *parameterised instantiation*. Examples, definitions and properties of this mechanism can be found in [79].

References

1. J. Adámek, H. Herrlich, G. Strecker (1990) *Abstract and Concrete Categories.* Wiley Interscience, New York
2. R. Allen, D. Garlan (1997) A formal basis for architectural connectors. *ACM TOSEM* 6(3):213–249
3. F. Arbab (1998) What do you mean, coordination? In: *Bulletin of the Dutch Association for Theoretical Computer Science (NVTI),* March 1998
4. M. Arbib, E. Manes (1975) *Arrows, Structures and Functors: The Categorical Imperative.* Academic Press, New York
5. M. Arrais, J. L. Fiadeiro (1996) Unifying theories in different institutions. In: M. Haveraaen, O. Owe, O.-J. Dahl (eds) *Recent Trends in Data Type Specification. LNCS, vol 1130.* Springer, Berlin Heidelberg New York, pp 81–101
6. A. Asperti, G. Longo (1991) *Categories, Types and Structures: An Introduction to Category Theory for the Working Computer Scientist.* MIT Press, Cambridge, MA
7. E. Astesiano, M. Broy, G. Reggio (1999) Algebraic specification of concurrent systems. In: [8]
8. E. Astesiano, B. Krieg-Brückner, H.-J. Kreowski (eds) (1999) *IFIP WG 1.3 Book on Algebraic Foundations of System Specification.* Springer, Berlin Heidelberg New York
9. R. J. Back (1990) Refinement calculus II: parallel and reactive programs. In: J. deBakker, W. deRoever, G. Rozenberg (eds) *Stepwise Refinement of Distributed Systems. LNCS, vol 430.* Springer, Berlin Heidelberg New York, pp 67–93
10. R. J. Back, K. Sere (1996) Superposition refinement of reactive systems. *Formal Aspects of Computing* 8(3):324–346
11. J. P. Banâtre, D. Le Métayer (1993) Programming by multiset transformation. *Communications ACM* 16(1):55–77
12. M. Barr, C. Wells (1990) *Category Theory for Computing Science.* Prentice Hall, London
13. H. Barringer (1987) The use of temporal logic in the compositional specification of concurrent systems. In: A. Galton (ed) *Temporal Logics and Their Applications.* Academic Press, New York
14. L. Bass, P. Clements, R. Kasman (1998) *Software Architecture in Practice.* Addison-Wesley, Reading, MA
15. M. Bednarczyk (1988) *Categories of Asynchronous Transition Systems.* PhD Thesis, University of Sussex

16. G. Berry, G. Boudol (1992) The chemical abstract machine. *Theoretical Computer Science* 96:217–248
17. R. Burstall, J. Goguen (1977) Putting theories together to make specifications. In: R. Reddy (ed) *Proc. Fifth International Joint Conference on Artificial Intelligence,* August 1977, Cambridge, MA, pp 1045–1058
18. R. Burstall, J. Goguen (1980) The semantics of CLEAR, a specification language. In: *Proc. Advanced Course on Abstract Software Specification. LNCS, vol 86*. Springer, Berlin Heidelberg New York, pp 292–332
19. K. Chandy, J. Misra (1988) *Parallel Program Design – A Foundation*. Addison-Wesley, Reading, MA
20. J. F. Costa, A. Sernadas, C. Sernadas (1993) Data encapsulation and modularity: three views of inheritance. In: A. Borzyszkowski, S. Sokolowski (eds) *Mathematical Foundations of Computer Science. LNCS, vol 711*. Springer, Berlin Heidelberg New York, pp 382–391
21. J. F. Costa, A. Sernadas, C. Sernadas, H.-D. Ehrich (1992) Object interaction. In: *Mathematical Foundations of Computer Science. LNCS, vol 629*. Springer, Berlin Heidelberg New York, pp 200–208
22. R. Crole (1993) *Categories for Types*. Cambridge University Press, Cambridge
23. E. Dijkstra (1976) *A Discipline of Programming*. Prentice Hall, London
24. H.-D. Ehrich, J. Goguen, A. Sernadas (1991) A categorial theory of objects as observed processes. In: J. deBakker, W. deRoever, G. Rozenberg (eds) *Foundations of Object-Oriented Languages. LNCS, vol 489*. Springer, Berlin Heidelberg New York, pp 203–228
25. H.-D. Ehrich, A. Sernadas, C. Sernadas (1989) Objects, object types and object identity. In: H. Ehrig et al. (eds) *Categorical Methods in Computer Science with Aspects from Topology. LNCS, vol 393*. Springer, Berlin Heidelberg New York, pp 142–156
26. H. Ehrig, M. Große-Rhode, U. Wolter (1998) Applications of category theory to the area of algebraic specification in computer science. *Applied Categorical Structures* 6:1–35
27. H. Ehrig, K.-D. Kiermeier, H.-J. Kreowski, W. Kuhnel (1974) *Universal Theory of Automata: A Categorical Approach*. B. G. Teubner, Stuttgart
28. H. Ehrig, B. Mahr (1985) *Fundamentals of Algebraic Specification 1: Equations and Initial Semantics. EATCS Monographs on Theoretical Computer Science, vol 6*. Springer, Berlin Heidelberg New York
29. H. Ehrig, B. Mahr (1990) *Fundamentals of Algebraic Specification 2: Module Specifications and Constraints. EATCS Monographs on Theoretical Computer Science, vol 21*. Springer, Berlin Heidelberg New York
30. E. Emerson, E. Clarke (1982) Using branching time logic to synthesize synchronisation skeletons. *Science of Computer Programming* 2:241–266
31. J. L. Fiadeiro (1996) On the emergence of properties in component-based systems. In: M. Wirsing, M. Nivat (eds) *Algebraic Methodology and Software Technology. LNCS, vol 1101*. Springer, Berlin Heidelberg New York, pp 421–443
32. J. L. Fiadeiro, J. F. Costa (1996) Mirror, mirror in my hand: a duality between specifications and models of process behaviour. *Mathematical Structures in Computer Science* 6:353–373

33. J. L. Fiadeiro, J. F. Costa (1995) Institutions for behaviour specification. In: E. Astesiano, G. Reggio, A. Tarlecki (eds) *Recent Trends in Data Type Specification. LNCS, vol 906*. Springer, Berlin Heidelberg New York, pp 273–289
34. J. L. Fiadeiro, A. Lopes (1997) Semantics of architectural connectors. In: M. Bidoit, M. Dauchet (eds) *Theory and Practice of Software Development. LNCS, vol 1214*. Springer, Berlin Heidelberg New York, pp 505–519
35. J. L. Fiadeiro, A. Lopes (1999) Algebraic semantics of coordination, or what is in a signature? In: A. Haeberer (ed) *Algebraic Methodology and Software Technology. LNCS, vol 1548*. Springer, Berlin Heidelberg New York, pp 293–307
36. J. L. Fiadeiro, A. Lopes, T. Maibaum (1997) Synthesising interconnections. In: R. Bird, L. Meertens (eds) *Algorithmic Languages and Calculi*. Chapman Hall, London, pp 240–264
37. J. L. Fiadeiro, A. Lopes, M. Wermelinger (2003) A mathematical semantics for architectural connectors. In: R. Backhouse, J. Gibbons (eds) *Generic Programming. LNCS, vol 2793*. Springer, Berlin Heidelberg New York, pp 90–234
38. J. L. Fiadeiro, T. Maibaum (1991) Describing, structuring, and implementing objects. In: J. deBakker, W. deRoever, G. Rozenberg (eds) *Foundations of Object-Oriented Languages. LNCS, vol 489*. Springer, Berlin Heidelberg New York, pp 274–310
39. J. L. Fiadeiro, T. Maibaum (1992) Temporal theories as modularisation units for concurrent system specification. *Formal Aspects of Computing* 4(3):239–272
40. J. L. Fiadeiro, T. Maibaum (1995) Verifying for reuse: foundations of object-oriented system verification. In: C. Hankin, I. Makie, R. Nagarajan (eds) *Theory and Formal Methods*. World Scientific Publishing Company, Singapore, pp 235–257
41. J. L. Fiadeiro, T. Maibaum (1995) Interconnecting formalisms: supporting modularity, reuse and incrementality. In: G. E. Kaiser (ed) *ACM SIGSOFT Software Engineering Notes* 20(4):72–80
42. J. L. Fiadeiro, T. Maibaum (1996) A mathematical toolbox for the software architect. In: *Proc. 8th Int. Workshop on Software Specification and Design*. IEEE Computer Society Press, Silver Spring, MD, pp 46–55
43. J. L. Fiadeiro, T. Maibaum (1996) Design structures for object-based systems. In: S. Goldsack, S. Kent (eds) *Formal Methods and Object Technology*. Springer, Berlin Heidelberg New York, pp 183–204
44. J. L. Fiadeiro, T. Maibaum (1997) Categorical semantics of parallel program design. *Science of Computer Programming* 28(2–3):111–138
45. J. L. Fiadeiro, A. Sernadas (1988) Structuring theories on consequence. In: D. Sannella, A. Tarlecki (eds) *Recent Trends in Data Type Specification. LNCS, vol 332*. Springer, Berlin Heidelberg New York, pp 44–72
46. J. L. Fiadeiro, A. Sernadas, C. Sernadas (1988) Knowledge bases as structured theories. In: K. Nori, S. Kumar (eds) *Foundations of Software Technology and Theoretical Computer Science. LNCS, vol 338*. Springer, Berlin Heidelberg New York, pp 469–486

47. P. Fingar (2000) Component-based frameworks for e-commerce. *Communications ACM* 43(10):61–66
48. N. Francez, I. Forman (1996) *Interacting Processes*. Addison-Wesley, Reading, MA
49. P. Freyd, A. Scedrov (1990) *Categories, Allegories*. North-Holland, Amsterdam
50. D. Garlan, D. Perry (1994) Software architecture: practice, potential, and pitfalls. In: *Proc. 16th International Conference on Software Engineering*. IEEE Computer Society Press, Silver Spring, MD, pp 363–364
51. D. Gelernter, N. Carriero (1992) Coordination languages and their significance. *Communications ACM* 35(2):97–107
52. J. Goguen (1971) Mathematical representation of hierarchically organised systems. In: E. Attinger (ed) *Global Systems Dynamics*. Krüger, New York, pp 112–128
53. J. Goguen (1973) Categorical foundations for general systems theory. In: F. Pichler, R. Trappl (eds) *Advances in Cybernetics and Systems Research*. Transcripta Books, New York, pp 121–130
54. J. Goguen (1975) Objects. *International Journal of General Systems* 1(4):237–243
55. J. Goguen (1986) Reusing and interconnecting software components. *IEEE Computer* 19(2):16–28
56. J. Goguen (1989) Principles of parametrised programming. In: T. J. Biggerstaff, A. J. Perlis (eds) *Software Reusability*. Addison-Wesley, Reading, MA, pp 159–225
57. J. Goguen (1991) A categorical manifesto. *Mathematical Structures in Computer Science* 1(1):49–67
58. J. Goguen (1991) Sheaf semantics of concurrent interacting objects. *Mathematical Structures in Computer Science* 1(2):159–191
59. J. Goguen (1996) Parametrised programming and software architecture. In: *Proc. 4th International Conference on Software Reuse*. IEEE Computer Society Press, Silver Spring, MD, pp 2-11
60. J. Goguen, R. Burstall (1984) Some fundamental algebraic tools for the semantics of computation, part 1: comma categories, colimits, signatures and theories. *Theoretical Computer Science* 31(2):175–209
61. J. Goguen, R. Burstall (1984) Some fundamental algebraic tools for the semantics of computation, part 2: signed and abstract theories. *Theoretical Computer Science* 31(3):263–295
62. J. Goguen, R. Burstall (1992) Institutions: abstract model theory for specification and programming. *Journal ACM* 39(1):95–146
63. J. Goguen, R. Burstall (1986) A study in the foundations of programming methodology: specfications, institutions, charters and parchments. In: D. Pitt et al. (eds) *Category Theory and Computer Programming*. *LNCS, vol 240*. Springer, Berlin Heidelberg New York, pp 313–333
64. J. Goguen, S. Ginali (1978) A categorical approach to general systems theory. In: G. Klir (ed) *Applied General Systems Research*. Plenum, New York, pp 257–270
65. J. Goguen, J. Thatcher, E. Wagner (1978) An initial algebra approach to the specification, correctness and implementation of abstract data types. In:

R. Yeh (ed) *Current Trends in Programming Methodology IV*. Prentice Hall, London, pp 80–149

66. J. Goguen, J. Thatcher E. Wagner, J. Wright (1973) A junction between computer science and category theory I: basic concepts and examples. Technical report, IBM Watson Research Center, Yorktown Heights NY, Report RC 4526 (part 1) and 5908 (part 2)
67. R. Goldblatt (1987) *Logics of Time and Computation*. CSLI, Stanford
68. R. Goldblatt (1989) *Topoi: The Categorial Analysis of Logic*. North-Holland, Amsterdam
69. C. A. R. Hoare (1985) *Communicating Sequential Processes*. Prentice Hall, London
70. B. Jacobs (2001) *Categorical Logic and Type Theory*. Elsevier, Amsterdam
71. S. Johnson (2001) *Emergence: The Connected Lives of Ants, Brains, Cities and Software*. Penguin, London
72. S. Katz (1993) A superimposition control construct for distributed systems. *ACM TOPLAS* 15(2):337–35
73. W. Kent (1993) Participants and performers: a basis for classifying object models. In: *Proc. OOPSLA 1993 Workshop on Specification of Behavioral Semantics in Object-Oriented Information Modeling*
74. J. Lambek, P. Scott (1986) *Introduction to Higher-Order Categorical Logic*. Cambridge University Press, Cambridge
75. F. W. Lawvere, S. H. Schanuel (1997) *Conceptual Mathematics – A First Introduction to Categories*. Cambridge University Press, Cambridge
76. J. Loeckx, H.-D. Ehrich, M. Wolf (1996) *Specification of Abstract Data Types*. Wiley, New York
77. A. Lopes, J. L. Fiadeiro (1999) Using explicit state to describe architectures. In: E. Astesiano (ed) *Fundamental Approaches to Software Engineering*. *LNCS, vol 1577*. Springer, Berlin Heidelberg New York, pp 144–160
78. A. Lopes, J. L. Fiadeiro (2004) Superposition: composition vs refinement of non-deterministic action-based systems. *Formal Aspects of Computing* 16(1):5–18
79. A. Lopes, M. Wermelinger, J. L. Fiadeiro (2003) Higher-order architectural connectors. *ACM TOSEM* 12(1):64–104
80. S. MacLane (1971) *Categories for the Working Mathematician*. Springer, Berlin Heidelberg New York
81. J. Magee, J. Kramer, M. Sloman (1989) Constructing distributed systems in Conic. *IEEE TOSE* 15(6):663–675
82. T. Maibaum, P. Veloso, M. Sadler (1985) A theory of abstract data types for program development: bridging the gap? In: H. Ehrig et al. (eds) *Formal Methods and Software Development*. *LNCS, vol 186*. Springer, Berlin Heidelberg New York, pp 214–230
83. Z. Manna, P. Wolper (1984) Synthesis of communicating processes from temporal logic specifications. In: *ACM TOPLAS* 6(1):68–93
84. N. Mehta, N. Medvidovic, S. Phadke (2000) Towards a taxonomy of software connectors. In: *Proc. 22nd International Conference on Software Engineering*. IEEE Computer Society Press, Silver Spring, MD, pp 178–187
85. J. Meseguer (1989) General logics. In: H.-D. Ebbinghaus et al. (eds) *Logic Colloquium 87*. North-Holland, Amsterdam, pp 275–329

86. B. Meyer (1992) *Object-Oriented Software Construction.* Addison-Wesley, Reading, MA
87. M. Moriconi, X. Qian (1994) Correctness and composition of software architectures. *ACM SIGSOFT Software Engineering Notes* 19(5):164–174
88. D. Perry, A. Wolf (1992) Foundations for the study of software architectures. *ACM SIGSOFT Software Engineering Notes* 17(4):40–52
89. B. Pierce (1991) *Basic Category Theory for Computer Scientists.* MIT Press, Cambridge, MA
90. D. Pitt, S. Abramsky, A. Poigné, D. Rydeheard (eds) (1985) *Category Theory and Computer Programming. LNCS, vol 240.* Springer, Berlin Heidelberg New York
91. A. Poigné (1992) Basic category theory. In: *Handbook of Logic in Computer Science, vol 1.* Oxford University Press, pp 413–640
92. H. Reichel (1987) *Initial Computability, Algebraic Specifications, and Partial Algebras.* Oxford University Press
93. G.-C. Roman, P. J. McCann, J. Y. Plun (1997) Mobile UNITY: reasoning and specification in mobile computing. *ACM TOSEM* 6(3):250–282
94. R. Rosen (2000) *Life Itself.* Columbia University Press, New York
95. D. Sannella, A. Tarlecki (1988) Building specifications in an arbitrary institution. *Information and Control* 76:165–210
96. D. Sannella, S. Sokolowski, A. Tarlecki (1992) Toward formal development of programs from algebraic specifications: parameterisation revisited. *Acta Informatica* 29:689–736
97. V. Sassone, M. Nielsen, G. Winskel (1996) Models for concurrency: towards a classification. *Theoretical Computer Science* 170:277–296
98. M. Shaw (1996) Procedure calls are the assembly language of software interconnection: connectors deserve first-class status. In: D. A. Lamb (ed) *Studies of Software Design. LNCS, vol 1078.* Springer, Berlin Heidelberg New York, pp 17–32
99. M. Shaw, D. Garlan (1996) *Software Architecture: Perspectives on an Emerging Discipline.* Prentice Hall, London
100. D. Smith (1993) Constructing specification morphisms. *Journal of Symbolic Computation* 15(5–6):571–606
101. Y. Srinivas, R. Jüllig (1995) Specware™: formal support for composing software. In: B. Möller (ed) *Mathematics of Program Construction. LNCS, vol 947.* Springer, Berlin Heidelberg New York, pp 399–422
102. A. Tarlecki, R. Burstall, J. Goguen (1991) Some fundamental algebraic tools for the semantics of computation, part 3: indexed categories. *Theoretical Computer Science* 91:239–264
103. P. Veloso, T. Maibaum, M. Sadler (1985) Program development and theory manipulation. *Proc. Third Int. Workshop on Software Specification and Design.* IEEE Computer Society Press, Silver Spring, MD, pp 228–232
104. P. Veloso, T. Maibaum (1995) On the modularisation theorem for logical specifications. *Information Processing Letters* 53:287–293
105. R. Walters (1991) *Categories and Computer Science.* Cambridge Computer Science Texts
106. G. Winskel (1984) Synchronization trees. *Theoretical Computer Science* 34:33-82

107. G. Winskel (1985) Categories of models for concurrency. In: S. D. Brookes et al. (eds) *Seminar on Concurrency. LNCS, vol 197.* Springer, Berlin Heidelberg New York, pp 246–267
108. G. Winskel (1987) Petri nets, algebras, morphisms and compositionality. In: *Information and Computation* 72:197–238
109. G. Winskel, M. Nielsen (1995) Models for concurrency. In: S. Abramsky, D. Gabbay, T. Maibaum (eds) *Handbook of Logic in Computer Science, vol 4,* Oxford University Press, pp 1–148
110. P. Wolper (1989) On the relation of programs and computations to models of temporal logic. In: B. Banieqbal, H. Barringer, A. Pnueli (eds) *Temporal Logic in Specification. LNCS, vol 398.* Springer, Berlin Heidelberg New York, pp 75–123
111. P. Zave (1993) Feature interactions and formal specifications in telecommunications. *IEEE Computer* XXVI(8):20–30
112. P. Zave, M. Jackson (1993) Conjunction as composition. *ACM TOSEM* 2(4):371–411

Index

A

$a{\downarrow}C$, 35, 125
$a{\downarrow}\varphi$, 111
action (CommUnity), 179
adjoint, 151
adjunction, 151
  (co)unit, 152
  dual of, 152
amalgamated sum. *See* pushout
amnestic concrete category, 96, 99
ANCESTOR, 22
architectural connector
  ASYNC, 200, 202
  complete, 217
  EXTENSION CORD, 209
  generalised, 216
  heterogeneous, 220
  in CommUnity, 199
  INHIBITION, 210
  instantiation, 200
  semantics, 199
  SUBSUMPTION, 207
  SYNC, 205
architectural school, 214
arrow category, 126
ASYNC, 202
AUTOM, 25
automata
  category of, ***AUTOM***, 25
  reachable, 39

B

base of a concrete category, 95

C

cable, 183, 188
Cartesian morphism, 100
CAT, 86
category
  ANCESTOR, 22
  arrow, 126
  AUTOM, 25
  CAT, 86
  c-DSGN, 186
  CLASS_SPEC, 45, 85, 141
  CLOS, 53
  cocomplete, 77
  comma category, 35, 111, 125
  concrete, 95
  coordinated, 172
  c-SIGN, 185
  definition, 20
  discrete, 32
  dual, 31
  equivalent, 144
  finitely cocomplete, 77
  functor structured, 117
  generated from a graph, 22
  GRAPH, 21
  indexed, 125
  isomorphic, 87
  monoids as categories, 33
  of categories, 86
  of partial functions, 37
  of power sets and inverse
    functions, 38
  of processes, 118

opposite, 31
PAR, 37
POWER, 38
preorder as a, 22
PRES, 54
$PRES_{FOLTL}$, 85
$PRES_{LTL}$, 49
product of, 32
REACH, 39
SET, 21
sets as categories, 32
$SET_{\perp}$, 33
spa(φ), 117
SPRES, 54
subcategory, 37
THEO, 54
$THEO_{LTL}$, 49
c-DSGN, 186
channel (CommUnity), 178
class inheritance hierarchies
as graphs, 16
CLASS_SPEC, 45, 85, 141
cleavage, 101
CLOS, 53
closure system, 53
coadjoint, 152
co-Cartesian morphism, 100
cocompleteness, 77
cocone, 75
base of, 75
category of, 77
colimit, 76
commutative, 76
edge of, 76
vertex of, 75
coequaliser, 72
quotients in *SET*, 71
cofibration, 101
colimit
concrete, 98
definition of, 76
in *spa(φ)*, 120
of theories and presentations in (π-)institutions, 136
of theories and presentations in LTL, 107
vs. (co)fibrations, 107
comma category, 35, 111, 125
CommUnity
cable, 188
compositionality, 196
configuration, 189
design, 182
design morphism, 186
program, 180
refinement morphism, 191
signature, 182
signature morphism, 184
commutative (diagram), 24
composition law, 20
compositionality
design formalisms, 213
in CommUnity, 196
of programs relative to specifications, 112
concrete
(co)limits, 98
(co)reflective subcategories, 98
amnestic category, 96, 99
category, 95
fibre-complete category, 96, 105
fibre-discrete category, 96
functor, 97
subcategories, 98
cone, 78
configuration, 189
refinement, 195
well-formed, 189
connector. *See* architectural connector
higher-order, 231
construct, 95. *See also concrete category*
contravariant
functor, 84
graph homomorphism, 19
coordinated category/functor, 172
coproduct, 62

coreflection
arrow, 40
coreflective subcategory, 40
REACH as a coreflective subcategory of ***AUTOM***, 39
creates
colimits, 92
c-SIGN, 185

D

D(g), 179
D(v), 179
design, 182
action, 206
buffer, 180
cart, 205
check-in, 205
counter, 206
extend, 209
gate, 205
inhibit, 211
monitored_cart, 207
morphism, 186
pipe, 218
printer, 190
receiver, 200
refinement morphism, 191
sender, 181, 183
subsume, 207
user, 190
design formalism, 211
compositionality, 213
diagram, 23
commutative, 24
shape, 23
discrete
category, 32
lift, 170
structures (functor/concrete category has), 170
dual
of a category, 31
of a functor, 83
of a natural transformation, 142
of an adjunction, 152

E

Eiffel class specification, 43
embedding, 87
epi (also epic and epimorphism), 30
equaliser, 73
equivalence of categories, 144
EXTENSION CORD, 209

F

faithful, 87
fibration, 100
cloven, 101
split, 104
theories(presentations) as, 136
universal constructions, 107
vs. indexed categories, 127
fibre
general definition, 99
of a concrete category, 96
fibre-complete concrete category, 96, 105
fibred product. *See* pullback
fibre-discrete concrete category, 96
fitting, 233
free functor, 152
full
embedding, 87
functor, 87
subcategory, 38
vs coreflections, 41
vs. reflections, 43
functor, 83
(co)reflective, 146
(co)reflector, 88, 147
composition law, 86
contravariant, 84
coordinated, 172
creates colimits, 92
dual of, 83
embedding, 87
faithful, 87
full, 87

has discrete structures, 170
identity, 83
inverse, 87
isomorphism, 87
lifts colimits, 92
nodes, 83
preserves colimits, 91
preserves isomorphisms, 88
reflects colimits, 92
reflects isomorphisms, 88
right/left adjoint, 151
functor structured category, 117

G

Gamma
morphisms, 173
programs, 173
graph, 21
category of, 21
definition, 15
dual, 19
homomorphism, 18
path, 20

H

higher-order connector, 231
fitting morphism, 233
instantiation, 234
hom-set, 20

I

in(v), 178
indexed category, 125
as a split fibration, 127
flattening of, 126
indiscrete. See discrete (dual)
INHIBITION, 210
initial object, 58
in ***ANCESTOR***, 60
in ***LOGI***, 58
in ***PAR***, 58
in ***SET***, 58
in $\boldsymbol{SET_{\bot}}$, 58
in ***spa(φ)***, 121
institution, 131
CTL*, 161
defined via a split (co)fibration, 132
generalised models, 134
initial/terminal semantics, 139
modal logics, 134
morphism, 165
the p-property, 138
interpretation between temporal theories. *See* temporal theory: morphism of
isomorphism, 27
functor, 87
inverse, 27
isomorphism-closed full subcategory, 39

L

L(g), 179
left adjoint, 152
lifts
colimits, 92
vs. fibrations, 105
vs. functor structured categories, 120
limit, 78
concrete, 98
in ***PROC***, 123
in ***spa(φ)***, 120
of theories and presentations in (π-)institutions, 136
of theories and presentations in LTL, 107
vs. (co)fibrations, 107
loc(v), 179

M

mono (also monic and monomorphism), 30
monoid, 32
morphism
composition law, 20
epi (or epic), 30

identity, 20
inverse, 27
isomorphism, 27
mono (or monic), 30
of graphs, 18
split (mono, epi), 30

N

natural transformation, 142
(co)unit of an adjunction, 152
composition law, 143
counit of a reflection, 148
dual of, 142
identity, 142
natural isomorphism, 144
unit of a reflection, 147
null object, 60

O

object
initial, 58
null, 60
terminal, 59
zero, 60
out(v), 179

P

PAR, 37
path in a graph, 20
pointed sets. *See* $SET_\perp$
POWER, 38
p-property, 138
PRES, 54, 136
presentation lemma, 51
preserves
colimits, 91
isomorphisms, 88
$PRES_{FOLTL}$, 85
$PRES_{LTL}$, 49
PROC, 118
universal constructions, 123
process. See ***PROC***
product, 65
in ***LOGI***, 65
in ***PROC***, 123
in ***SET***, 65
in $SET_\perp$, 65
of categories, 32, 86
of functors, 86
proof systems
as graphs, 18
prv(v), 179
pullback, 73
in ***PROC***, 123
in $SET_\perp$, 73
pushout, 68
in ***CLASS_SPEC***, 78
in ***SET***, 68
vs. multiple inheritance, 69
vs. the 'Join Semantics rule', 78

R

R(g), 180
REACH, 39
realisation
of configurations (diagrams), 112
of specifications by programs, 111
reduct
for temporal logic, 50
refinement
of configurations, 195
of designs (morphism), 191
reflection
(co)reflective functor, 146
arrow, 42
for a functor, 146
reflective subcategory, 43
THEO as a reflective subcategory of ***PRES*** and ***SPRES***, 54
reflector
for a functor, 147
for a subcategory, 88
reflects
colimits, 92
isomorphisms, 88
right adjoint, 151

S

satisfaction condition
  in CTL*, 162
  in institutions, 131
  in temporal logic (LTL), 50
SET, 21
SET$_{\perp}$, 33
shape (of a diagram), 23
skip, 180
slice category. *See* comma category
spa(φ)
  as a (co)fibration, 118
  functor structured categories, 117
  universal constructions, 120
split (mono, epi), 30
split fibration, 104
SPRES, 54, 136
strict theory presentation
  in a (π-)institution, 136
  in a closure system, 54
subcategory, 37
  coreflective, 40
  full, 38
  isomorphism-closed full, 39
  reflective, 43
SUBSUMPTION, 207
sum, 62
  in ***LOGI***, 63
  in ***PROOF***, 63
  in ***SET***, 63
superposition (or superimposition), 187
SYNC, 205

T

temporal logic
  presentations, 48
  propositions, 47
  reducts, 50
  semantics, 47
  signatures, 46
  theories, 48
temporal theory
  as a concrete category, 96
  category of, 49
  morphism of, 49
  presentation of, 49
terminal object, 59
  in ***ANCESTOR***, 60
  in ***LOGI***, 60
  in ***PAR***, 60
  in ***PROC***, 123
  in ***SET***, 59
  in ***SET***$_{\perp}$, 60
  in ***spa(φ)***, 121
THEO, 54, 136
THEO$_{LTL}$, 49
theory(presentation)
  as a split (co)fibration, 136
  in a (π-)institution, 136
  in a closure system, 54
  in temporal logic, 49
  universal constructions, 107, 136
transition systems
  as graphs, 17

U

U(g), 179
unit
  of an adjunction, 152

Z

zero object, 60

–

π-institution, 130
  presented by an institution, 131, 137

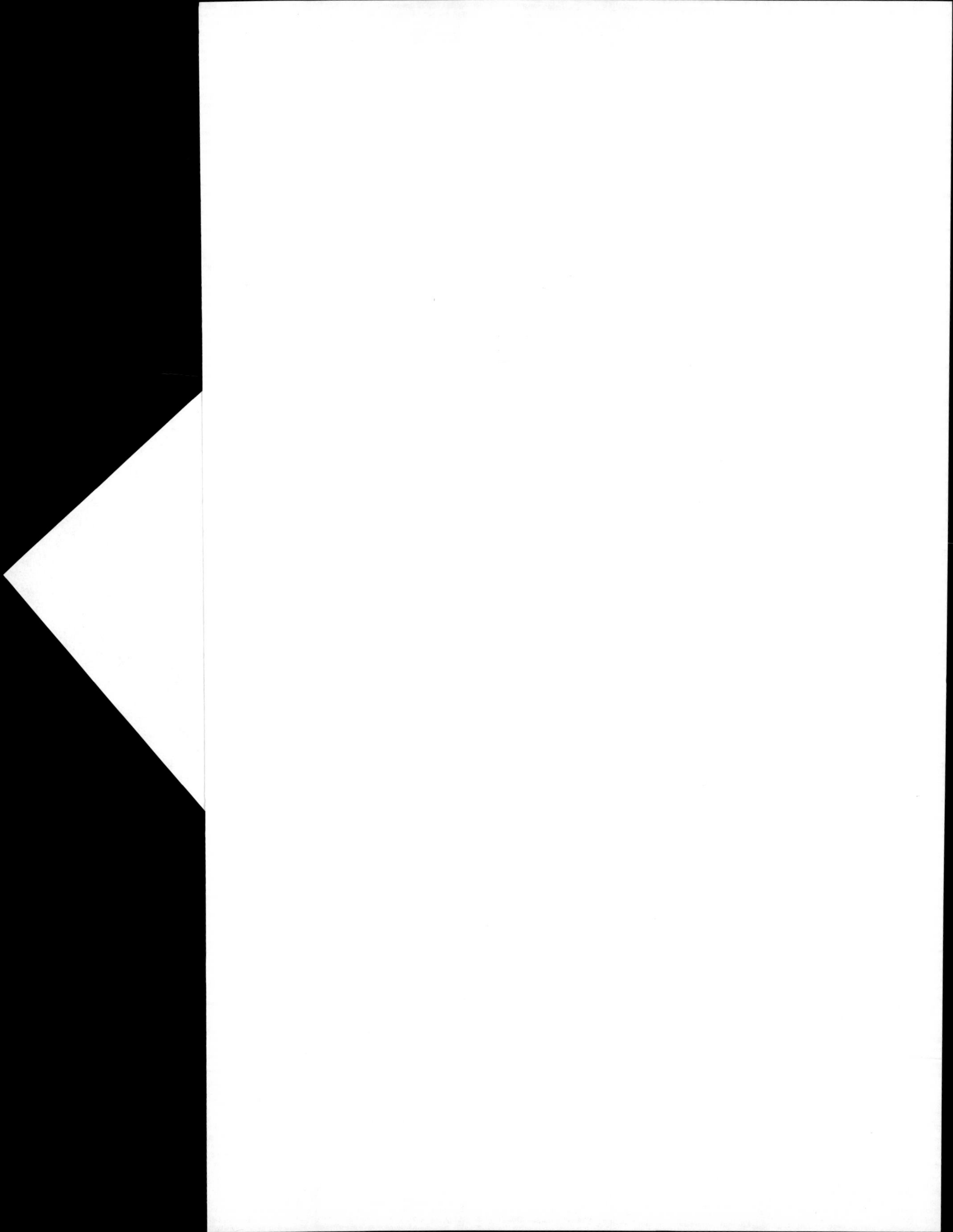

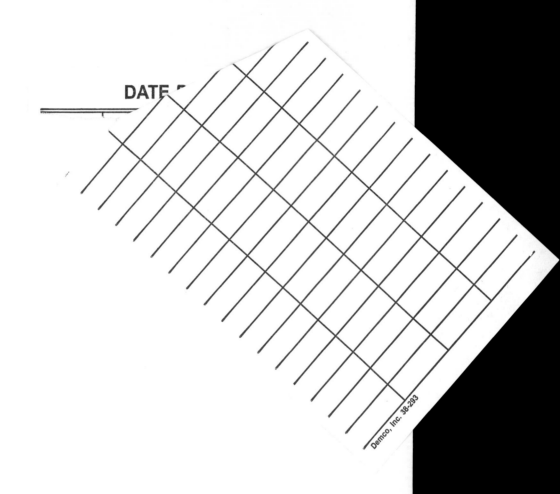